Lean Tools in Apparel Manufacturing

The Textile Institute Book Series

Incorporated by Royal Charter in 1925, The Textile Institute was established as the professional body for the textile industry to provide support to businesses, practitioners, and academics involved with textiles and to provide routes to professional qualifications through which Institute Members can demonstrate their professional competence. The Institute's aim is to encourage learning, recognize achievement, reward excellence, and disseminate information about the textiles, clothing and footwear industries and the associated science, design and technology; it has a global reach with individual and corporate members in over 80 countries.

The Textile Institute Book Series supersedes the former "Woodhead Publishing Series in Textiles" and represents a collaboration between The Textile Institute and Elsevier aimed at ensuring that Institute Members and the textile industry continue to have access to high caliber titles on textile science and technology.

Books published in The Textile Institute Book Series are offered on the Elsevier website at: store. elsevier.com and are available to Textile Institute Members at a substantial discount. Textile Institute books still in print are also available directly from the Institute's website at: www.textileinstitute.org

To place an order, or if you are interested in writing a book for this series, please contact Matthew Deans, Senior Publisher: m.deans@elsevier.com

Recently Published and Upcoming Titles in the Textile Institute Book Series:

Handbook of Natural Fibres: Volume 1: Types, Properties and Factors Affecting Breeding and Cultivation, 2nd Edition, Ryszard Kozlowski Maria Mackiewicz-Talarczyk, 978-0-12-818398-4
Handbook of Natural Fibres: Volume 2: Processing and Applications, 2nd Edition, Ryszard Kozlowski Maria Mackiewicz-Talarczyk, 978-0-12-818782-1
Advances in Textile Biotechnology, Artur Cavaco-Paulo, 978-0-08-102632-8
Woven Textiles: Principles, Technologies and Applications, 2nd Edition, Kim Gandhi, 978-0-08-102497-3
Auxetic Textiles, Hong Hu, 978-0-08-102211-5
Carbon Nanotube Fibres and Yarns: Production, Properties and Applications in Smart Textiles, Menghe Miao, 978-0-08-102722-6
Sustainable Technologies for Fashion and Textiles, Rajkishore Nayak, 978-0-08-102867-4
Structure and Mechanics of Textile Fibre Assemblies, Peter Schwartz, 978-0-08-102619-9
Silk: Materials, Processes, and Applications, Narendra Reddy, 978-0-12-818495-0
Anthropometry, Apparel Sizing and Design, 2nd Edition, Norsaadah Zakaria, 978-0-08-102604-5
Engineering Textiles: Integrating the Design and Manufacture of Textile Products, 2nd Edition, Yehia Elmogahzy, 978-0-08-102488-1
New Trends in Natural Dyes for Textiles, Padma Vankar Dhara Shukla, 978-0-08-102686-1
Smart Textile Coatings and Laminates, 2nd Edition, William C. Smith, 978-0-08-102428-7
Advanced Textiles for Wound Care, 2nd Edition, S. Rajendran, 978-0-08-102192-7
Manikins for Textile Evaluation, Rajkishore Nayak Rajiv Padhye, 978-0-08-100909-3
Automation in Garment Manufacturing, Rajkishore Nayak and Rajiv Padhye, 978-0-08-101211-6
Sustainable Fibres and Textiles, Subramanian Senthilkannan Muthu, 978-0-08-102041-8
Sustainability in Denim, Subramanian Senthilkannan Muthu, 978-0-08-102043-2
Circular Economy in Textiles and Apparel, Subramanian Senthilkannan Muthu, 978-0-08-102630-4
Nanofinishing of Textile Materials, Majid Montazer Tina Harifi, 978-0-08-101214-7
Nanotechnology in Textiles, Rajesh Mishra Jiri Militky, 978-0-08-102609-0
Inorganic and Composite Fibers, Boris Mahltig Yordan Kyosev, 978-0-08-102228-3
Smart Textiles for In Situ Monitoring of Composites, Vladan Koncar, 978-0-08-102308-2
Handbook of Properties of Textile and Technical Fibres, 2nd Edition, A. R. Bunsell, 978-0-08-101272-7
Silk, 2nd Edition, K. Murugesh Babu, 978-0-08-102540-6

The Textile Institute Book Series

Lean Tools in Apparel Manufacturing

Edited by

Prabir Jana

Manoj Tiwari

Woodhead Publishing is an imprint of Elsevier
The Officers' Mess Business Centre, Royston Road, Duxford, CB22 4QH, United Kingdom
50 Hampshire Street, 5th Floor, Cambridge, MA 02139, United States
The Boulevard, Langford Lane, Kidlington, OX5 1GB, United Kingdom

Copyright © 2021 Elsevier Ltd. All rights reserved.

No part of this publication may be reproduced or transmitted in any form or by any means, electronic or mechanical, including photocopying, recording, or any information storage and retrieval system, without permission in writing from the publisher. Details on how to seek permission, further information about the Publisher's permissions policies and our arrangements with organizations such as the Copyright Clearance Center and the Copyright Licensing Agency, can be found at our website: www.elsevier.com/permissions.

This book and the individual contributions contained in it are protected under copyright by the Publisher (other than as may be noted herein).

Notices
Knowledge and best practice in this field are constantly changing. As new research and experience broaden our understanding, changes in research methods, professional practices, or medical treatment may become necessary.

Practitioners and researchers must always rely on their own experience and knowledge in evaluating and using any information, methods, compounds, or experiments described herein. In using such information or methods they should be mindful of their own safety and the safety of others, including parties for whom they have a professional responsibility.

To the fullest extent of the law, neither the Publisher nor the authors, contributors, or editors, assume any liability for any injury and/or damage to persons or property as a matter of products liability, negligence or otherwise, or from any use or operation of any methods, products, instructions, or ideas contained in the material herein.

British Library Cataloguing-in-Publication Data
A catalogue record for this book is available from the British Library

Library of Congress Cataloging-in-Publication Data
A catalog record for this book is available from the Library of Congress

ISBN: 978-0-12-819426-3 (print)

ISBN: 978-0-12-819427-0 (online)

For information on all Woodhead Publishing publications
visit our website at https://www.elsevier.com/books-and-journals

Publisher: Matthew Deans
Acquisition Editor: Brian Guerin
Editorial Project Manager: Chiara Giglio
Production Project Manager: Anitha Sivaraj
Cover Designer: Victoria Pearson

Typeset by MPS Limited, Chennai, India

Contents

List of contributors	**xiii**
Foreword	**xv**
Preface	**xix**

1. Lean management in apparel manufacturing   **1**
Prabir Jana and Manoj Tiwari
1.1	Introduction	1
1.2	Lean: definition, philosophy, and its evolution	2
1.3	Lean and different approaches	3
1.4	Apparel manufacturing and lean	8
	1.4.1 Lean implementation in apparel manufacturing: challenges	12
1.5	Conclusion	15
	References	15

2. Lean terms in apparel manufacturing   **17**
Prabir Jana and Manoj Tiwari
2.1	Introduction	17
2.2	Lean Terms	17

3. Fundamentals of lean journey   **47**
Manoj Tiwari
3.1	Introduction	47
3.2	Value addition	48
	3.2.1 Criteria for value addition	49
3.3	3M of lean manufacturing	53
3.4	Types of wastes	54
	3.4.1 Muda of transportation	57
	3.4.2 Muda of motion	57
	3.4.3 Muda of waiting	57
	3.4.4 Muda of rework	57
	3.4.5 Muda of overprocessing	58
	3.4.6 Muda of inventory	58
	3.4.7 Muda of overproduction	59
3.5	The lean way for improvement and its challenges	60
3.6	Lean way of manufacturing	62
3.7	A framework to initiate lean interventions	63

	3.7.1	How to start	65
	3.7.2	Continuous improvements	67
3.8	Step-by-step implementation of lean tools		72
3.9	Conclusion		76
References			76

4. Lean problem-solving   **81**

Manoj Tiwari and Yuvraj Garg

4.1	Introduction		81
4.2	Kaizen		81
4.3	Value stream mapping		87
	4.3.1	Preparation for value stream mapping	87
	4.3.2	Conducting value stream mapping in a practical environment	91
4.4	A3 problem-solving		107
4.5	Standardized work		112
4.6	Plan—do—check—act cycle		117
4.7	Look—ask—model—discuss—act cycle		118
4.8	Define—measure—analyze—improve—control		119
	4.8.1	Define	119
	4.8.2	Measure	120
	4.8.3	Analyze	120
	4.8.4	Improve	120
	4.8.5	Control	121
4.9	Root-cause analysis		121
	4.9.1	Ishikawa diagram	122
	4.9.2	5-Why analysis	124
4.10	Conclusion		127
References			127

5. Visual management   **131**

Ankur Makhija, Chandrajith Wickramasinghe and Manoj Tiwari

5.1	Introduction		131
5.2	Gemba		133
	5.2.1	Gemba walk	134
	5.2.2	Gemba centric management	135
	5.2.3	Rules of Gemba management	136
	5.2.4	Best practices for successful Gemba walks	136
5.3	Visual factory		137
	5.3.1	Characteristics of visual tools	138
	5.3.2	Classification of visual tools	139
	5.3.3	Basics of visual factory system	139
	5.3.4	Components of visual management systems	139
	5.3.5	Steps in creating a visual factory	147
	5.3.6	Visual factory: zoning approach for plant layout	151

	5.3.7	Visual garment factory	152
	5.3.8	Benefits of visual factory	154
5.4	5S		156
	5.4.1	Benefits of 5S	164
	5.4.2	Implementing 5S	166
5.5	Andon		182
	5.5.1	Types of Andon systems	183
	5.5.2	Implementing the light and sound in visual factory	185
	5.5.3	Working of Andon system	186
	5.5.4	Benefits of Andon system	188
5.6	Yamazumi charts		188
	5.6.1	How to make a Yamazumi chart?	189
	5.6.2	Benefits of Yamazumi chart	190
5.7	Turtle diagram		191
	5.7.1	Components of a Turtle diagram	191
	5.7.2	Benefits of Turtle diagram	193
	5.7.3	Applications of Turtle diagram	194
5.8	Affinity diagram		195
	5.8.1	Creating an affinity diagram	196
	5.8.2	Advantages of the affinity diagram	198
5.9	Dashboard		198
	5.9.1	Types of dashboards	201
	5.9.2	How to make a dashboard?	201
	5.9.3	Benefits of dashboard	203
	5.9.4	Application of dashboards in apparel industry	204
5.10	Conclusion		204
	References		205

6. Rapid setup   **209**

Chandrark Karekatti and Chandrajith Wickramasinghe

6.1	Introduction		209
6.2	Tool changeover		211
	6.2.1	What is tool changeover time	211
	6.2.2	Design for changeover	212
	6.2.3	Changeover methodology	213
	6.2.4	Prerequisites for SMED implementation	216
6.3	Style changeover in apparel manufacturing		216
6.4	Single Minute Exchange of Die in apparel manufacturing		218
	6.4.1	Organizational improvement	219
	6.4.2	Design improvement	222
	6.4.3	SMED examples in apparel manufacturing	222
6.5	Impact of rapid changeover		228
6.6	Conclusion		229
	References		229

7.	**Autonomation**	**233**

Chandrark Karekatti and Prabir Jana

7.1	Introduction	233
7.2	Jidoka	234
	7.2.1 Jidoka methodology	235
	7.2.2 Jidoka functions	237
	7.2.3 Jidoka in apparel manufacturing machinery and equipment	238
	7.2.4 Jidoka system in the apparel manufacturing process	240
7.3	Poka-Yoke	242
	7.3.1 Poka-Yoke categories	243
	7.3.2 Poka-Yoke implementation in the apparel industry	247
7.4	Conclusion	254
	References	254

8.	**Process balancing**	**257**

Chandrark Karekatti and Manoj Tiwari

8.1	Introduction	257
8.2	Challenge of high inventory levels in apparel manufacturing	258
8.3	Continuous flow manufacturing	259
	8.3.1 Key traits of flow manufacturing	261
	8.3.2 Key time-related definitions in flow manufacturing	261
	8.3.3 Implementing single-piece flow in apparel manufacturing	266
	8.3.4 Challenges in flow manufacturing	271
8.4	Constraint management	273
	8.4.1 Drum−buffer−rope system	273
	8.4.2 Implementing drum−buffer−rope system	274
8.5	Production leveling	275
	8.5.1 Heijunka approach for production leveling	276
	8.5.2 Heijunka approach for product leveling	277
	8.5.3 Heijunka box	279
8.6	Kanban	279
	8.6.1 Kanban card	280
	8.6.2 Types of Kanban	280
	8.6.3 Supermarket	282
	8.6.4 Implementing Kanban	283
	8.6.5 Determining the number of Kanban cards	286
8.7	Alternatives to Kanban	287
8.8	Conclusion	288
	References	288

9.	**Apparel manufacturing systems**	**291**

Manoj Tiwari and Prabir Jana

9.1	Introduction	291
9.2	Manufacturing systems	292
9.3	Evolution of apparel manufacturing systems	293

	9.4	Apparel manufacturing systems	294
		9.4.1 Make-through	295
		9.4.2 Assembly line	295
		9.4.3 Modular manufacturing	297
	9.5	Success parameters of a manufacturing system	307
		9.5.1 Flexibility	307
		9.5.2 Cost-effectiveness	307
		9.5.3 Quality	307
		9.5.4 Speed	307
	9.6	Conclusion	307
	References		308
10.	**Lean quality management**		**311**
	Rajesh Bheda		
	10.1	Introduction	311
		10.1.1 The meaning of quality and quality improvement	312
	10.2	Lean philosophy and quality management	314
	10.3	A brief history of statistical methods in quality	317
	10.4	What is statistical process control?	317
		10.4.1 Types of charts	317
		10.4.2 Statistical process control implementation	320
	10.5	Acceptable quality level	322
		10.5.1 Working of the acceptance sampling plans	323
		10.5.2 Ensuring success of acceptance sampling plans	325
	10.6	Six Sigma	326
	10.7	Conclusion	327
	References		328
11.	**Lean human resources**		**331**
	Chandrark Karekatti		
	11.1	Introduction	331
	11.2	Lean human resource management	332
	11.3	Lean and downsizing	338
	11.4	The role of HR in lean implementations in the apparel industry	339
		11.4.1 Organization-wide lean implementation initiatives	339
		11.4.2 Engagement in employee participation initiatives	344
	11.5	Harada method	346
	11.6	Conclusion	351
	References		351
12.	**Total productive maintenance**		**355**
	Rashmi Thakur and Deepak Panghal		
	12.1	Introduction	355
		12.1.1 History and evolution	355
		12.1.2 Definition	356

	12.2	Basic concept of TPM	356
		12.2.1 Pillars of TPM	357
		12.2.2 Types of maintenance	358
		12.2.3 Six major losses in the absence of TPM	359
		12.2.4 Maintenance performance indicator	359
		12.2.5 Overall Equipment Effectiveness (OEE) and its calculation	360
	12.3	Benefits and challenges of TPM	362
		12.3.1 Benefits of TPM	362
		12.3.2 Challenges in implementation of TPM	363
	12.4	TPM in the textile and apparel manufacturing industry	363
	12.5	Smart maintenance	366
		12.5.1 Technological intervention	370
		12.5.2 Challenges of smart maintenance	375
	12.6	Conclusion	376
	References		377

13. Lean supply chain management — **381**
Prabir Jana

	13.1	Introduction	381
	13.2	Evolution of lean supply chain	382
		13.2.1 Demand amplification and bullwhip effect	382
		13.2.2 Product development	384
		13.2.3 Inventory management	386
		13.2.4 Supply chain structure	390
	13.3	Lean applications in retail supply chain	392
		13.3.1 Lean warehousing and transport	392
		13.3.2 Lean distribution	393
		13.3.3 Lean in e-commerce	394
	13.4	Conclusion	395
	References		395

14. Agile manufacturing — **399**
Riddhi Malviya

	14.1	Introduction	399
		14.1.1 Lean manufacturing: risks and mitigations	400
		14.1.2 Lean paves the way for agility	401
	14.2	Agile manufacturing	401
		14.2.1 Agile	402
		14.2.2 Agile manifest	402
		14.2.3 Agile manufacturing system	402
		14.2.4 The growing popularity of agile in manufacturing	403
	14.3	The symbiotic relationship between lean and agile	404
	14.4	Key advancements needed to support agile in the future	405

	14.4.1	The digital thread	405
	14.4.2	Data management and sharing	405
	14.4.3	Changing the mindset	405
	14.4.4	Checking the feasibility of agile	406
14.5	Lean versus agile		406
	14.5.1	Prime integrant	406
	14.5.2	Cornerstones	407
	14.5.3	External vendors	407
	14.5.4	Corporate structure and values	408
	14.5.5	Product	408
14.6	Agile organization		408
	14.6.1	Reasons for a low emphasis on collaboration	409
	14.6.2	Ways agile can help a company improve its culture and be more innovative	410
	14.6.3	Is bringing outside help imperative to apply agile concepts?	410
14.7	Strategic tools for future agile firms		411
	14.7.1	Understand that clients are more important to the business than investors	411
	14.7.2	Performance metrics centered around clients	411
	14.7.3	Activities report to the association, not the immediate department	411
	14.7.4	Budgeting is performed by the association, not at the single department level	411
	14.7.5	Work should be done in small groups working short cycles	412
	14.7.6	Connection to HR	412
14.8	New horizons for lean manufacturing		412
14.9	Case study		418
	14.9.1	Agility: an unlikely savior in trying times	418
14.10	Conclusion		419
References			420

Index 423

List of contributors

Rajesh Bheda Rajesh Bheda Consulting Pvt. Ltd., Gurgaon, India

Yuvraj Garg Department of Fashion Technology, National Institute of Fashion Technology, Jodhpur, India

Prabir Jana Department of Fashion Technology, National Institute of Fashion Technology, New Delhi, India

Chandrark Karekatti Ananta Garments Ltd., Dhaka, Bangladesh

Ankur Makhija Department of Fashion Technology, National Institute of Fashion Technology, Gandhinagar, India

Riddhi Malviya VF Corporation, Greensboro/Winston-Salem, North Carolina, United States

Deepak Panghal Department of Fashion Technology, National Institute of Fashion Technology, New Delhi, India

Rashmi Thakur Department of Fashion Teachnology, National Institute of Fashion Teachnology, Mumbai, India

Manoj Tiwari Department of Fashion Technology, National Institute of Fashion Technology, Jodhpur, India

Chandrajith Wickramasinghe Corrigo Consultancy Pvt. Ltd., Noida, India

Foreword

Textiles are everywhere. Everybody owns at least some pieces of clothing. You can go to the farthest corners in the Amazon rainforest and find a tribal member wearing a T-shirt. This widespread need and the resulting huge market for clothing made the apparel industry a significant catalyst in multiple ways.

Textiles were the starting point of the industrial revolution. In medieval times, the constraint was spinning. It needed around 10 spinners to keep one weaver busy, and spinning was a common task in most households. The first spinning mill by Thomas Cotchett around 1702 in Derby, UK, was a major failure. Only in 1722, after some serious industrial espionage of other machines in Italy by John Lombe was this spinning mill able to spin yarn in adequate quality. Machines were now able to spin yarn much cheaper and faster than humans could. However, the mill produced silk yarn, which back then like nowadays was a luxury product that few could afford.

The real money was in the much cheaper cotton, which even nowadays makes up most of our clothes. Unfortunately, cotton is a much shorter fiber than silk and therefore much more difficult to spin. The desire to spin cotton yarn mechanically leads to a quick succession of inventions. Richard Arkwright and John Kaye invented the water frame, a machine to stretch cotton fibers in 1769. In 1770 James Hargreaves invented the spinning jenny to twist the fibers. The first major cotton spinning mills was opened by Richard Arkwright in 1772 in Cromford, UK—and it was a smashing success. Arkwright opened multiple mills and received the knighthood for his work. His son continued the spinning business and became the richest man in Europe.

Multiple other entrepreneurs also established spinning mills. Since spinning was now no longer the constraint in the value chain, attention moved to weaving. Soon, mechanized looms were used in large weaving factories. At one point, half of the cotton production of the entire world was spun in the United Kingdom. It was cheaper to ship cotton from India to England, process it, and then ship the textiles back to India rather than processing these in India. Combined with other key technologies like the steam engine and railways, *textiles were the products that started the industrial revolution, transforming our world.*

Textiles were also a catalyst in lean manufacturing. The archetype of lean manufacturing is the Toyota Motor Corporation with its Toyota production system. However, Toyota originated as a spin off from another company, Toyoda Automatic Loom. Its founder, Sakichi Toyoda, started numerous businesses and was the source of many inventions in weaving. His Model G automatic loom is most famous, which is also the origin of lean manufacturing. Different mistake proofing devices (in Japanese called *poka yoke*) stopped the machine automatically when a yarn broke. Another gadget automatically switched shuttles when the yarn was used up. Altogether this is known in Japanese and in the Lean world as *jidoka*, usually translated as *autonomation*.

The most famous concept in lean manufacturing, the reduction of material using methods like kanban, was invented by the Toyota Motor employee Taiichi Ohno. But even here the inspiration springs from textiles. When Ohno was still working for Toyoda Automatic Loom, one of their competitors *Nichibo* performed much better. Quality was much better while cost was also lower. Ohno was tasked to learn about the secret to the success of *Nichibo*. His main finding was that *Nichibo* used much smaller lot sized and in general had much less inventory on the shop floor. This is probably the inspiration for Ohno to do the same at Toyota. Hence, *textiles were also the origin of lean manufacturing at Toyota*.

Finally, textiles also have a major impact on the economies of individual countries. While spinning and weaving are nowadays easy to automate, the handling of the fabric is much trickier. Picking up a steel or aluminum part usually does not change the shape of the part, and robots can easily position and work these parts. Picking up a half-finished shirt, however, changes the shape of the product. The textile easily bends under gravity and shifts dynamically when moved. Hence, modern robots still have difficulties handling textiles, and manual work is commonly employed.

Almost all countries started out as agricultural societies. As their technical skills progress, they turn into industrial societies, while agriculture became less important. The final step is to transform from an industrial economy to a service economy to increase prosperity even more. Manufacturing is less significant, and the relevance of agriculture is marginal. Pretty much all countries go through these stages, where a peak in industry is needed to transform from a poor agricultural based economy to a wealthy service-based economy.

Here, too, textile and apparel manufacturing as well as lean manufacturing have a key role. Since the production of apparel needs a lot of manual labor, it is too expensive for advanced economies with their high labor cost. This creates an opportunity for poorer countries to industrialize. The country can compete with advanced economies on labor cost, and it is difficult nowadays to find textile processing in advanced economies. By producing apparel they will learn about technology, supply chains, lean manufacturing, and management through processing textiles. Lean manufacturing can be of significant help with these high labor apparel manufacturing. Using this knowledge as a stepping stone, the country can move up towards more difficult technologies and eventually become a wealthy service-based economy.

Almost all rich nations have moved through this process, usually including a lot of textile and apparel manufacturing. Hence, even if a wealthy country no longer has a significant apparel industry, *textiles are also a key stepping stone to lift a country out of poverty and towards prosperity*.

Overall, the apparel industry is of major significance for the prosperity of the world, both historically and current. Hence, this book will be very helpful in

Foreword xvii

improving the apparel industry, especially under the constant pressure to reduce lead times and cost, requiring reduction of waste (muda), fluctuations (mura), and overburden (muri).

The book itself looks at the fundamentals of lean in apparel in Chapter 1, Lean Management in Apparel Manufacturing, followed by an in-depth glossary in Chapter 2, Lean Terms in Apparel Manufacturing. It then sets out on a lean journey and shows how to approach lean within the apparel industry. It goes into significant detail on the most important lean techniques including problem-solving, visual management, quick changeovers, autonomation (Jidoka), balancing and leveling, quality management, maintenance, supply chain management, and finally agile manufacturing. I sincerely hope that this book will help you on the path towards success, for your country, for your company, and also for yourself.

Christoph Roser

www.AllAboutLean.com

Professor of Production Management,

Karlsruhe University of Applied Sciences, Karlsruhe, Germany

Preface

Congratulations on picking up this unique book that will help you go through the Lean tools and techniques in the context of Apparel and Textiles!!! You may be curious about why there are images of Beehives, Weaver Birds, Ants, and Termite mounds over the cover page! Well, before we learn from books and fellow human beings, there are learnings from nature.

Honeybees expend the least amount of energy to build the hexagonal honeycomb cells that gives the highest structural stability, zero space between cells (most shared walls), and maximizes honey storage capacity without spilling out. That means, maximizing output with minimum resources... that is the gist of being Lean!!! No wonder, beehives are considered as "Nature's Best Zero-Waste Structure."

Baya Weavers are tiny birds with high engineering skills. The beautifully, engineered nests are "Nature's Best Poka-Yoke" *due to its* safety (in terms of height and location), stability in high winds, upside down tubular entrance (resistant to rain), and clever compartments ensuring eggs and kids are not falling off accidentally.

Ants are small creatures but known for their active, persistent, perseverance, diligent, energetic, hard-working, industrious, and lively behavior. One very important quality of ants is the carefully efficient management of resources without wastage. Other notable qualities of ants include well-organized nature, flexibility, commitment, resource sharing, well-disciplined behavior, cooperation, and coordination (teamwork), leadership, decision-making capacity, good planning ability, and effective communication. These characteristics are key traits of the lean philosophy. We call them "Nature's Best Team-workers."

Termites are also social insects just like ants, bees, and other social insects, where the synergic bonding and unity of the colony greatly surpasses that of the individual termite. This makes the tiny termite superpowers. Also, a termite mound (that is made of mud) is a structure or facility without a foreman, which means no one termite is in-charge of the mound. This is also a quality of lean culture with a message "Contribution from All... Ownership of All...!!!"

Before we begin our journey into the industrial applications of lean, we salute all these intelligent species and their lean applications that are found in Mother Nature.

With the advancement of technology, consumers have access to everything, everywhere, and at any time. Physical distances and locations are not at all relevant as the products are just a click away on mobile phones, that too with the best features and at the best prices, and of course inline of what a customer wants, even customizations are possible.

This has given a new dimension to the already fierce competition, where everybody wants to tap the business potential through reaching to the customer at the earliest, with the best product at the best price. With this objective to survive and thrive in the cut-throat business environment, the manufacturers and service providers are pushed for improved resource utilization. Changing the manufacturing bases from one country to another in search for cheaper labor and other resources is just one part of the story, the key is effective and efficient resource utilization and achieving operational excellence through continuous improvement. Textiles and Apparel are also no exceptions to this global competition increasing with each passing day. Shorter product life cycles, product diversifications, and consumers' never-ending expectations to get the customized products are pushing the manufacturers to adopt the Best as well as the Next practices.

Most probably, Lean philosophy is one of the greatest gifts ever, to the manufacturing world from the Toyota Motor Corporation. In the last 70 years, there are several success stories across the globe and across the industries, where lean transformations have changed the fortunes. Though, lean concepts were first developed and practiced by the automobile industry and quickly adopted by other industries as well. As rightly stated by Womack (1990), "lean production is a superior way for humans to make things ... It follows that the whole world should adopt lean production, and as quickly as possible." Apparel manufacturing is a labor-intensive environment with relatively lesser automation. Most of the typical apparel manufacturing environments are full of inefficiencies (to name a few, excess inventories, waiting time, overproduction, overprocessing, process imbalances, product quality-related issues, lesser operator utilization, poor information, and material flow, high material consumptions, and longer lead times). The vary reason for same is not adopting to the lean practices, due to the "cultural inertia," people still feel safe and comfortable in their traditional approaches of hiding the problems under a high level of inventories. This is a typical Mass manufacturing mentality, and here again, the words of Womack (1996) fit perfectly well, "What life has taught many employees is that one of the best features of mass production is that problems are always a mystery and therefore no-one's fault. Exposing problems, by contrast, suggests that someone will be assigned the blame and punished."

In recent years, being lean has become a necessity and not just a choice. This has resulted in an increased momentum towards adopting lean culture in the textiles and apparel industry, though it is still way behind when compared to the automobiles industry. We have great literature and research conducted by the authorities in lean manufacturing, but most of them have been written in the context of other industries but the textiles and apparel industry. Also, the practical implementations-related aspects of lean tools and textile and apparel-specific examples are missing in such available literature. Hence, the people from the textiles and apparel industry face difficulties in technical know-how of lean transformation. Our experience shows that most of the lean initiatives in textile and apparel organizations are either once bitten twice shy or "me too" approach without a clear goal or purpose.

Preface xxi

The reasons are multifold: first, most lean projects are initiated through external consultants or exploration projects through student internees, thereby lacks continuity; secondly, the consultants have often overcomplicated the lean applications drawing a negative vibe from organizations; last but not least, the hurry of textiles and apparel organizations to get the result.

Given the same, this book may be a handy resource guide for conceptual clarity and practical aspects of lean implementation in the textile and apparel industry environments. Wide coverage of different areas of lean makes this book equally useful to the lean practitioners, consultants, academicians, and students as well as to people who are dealing in the fields of textile and apparel. This book aims to provide readers a seamless, easy, and gradual progression for a lean transformation where different relevant tools (*not all the lean tools and techniques, as there are many lean solutions practiced by the automobile industry*) are discussed with its practical implications in the textile and apparel industry. Authors have tried to ensure that the readers can use this book without jumping forward and backward from one chapter to another. At the same time, this book can be referred by the users for specific chapters or topics as well.

Chapter 1: This chapter discusses the brief history and evolution of lean principles, different approaches, and stages of lean implementation. The challenges in lean implementation in the context of apparel manufacturing have also been discussed. The objective of this chapter is to introduce and orient readers to provide an overview of lean management practices in apparel manufacturing.

Chapter 2: Lean is not a single term or technique but a combination of several tools and techniques. Lean philosophy involves different approaches as well. Lean experts have explained them through different elements, principles, as well as tools. Over the years, different academicians, consultants, and industry practitioners have also combined multiple approaches and tools, and given a different name to the combined process. In this chapter, each term is categorized to explain if it is a tool, a philosophy, a manufacturing system, or a simple measure. The nature of the tool is further categorized into diagnostic, improvement, training, or monitoring and control tool. The improvement tool is further categorized into whether it follows waste minimization or production leveling approach. The authors have tried to list those lean terms which are popularly used or have potential application in apparel manufacturing.

Chapter 3: This chapter discusses the step-by-step methodology to kick-start lean transformations in the context of apparel manufacturing. Value-addition, Muda, Mura, Muri (3M), lean waste (7W), challenges to lean, Kaizen event, Hoshin Kanri, Lean framework have been discussed in the context of apparel manufacturing.

Chapter 4: This chapter discusses the key tools to find process waste and problem-solving in the context of apparel manufacturing. The managers and decision makers in conventional apparel manufacturing are essentially involved in activities, such as communicating necessary information, increasing efficiencies or productivity, and troubleshooting the day-to-day problems. In this chapter, lean-based alternatives paths, tools, and techniques to traditional problem-solving have been discussed. Apart from value stream mapping, the other problem-solving techniques, such as A3, DMAIC, LAMDA, standardized work, plan do check act cycle, root-cause analysis, Ishikawa diagram, and 5-why analysis are also explained.

Chapter 5: This chapter discusses the philosophies of 5S and visual management in the context of apparel manufacturing. Several visual management tools and techniques, such as Gemba, Visual factory, 6S audit, Andon, a different type of charts & diagrams, Dashboard have been discussed in this chapter.

Chapter 6: Developing rapid setup or quick style changeover capability is important for the successful implementation of lean, flexible, or small-order manufacturing systems. It has become even more important for apparel manufacturers catering to high fashion garments, with shorter lead times, smaller order quantities with diversified products. This chapter discusses the rapid setup implementation process in apparel manufacturing. Other related areas such as prerequisites for SMED, changeover approaches, and changeover methodology have been discussed with practical success stories in apparel manufacturing.

Chapter 7: Autonomation, in simple terms, is described as automation with human intelligence. Autonomation is a self-reliant work system. that is designed in such a way that it either prevents the user/work system from making errors or triggers a rectification mechanism to correct the error/variation by detecting abnormality/error and raises alarm or on detecting abnormality. This chapter is an attempt to document autonomation and its allied tools with the methodology of the application of these tools in the apparel industry.

Chapter 8: The competitive business environment pushes organizations to adopt the just-in-time (JIT) approach to survive. Resource balancing which is carefully controlled by the demand is the need of the hour. Lean provides some effective tools such as load-leveling (Heijunka), pull production with controlled WIP (Kanban), and flow management using theory of constraint suitable for such challenging environments. Process balancing using different lean approaches in the context of apparel manufacturing has been discussed in this chapter.

Chapter 9: With the advancement of technology, the business environment is becoming dynamic and competitive than ever. The organizations have to adapt to the changes to survive and maintain a competitive advantage over others. The ever-increasing customer expectations have always pushed for the transformation of manufacturing systems, and as a result, the manufacturing systems have undergone many evolutions and paradigm shifts. This evolution can be witnessed clearly as craftsmanship-based make-through systems to the modern lean philosophy-based agile manufacturing systems. Lean philosophy-based manufacturing systems helps immensely in achieving such objectives to achieve operational excellence. This chapter discusses various apparel manufacturing systems in the context of lean manufacturing.

Chapter 10: This chapter elucidates the application of principles of quality management for lean implementation in the apparel industry. It covers the importance of quality, in terms of quality of design and quality of conformance. It focuses on how to control the processes through the statistical techniques like statistical process control, how to make a judgment about the quality of the produced lots using acceptance sampling and introduction to concepts of six sigma. The use of various control charts and their implementation have also been discussed. The chapter focuses to make us understand that building quality into the design and managing it at the various process levels have to be seen as an integral part of one of the several strategic initiatives taken by the top management.

Chapter 11: Any improvement is all about people. The magic is not there in the philosophies, tools, and techniques, but it is in the execution by the concerned people. And the human resource is the one who is going to initiate, implement, and making any improvement sustainable. This chapter discusses this important aspect of human resource management in a lean journey. How lean and human resources mutually support each other and helps in empowerment has been also discussed.

Chapter 12: With technological advancements, production has mainly turned out to be unmanned, owing to robotics and automation in industry. Equipment in production has become complex owing to automation. This further affirms that production output as well as the quality depends on the equipment. However, the rate at which production is getting automated and thus unmanned, maintenance is not. This further requires maintenance organization and philosophy to be upgraded for the facilitation of modification in maintenance methodology and enhancement of maintenance skills as well, thus leading to total productive maintenance (TPM). This chapter briefly reviews the basic concepts of TPM and majorly focuses on the evolution of TPM over time especially in the form of smart maintenance which is the need of the hour across all the industry to cope up with Industry 4.0.

Chapter 13: The apparel and textile supply chain is explained with its evolution from industrial dynamics and logistics functions. The chapter discussed the reason behind the bullwhip effect in the apparel supply chain and tools to minimize it, the use of critical chain, virtual prototyping, and collaborative product development to reduce lead time and increase efficiency in product development. The chapter also discusses the evolution of and role of postponement, vendor-managed inventory, and JIT to manage inventory in the apparel supply chain. The different supply chain structure and it's bearing on lean apparel manufacturing is explained. Examples of lean principles in logistics, warehousing, distribution, and e-commerce operations are also explained to expand the domain of lean in supply chain management.

Chapter 14: The rise of consumerism and the growing desire for unique fashion have paved the way for countless clothing brands promoting creativity and innovation in design. For such brands, change and adaptability are key, rather than value or process perfection. Thus a methodology like agile manufacturing—which places the main focus on rapid response to the customer—is fitting. This chapter discusses the concepts of Agile and Leagile, which can be considered as an extension to a lean approach.

This book is a result of constant rigorous efforts of more than 3 years, and it could have not been completed without contributions from several people. It is our privilege to thank them for their contributions. First and foremost, we are thankful to our publisher Elsevier for publishing this book under the *Textile Institute Book Series*. We are highly thankful to Brian Guerin, Senior Acquisitions Editor for considering the request and taking lead to making it happen. We are also thankful to Editorial Project Managers Ana Claudia Abad Garcia and Chiara Giglio for all the support and addressing our queries and updates in a highly professional manner. They were instrumental in handling all the related issues and prompt redressal always. We are also thankful to the Copyrights Coordinators Ashwathi Aravindakshan and Srinivasan Bhaskaran for their assistance in getting permissions as and when required.

We feel thankful and fortunate to have Prof. Roser Christoph, Professor of Production Management, Karlsruhe University of Applied Sciences, Germany, for his motivating and insightful foreword to this book. In the last many years, Prof. Christoph has been instrumental in sharing knowledge and experience through his clear, crisp, and impactful discussions and writings on different aspects of lean manufacturing. Needless to say that his blog "AllAboutLean" is a great platform for lean discussions.

We are grateful to Deepak Mohindra, Editor-in-Chief, Apparel Resources Pvt. Ltd., Chandrajith Wickramasinghe, Corrigo Consultancy Pvt. Ltd., and Sunaina Khanna, Methods Apparel Consultancy India Pvt. Ltd., for their constant support and allowing us to use relevant value-added content in this book. We must be thankful to all the authors who have contributed to this book and make it a useful and value-added resource.

We are also thankful to two young and creative brains Deepthi Baskar and Anushree Chand of National Institute of Fashion Technology, India for their inputs on English language corrections and proofreading.

We wish that this book shall serve the purpose and help in enriching the readers with conceptual and practical knowledge of lean concepts and help them in making informed decisions about the selection and implementation of appropriate lean tools.

To conclude, we have a wonderful quote by Dominick Coniguliaro

"If you want to be where you've never been, you have to go the way you had never walked. If you want to achieve what you never had, you have to do what you never did."

We wish you all the very best for your Lean journey.

Prabir Jana[1] and Manoj Tiwari[2]
[1]Department of Fashion Technology, National Institute of Fashion Technology, New Delhi, India, [2]Department of Fashion Technology, National Institute of Fashion Technology, Jodhpur, India

Lean management in apparel manufacturing

Prabir Jana[1] and Manoj Tiwari[2]

[1]Department of Fashion Technology, National Institute of Fashion Technology, New Delhi, India, [2]Department of Fashion Technology, National Institute of Fashion Technology, Jodhpur, India

> *There are four purposes of improvement: easier, better, faster, and cheaper. These four goals appear in the order of priority.* — *Shigeo Shingo*

1.1 Introduction

"The lesser is better" is the manufacturing mantra in the current times; this enables in reaching developing products and services in the right quantity, right quality, and at right time. Due to the unstable and ambitious demand patterns of the consumers, the days with big orders quantities and set designs are no more. The order quantities are getting smaller with more variations, and delivery timelines are shrinking. The global competition is so fierce that one can no longer work in the set pattern of traditional business; in true sense, the business environment has become volatile, uncertain, complex, and ambiguous (collectively termed as VUCA) (Ebrahim, Krishnakanthan, & Thaker, 2018). This has led to a transformation of organizations adopting newer and better practices in all the areas of business including manufacturing. Cost, quality, and delivery are three key control parameters of today's manufacturing and the same is true with apparel manufacturing as well. Any kind of value addition in cost, quality, and delivery may take the burden off from the management system (Technopak Advisors, 2011).

Lean management, a Japanese philosophy, brings solutions to achieve operational excellence and provides a vital push to survive and compete in today's competitive business environment. However, lean concepts are not new to the world but it has been constantly proving its worth and relevance. In fact, in recent years a huge number of organizations have adopted a lean way of working. Lean manufacturing tools are applied world-wide and it provides an edge in cost, quality, productivity, flexibility, and quick response (Schonberger, 2007). The possible reason for the same is that the lean approach focuses on using less of everything compared to the traditional way of working (Bhasin & Burcher, 2006; Lathin & Mitchell, 2001). It focuses on productivity improvement through reducing waste (Bowes, 2009) by applying lesser resources including (lesser human effort,

Lean Tools in Apparel Manufacturing. DOI: https://doi.org/10.1016/B978-0-12-819426-3.00014-X
© 2021 Elsevier Ltd. All rights reserved.

manufacturing space, setup, and lesser processing time in developing the product) (Wang, 2011). We shall be discussing lean management principles and their practical applications in this chapter. This chapter aims to introduce readers about the holistic view of the lean approach.

1.2 Lean: definition, philosophy, and its evolution

In simple words, lean philosophy may be summarized as identification, reduction, or elimination wastages in the system wherever possible. This looks simple but difficult to implement and execute as this is not just a concept but a lifestyle that needs to be followed religiously. Lean is not time bounded but a continuous improvement journey where there is no limit to achieving excellence. The world is highly grateful to Japan (Toyota Corporation) to introduce this philosophy of Kaizen, known as synonymous to lean world-wide. The term lean production was coined by John Krafcik because it uses lesser resources as compared to mass production (Womack, Jones, & Roos, 2007). Lean is not a single concept of technique but an amalgam of theories applied constantly to achieve operational excellence.

Lean management which is a generic process management philosophy (Wang, 2011) has its roots in the Toyota Production System (TPS) which was developed and practiced by Taiichi Ohno in the 1950s at the Toyota Corporation, Japan (Jana & Tiwari, 2018). TPS was a combination of practices followed by employees of the Toyota Motor Company in the form of their day-to-day efforts to achieve continual improvements. Lean ideas provide newer ways to identify and root out the waste in the process (Hammer & Somers, 2015). Kaizen, a term referred to as a Japanese approach for continuous improvement, is the genesis of the Lean way of working. In the 1980s a term named Lean manufacturing was first coined by researchers at MIT under the guidance of Dr. James Womack. In the 1980s a term named Lean manufacturing was first coined by researchers at MIT under the guidance of Dr. James Womack, who termed the lean approach as a solution to eliminating the process of waste (Womack & Daniel, 2003). Accordingly to the Lean Enterprise Institute, lean is a demand-driven approach based on the pull concept (LEI, 2000) which is in contrast to the traditional thinking of producing more or push concepts (Womack et al., 2007). Richard Schonberger played an important role in popularising "just-in-time" concepts in the United States in the 1980s, which emphasized the material flow aspects of lean (Wilson, Hill, & Glazer, 2013).

Around the 1930s Ford and General Motors (GM) of the United States were the leaders of automobile mass production, thanks to Henry Ford for introducing the assembly line system in the early 1900s. The assembly line manufacturing brought much-needed speed, accuracy, and consistency in manufacturing, and they were able to meet the global demand. On the other side, Toyota was a very small Japanese player in the domain in comparison to Ford and GM and facing stiff competition to survive. Toyota was not in a position to invest heavily in the infrastructure for mass manufacturing. So to survive, Toyota decided to have its way of manufacturing with a focus on smaller quantities, more variations, more flexibility, improved quality, and lower cost. In the initial years, Toyota was catering to local demands, but in a few years due to its solid foundation and practices, it started competing with the global players. Following their manufacturing principles religiously they became the global leader in the automobile market and left their initial competitors Ford and GM far behind (Technopak Advisors, 2011). According to Womack et al. (2007) after the Second World War, lean-based manufacturing principles and practices were pioneered by Eiji Toyoda and Taiichi Ohno. This has resulted in the rise of Japan as a global economic power as the systems developed and followed by Toyota were soon adopted by the other Japanese companies and industries (Womack et al., 2007).

Toyota's outstanding performance has mesmerized the world, and the companies from different domains tried to adopt and follow the practices and principles developed by Toyota, but only very few of them were able to succeed fully. There have been thousands of professionals and engineers who visited the Toyota manufacturing plants in Japan and in the United States to learn these practices but were unable to understand fully. Some people feel that the success mantra of Toyota lies in the cultural roots of Japan but that is not true as there are few Japanese companies (such as Honda and Nissan) that could not replicate the same principles in their practice. It seems that the tools and practices have not been fully understood by the researchers. The systems need to be understood holistically; the activities, connections, and production flow in the Toyota systems are rigidly scripted, yet their operations are flexible and adaptable to accommodate variations. Rigid specifications are the very thing resulting in flexibility, adaptability, and creativity in the system (Spear & Bowen, 1999). Lean thinking can be summarized in just five words (Hill, 2012):

Simple > Visual > Error-proof > Wasteless > Standard

Hill (2012) further explains lean is about making processes simple and visual. Once processes are simple and visual, it becomes easier for people in the "Gemba" to make them error-proof and to eliminate waste. Once processes are simple, visual, error-proof, and wasteless, it is important to make them standard so that the benefits are sustainable. Lastly, respect for people before, during, and after the process improvement activity is critical to success (Hill, 2012).

1.3   Lean and different approaches

Lean manufacturing, lean principles, lean tools, lean philosophy, lean management, lean thinking different phrases are being used by different authors and practitioners often contradicting each other, however most of the time with the same objective of removing waste and/or production leveling. Lean manufacturing is described as an interconnected and interdependent set of five key elements: manufacturing flow, organization, process control, metrics, and logistics (Feld, 2001, pp. 3−4). Each of these elements contains a set of lean principles that, when working together, all contribute to the development of a world-class manufacturing environment.

Manufacturing flow
1. Product/quantity assessment (product group)
2. Process mapping
3. Routing analysis (process, work, content, volume)
4. Takt calculations
5. Workload balancing
6. Kanban sizing
7. Cell layout
8. Standard work
9. One-piece flow

Process control
1. Total productive maintenance
2. Poka yoke
3. SMED
4. Graphical work instructions
5. Visual control
6. Continuous improvement
7. Line stop
8. SPC
9. 5S housekeeping

Metrics
1. On-time delivery
2. Process lead time
3. Total cost
4. Quality yield
5. Inventory (turns)
6. Space utilization
7. Travel distance
8. Productivity

Organization
1. Product-focused, multi disciplined team
2. Lean manager development
3. Touch labor cross-training skill matrix
4. Training (lean awareness, cell control, metrics, SPC, continuous improvement)
5. Communication plan
6. Roles and responsibility

Logistics
1. Forward plan
2. Mix-model manufacturing
3. Level loading
4. Workable work
5. Kanban pull signal
6. A,B,C parts handling
7. Service cell agreements
8. Customer/supplier alignment
9. Operational rules

Figure 1.1 Elements of lean manufacturing (Feld, 2001).

Manufacturing flow addresses the visible physical change; the organization addresses less visible changes like new roles and functions, new ways of working and multifunction teaming, etc. Process control is all about monitoring, controlling, stabilizing, and pursuing ways to improve the process. Metrics addresses result-based performance measures and logistics provides operating rules, mechanisms for planning, and controlling the flow of material. Fig. 1.1 lists the lean principles under five elements of lean manufacturing.

Another important aspect of implementing lean is whether different tools are to be implemented in any order of sequence. The comprehensive guideline suggested by Womack and Jones (Womack et al., 2007) and (Feld, 2001) suggests a three-stage process where each stage has specific lean principles and subsequent measures (refer Fig. 1.2).

Lean tools are differently categorized by different researchers based on their application. Rahman, Laosirihongthong, and Sohal (2010) grouped the 13 lean tools into three broad constructs: JIT (just-in-time), WE (waste elimination), and FM (flow management). JIT consists of six tools, reducing setup time, preventive maintenance, cycle time reduction, reducing inventory, new process equipment, and technology and quick change over techniques. WE consist of four tools, kanban, removing bottlenecks, poka yoke, and waste minimization. Finally, the last three practices, reducing production lot sizes, focusing on a single supplier and one-piece flow were identified under flow management practices (Gunarathne & Kumarasiri, 2017).

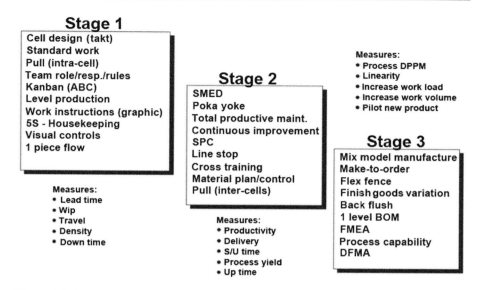

Figure 1.2 Stages of lean implementation (Feld, 2001; Womack et al., 2007).

Based on popularity and ease of use by the apparel manufacturers, the lean practices can be categorized as waste (Muda) elimination or control, and elimination or control of unevenness (mura). With this way of classification, lean tools such as kanban, Andon, value stream mapping (VSM), Yamazumi chart assist in the identification and steady elimination of waste in the process. On the other hand, lean tools such as takt time and Heijunka box are applied to smoothen the production flow, thereby steadily eliminating unevenness (Jana & Tiwari, 2018).

Further, the lean tools may also be categorized as either exploration tools or improvement tools. The lean tools (or techniques) that are applied to collect facts and figures systematically to assess the scope of improvement fall under exploration tools; lean tools such as Andon, VSM, overall equipment effectiveness, Yamazumi chart may be considered as exploration tools. The lean tools (or techniques that are applied to change or improve upon the product or process) fall under improvement tools; tools such as 5S, heijunka, jidoka, kanban, poka yoke, single minute exchange of die fall under the category of improvement tools (Jana & Tiwari, 2018). Once the exploration tools are implemented to understand the as-is scenario, the improvement tools are applied to improve the condition. The categorization of the lean tool as exploration and improvement further justifies the implementation order of lean tools found in organizations. A study in Sri Lanka (Silva, Perera, & Samarasinghe, 2011a) found that most of the industries (90%) tend to implement 5S and other visual management tools (like VSM) initially, which can easily make visible shop floor changes and quickly influence the financial status of the organization. Next, operation stability and continuous flow are achieved via takt time, one-piece flow, cellular manufacturing, etc. (refer Table 1.1). Heijunka is implemented at last which requires stability, standardization, and pull production.

Table 1.1 Stages of implementation of lean tools (Silva et al., 2011b).

Stages of implementation	Lean tools implemented
1	5S, VSM
2	Takt time, one-piece flow, cellular manufacturing, SMED, team work
3	Kaizen, pull system, standard work, kanban, visual displays and controls
4	TPM, JIT, supplier integration
5	Jidoka, poka yoke, problem-solving
6	Heijunka

JIT, Just-in-time; *SMED*, single minute exchange of die; *TPM*, total productive maintenance; *VSM*, value stream mapping.

Lean Sigma is another common phrase in lean terminology. Zhang, Hill, and Gilbreath identified five different views of lean sigma programs. From the metric point of view, lean sigma is about maximizing the sigma level (minimizing the defect rate or some other process capabilities metrics), often with the target of six sigma (or 3.4 defects per million opportunities). From the tool point of view, lean sigma is about applying managerial tools (e.g., brainstorming) and statistical tools (e.g., design of experiments) to problem-solving. From the project point of view, lean sigma is about defining and executing lean sigma projects with a black belt or green belt project leaders using the define, measure, analyze, improve, and control (DMAIC) five-step problem-solving methodology. From the program point of view, lean sigma is about a program management office that finds and prioritizes problems that need to be addressed and then charters and resources project to address those problems. From the philosophical point of view, it is about building a sustainable culture of leaders who are relentless about continuously improving processes to eliminate waste and defect (Wilson et al., 2013).

Lean philosophy results in enormous benefits to the organization. The biggest and the most significant improvement is a change in the attitude of taking up the issues and challenges. "The way people look at the things" is very vital for imbibing competitiveness in the system. With a lean approach, people seek betterment in the existing product or the process by considering issues, not as a problem but an opportunity for improvement. According to Conner (1993), "People can only change when they can do so. Ability means having the necessary skills and knowing how to use them. Willingness is the motivation to apply those skills to a particular situation." Adopting a lean approach in a system should satisfy the need for ability and willingness (Feld, 2001).

Lean-based set-ups are capable of creating an interdependent support system for various components of manufacturing, which results in reduced operating and administrative expenses. The other impact of the same is effective control of the day-to-day activities by focusing on point(s) of waste generation in the process. But, this is only possible where there is a when the system is robust in terms of its flexibility, responsiveness, ability to predict, and its consistency. Any lean approach-based process design should aim to achieve all these qualities simultaneously.

It is very important to understand that being lean is not a short-term goal but a long-term strategy or mission for continuous improvement. There may be situations where short-term results are not promising as expected but it does not mean the failure of the lean approach. The initial setback should be accepted in a long journey of improvements to achieve larger key benefits. Such setback should not be considered as a reason to scale-down the lean-based mission of overall improvements. Some of the key considerations should be ensured whenever there is a lean initiative:

Commitment from the top management: Constant and continuous commitment from the top management of the organization is one vital factor for the success of lean initiatives (Bowes, 2009). At many times, it is observed that the top management is only proactive at the time of introduction to lean initiatives and pushing for improvement only in the beginning and later their focus is shifted to other things. This results in an unattended, direction-less, incomplete mission that is just handed over to the subordinates. In a long journey of lean improvement, such an approach from the top management is just undesirable as it severely affects the morale of the employees involved and may fail due to lack of proper monitoring and decision making for further improvement (Ortiz, 2006).

Involvement of all the stakeholders: As the lean initiatives are very much interconnected with each other, and performance of one initiative (maybe in respective to any product or process) affects the outcome of other initiatives. In such a scenario to get the desired results with overall improvement, it becomes very important to involve the extended network of people such as suppliers and vendors, subcontractors, different departments of the organization, all the levels of management, and the buyers.

Training and education: Training and education play a vital role in transforming the workforce into a bunch of result-oriented teams. This can be very well understood from the quote from Fujio Cho, Honorary Chairman of Toyota Motor Corporation, "We get brilliant results from average people managing brilliant processes — while our competitors get average or worse results from brilliant people managing broken processes." The secret of success lies in the training, education, and coaching of the people (Technopak Advisors, 2011). The money, time, and efforts consumed in training are not an expense but an investment for a brighter future. Training and skill development of newer methods and practices, machine and equipment, newer domains of expertise make workforce better prepared and competent (by selecting the most appropriate lean tool or tools which can yield the best results) to future challenges in the dynamic world. With appropriate training, employees not only can handle the pressure situations more effectively and efficiently, but they are also mentally more open (welcoming) to adapt and excel.

Step-by-step assessment in a phased manner: Regular assessment of the progress done so far against the targeted goal and taking the necessary course of action is an essential requirement for any improvement process. It is very important to check if the changes are going on in the right direction or not. A step-by-step assessment may reveal insights indicating the need for fine-tuning or readjustment if any. Following the PDCA (plan—do—check—act) approach is one of the key recommendations of any lean-based initiative, as it gives a much-needed opportunity for timely assessment and opens up the ways for further improvement.

1.4 Apparel manufacturing and lean

Apparel manufacturing is a labor-intensive domain with a complex manufacturing environment. Right from the sourcing of raw material to the final shipment, several factors affect the process outcome. Despite all kinds of technological advancements, apparel manufacturing is still primarily driven by human resources only, which is in contrast to other industries (e.g., automobile manufacturing) where the application of lean practices is an established practice. This challenging environment presents a huge scope of improvement in many of the areas of an apparel business, which makes lean thinking and its tools even more relevant. This has pushed for transformation in the manufacturing systems and practices even further. Lean-based manufacturing set-ups with prime focus on waste minimization. Such models are sustainable aiming for continuous value addition in the final product.

As far as manufacturing is concerned, fundamentals in the newer lean-based systems are more or less the same only. It starts from creating a product with a value that customers are keen to buy at a higher price than of its manufacturing cost. But the difference in approach is on cutting down the cost by maintaining or even increasing the value of the product (Beaumont, Thibert, & Tilley, 2017).

Apparel manufacturing can be broadly classified into three stages: preproduction, production, and postproduction. Preproduction activities involve order confirmation, sample development, material procurement, planning of the order quantities, etc., while production is generally considered from the bulk spreading and cutting of the fabrics to the product output in form of semifinished or finished product. Postproduction activities involve finishing, packing, and shipment of the products. In a typical apparel manufacturing environment, most of the resources are involved in production and postproduction activities, and it is generally considered as core apparel manufacturing. This may be a probable reason that most of the lean tools and techniques are applied in these areas only. Another reason for the same may be that the actual conversion of raw materials into finished products happens in the production and postproduction only; hence, these areas are of the key focus as far as manufacturing is concerned.

Until the 1990s, it was only the automotive industry that had adopted TPS. Since then, it has spread into aerospace, general manufacturing, consumer electronics, healthcare, construction, and more recently to food manufacturing and meat processing (Silva, Perera, & Samarasinghe, 2011b). Lean initiatives at industry level can be traced in two ways: first, studies conducted by researchers which are published in journals and second, news items and articles published in trade journals. If we trace back the beginning of lean implementation in the Asian apparel manufacturing scenario, Sri Lankan industry took the initiative, lean was first implemented as an improvement method at MAS Intimates Linea Clothing Pallekale in 1996 (Gunarathne & Kumarasiri, 2017). Japan Sri Lanka Technical and Cultural Association (JASTECA) organizes and conducts national competitions in 5S, Kaizen, and corporate social responsibility culminating in the annual "JASTECA Awards." Sri Lanka's apparel manufacturer's dominance and passion for lean is obvious from the fact that between 1996 and 2018 apparel manufacturers from Sri Lanka have

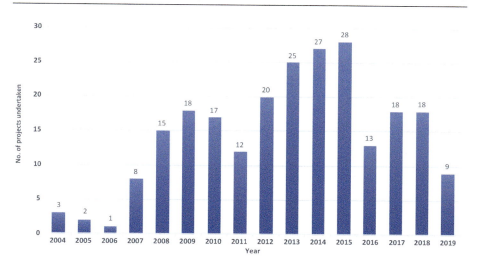

Figure 1.3 Projects undertaken in Lean Manufacturing at NIFT India. *NIFT*, National Institute of Fashion Technology.

bagged 35 out of 78 (or 44%) of the Taiki Akimoto 5S Awards amongst all industry sectors [Japan Sri Lanka Technical and Cultural Association (JASTECA), 2020].

Around the same time, the Indian industry also started implementing lean. National Institute of Fashion Technology (NIFT) is a premier institute in Southeast Asia, offering Bachelor's and Master's program in Fashion Technology. Students undertake dissertation projects mostly in association with the industry, researching on the topical issues faced by the industry. An analysis of lean projects undertaken by the Bachelor's and Master's students between the years 2000 and 2019 reveals interesting insight. Indian apparel manufacturing organizations started exploring lean only from the year 2004–05 (following neighboring Sri Lanka) and peaked between 2012 and 2015 and then gradually declined. The graph shown in Fig. 1.3 explains the trend.

There have been several lean tools applied in apparel manufacturing, though the number of lean applications practiced in this field is not as exhaustive as it is applied in other industries, especially automobile manufacturing. The reason for this limited application of lean practices may be that automobile manufacturing was the very first industry where lean practices were first applied, and the apparel manufacturing was never considered as a serious sector. But surely, in recent years, there has been a significant rise in adopting and implementing lean practices, though there is still a long path to travel. The lean movement is more consultant-driven and less academic-driven. Different organizations list down varying number and name of lean tools, principles, elements (Jana & Tiwari, 2018). The table (Table 1.2) lists the commonly used lean tools and their use in apparel manufacturing.

An analysis of news items published at www.apparelresources.com from the year 2011 till 2019 reveals that just 103 apparel manufacturing organizations

Table 1.2 Key lean tools used in apparel manufacturing (Jana & Tiwari, 2018).

Lean tool	Application
5S	Organizing the workplace
Andon	Visual indicator signaling an abnormal situation
Heijunka (level scheduling)	Load leveling or production leveling to eliminate/control unevenness (mura)
Jidoka	Automation with human intelligence also referred to as "intelligent automation" or "humanized automation"
JIT	Reducing flow times within production as well as response times from suppliers and to customers by adopting pull based mechanisms
Kanban (pull system)	Pull based (demand-driven) inventory replenishment approach
KPI	Measuring and monitoring performance (achieving targeted results) to evaluate the success
OEE	Overall performance measurement in context of the availability of the process or equipment to its the performance and quality
PDCA	A systematic and repetitive strategy (holding on, assessing the improvement, monitoring, and taking informed decisions for further improvement) for the continual (incremental) improvement of a product or process
Poka yoke (error proofing)	Error proofing or mistake proofing in a product or process
Root-cause analysis	Investigating the problems or issues to trace out the core reason(s)
SMED	Quick change over
Takt time	Final output rate of the end product to meet customer demand
TPM	Human resource based management approach to improve the overall effectiveness of the facility
Value stream mapping	Visual representation of an individual process in terms of its material and information flow according to specific production path from the start point to the end point
Yamazumi charts	Effective communication tool of a process in the form of stacked up bars or lad charts

JIT, Just-in-time; *KPI*, key performance indicator; *OEE*, overall equipment effectiveness; *PDCA*, plan−do−check−act; *SMED*, single minute exchange of die; *TPM*, total productive maintenance.

reported the use of lean tools and techniques for competitive advantage, to improve productivity and flexibility, to reduce lead time and change over time, and creating a unique value proposition. Although most of the organizations mentioned the use of lean tools in general without mentioning what specific tools were being used, the three mostly used lean tools were found to be 5S, 7Waste, and VSM. The heat map diagram (refer Fig. 1.4) shows the number of companies from Southeast Asia which reported the use of lean tools.

It is important to note that although the large organized manufacturing is concentrated in Southern India, the bulk of companies reported are from Northern India,

	2011	2012	2013	2014	2015	2016	2017	2018	2019
Bangladesh	0	6	0	1	1	1	2	2	1
Cambodia	0	0	0	0	0	1	1	0	0
India (north)	0	14	14	0	17	9	0	2	0
India (rest)	0	0	0	0	2	0	0	0	0
India (south)	1	5	0	0	11	2	0	3	0
Myanmar	0	0	0	0	0	0	0	1	0
Pakistan	0	1	0	0	0	0	0	0	0
Sri Lanka	0	0	0	0	0	1	0	0	0
Vietnam	0	0	0	0	0	2	0	1	1

Figure 1.4 Number of companies from Southeast Asia reported use of lean tools (2011–19).

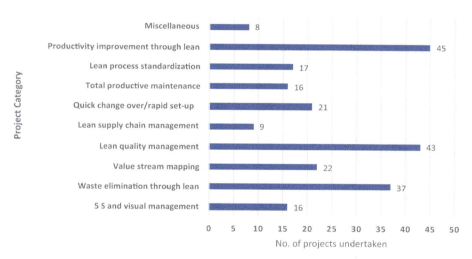

Figure 1.5 Types of lean projects undertaken at NIFT India. *NIFT*, National Institute of Fashion Technology.

this may be attributed to extrovert nature of companies in Northern India, presence of more number of MSME in the north (in comparison to few large enterprises in South) as well as geographical proximity of the media to the region.

If we correlate the number of projects undertaken by Indian organizations from 2011 to 2019, 170 projects were initiated through students, but only 80 were reported in media (www.apparelresources.com). Although there are organizations that are mutually exclusive between two sets, it can be concluded that nearly 50% of the lean projects are abandoned after initial exploration. Globally reported reasons behind barriers or challenges in implementing lean are backsliding, lack of implementation know-how, supervisor, and middle management resistance (Silva et al., 2011a).

Further analysis (refer Fig. 1.5) of lean projects undertaken by Indian apparel manufacturing organizations (through students' projects at NIFT India) reveals the propensity of the use of different lean tools by the organizations. It is clear from Fig. 1.5 that although productivity and quality improvement still the main agenda for lean implementation, three specific tools—waste minimization, VSM, and quick change over (during style change)—were majorly explored by the Indian industry.

1.4.1 Lean implementation in apparel manufacturing: challenges

Lean is all about producing more with lesser resources. In the cutthroat competitive environment, where it is not only a matter of doing business and survive but to strive continuously for achieving operational excellence, optimum utilization of resources (manpower, material, machines, information, space, etc.) to get the desired on-time quality output is the only way, and being lean is the most appropriate approach. Lean philosophy is very much capable of solving out the issues and challenges in a practical environment and that has been witnessed again and again in several industries. According to Locher (2008), successful adaption to lean practices results in a 40%−90% reduction in lean times, 30%−50% reduction in process times, and 30%−70% enhancement in the quality performance (Locher, 2008). Despite such promising improvement, the adapting lean approach and its applications are not truly observed in the apparel manufacturing where lean initiatives are yet to be accepted whole-heartedly. In apparel manufacturing, there are several myths and misconceptions related to adopting the lean way, and that works as challenges in successful lean implementations. Most of the apparel manufacturing plants still operate with the traditional approach of hiring and firing. Cutting down the costs by firing people, downsizing, and working with higher work-in-progress (WIP) are still considered as key tricks of being economical. Apart from it, typical apparel manufacturing set-ups struggle with additional issues such as poor quality, lower productivity levels, poor space utilization, longer assembly lines with higher levels of WIP, higher machine downtimes, and poor work allocation due to lack of work standardization. This results in longer lead times and higher operating costs. Some of the majorly observed challenges to lean implementation in apparel manufacturing are discussed as follows:

Lean is for standardized work: Many people believe that lean tools and techniques are only useful for repetitive and standardized work where the level of automation is higher, while apparel manufacturing is primarily human driven with relatively lesser/minimal automation. This brings the variability in doing the things and a possible source of errors. Here, it is important to understand the human angle of lean, which encourages human intervention but in a systematic manner. It also enhances the knowledge level and self-discipline with self-efficacy that results in output consistency (de Treville, Antonakis, & Edelson, 2005). A lean approach in the apparel manufacturing environment may help in identifying the root causes of such variability and sources of errors.

Apparel manufacturing is a different ball game: Many people dealing in apparel manufacturing believe that the nature of apparel manufacturing is different from other industries. Up to some extent it is true as well, considering the process complexities (different buyers and products have different specifications), and multiple factors involved (many of which are related to raw material suppliers and buyers) affecting the performance. One common example is the late arrival of fabric and trims from the suppliers, and delays in approval for preproduction activities at the sampling stage. Many times, such issues are beyond the control of apparel manufacturers; however, these inefficiencies severely affect the manufacturer's performance. This also makes people demoralized and they develop a feeling that no improvement shall work as the reason for all the failures are beyond their control. People consider it as one of the key reasons of nonsuitability of the lean approach in apparel manufacturing, but if observed closely, it can be realized that for most of the buyer and products the basic road-map of manufacturing is more or less same; of course, there are some deviations and that need to be addressed accordingly. Lean may help in bringing such variations on the same page (by effective measurement of failures and related factors) by streamlining and standardization of the activities.

Lean is a flavor of the month: Diffusion of lean initiative in an organization and sustainability of lean improvements for a longer duration is an important issue (Wickramasinghe & Wickramasinghe, 2016). There is research evidence where the lean initiative was not sustainable up to the satisfactory level; hence, these improvements were faded away or simply died after a few years from the introduction (Sohal & Egglestone, 1994). In apparel manufacturing, lean initiatives are not taken on a serious note. People generally consider it a short time activity of implementing some specific tools for a particular period, and after that, they bring back to their earlier work style. It is imperative to understand that lean is a never-ending mission, and it is a philosophy that is beyond just implementing a few tools and techniques. Considering lean as a flavor of a short time eventually result in unaccomplished goals, and then people start feeling that lean is not a remedy to their problems (Ortiz, 2006).

The lean mission is to be carried out by someone else: Change is always painful and difficult. Most of the people prefer working in their comfort zones (Technopak Advisors, 2011). Also, their mental blocks do not allow them to adapt and accept newer things. Fear of change is one easily observed scenario, whenever there is a question of embracing change for improvement (Ortiz, 2006). They feel that they are already doing their part in a good enough manner and there is no need for improvement. Hence, they take lean initiatives as an exercise to be undertaken by others. And the irony is that people from other departments and processes also think in the same manner. This results in a situation with a lesser commitment to improvements through lean initiatives.

We know the problems and also know its solutions: Most of the apparel manufacturing operations are run by the traditional approach, where scientific and logical approaches are missed out many times. People are generally aware of the problems and challenges but due to the tight delivery deadlines and work pressure, they generally end up in firefighting or to the quick-fix or Band-Aid to the problems. Such solutions are most of the time are instant results and generally not sustainable (Ortiz, 2006). This is a reactive approach to the issues where the root-cause of problems is never searched out. Problems remained unresolved and it pops up from time to time and the organizations waste a huge amount of resources in fixing up with such frequently occurring problems. With longer exposure to such firefighting mode, people start feeling it as "normal" and treat it as is the way the apparel industry works (Ortiz, 2006).

Why go for lean when we are already doing good: This is something very common with the apparel manufacturing organizations doing good business. People are generally not keen on any kind of intervention till they observe any problem, and they believe in the "Why to fix it, if it is not broken?" approach (Ortiz, 2006). People (especially those who are doing good or satisfactory) generally do not want to be questioned or reviewed on their work, hence they are reluctant to change though it is for improvement. Further, most of the organizations plan for expansions (in terms of increasing their production capacities or dealing with different products) rather focusing on the improving utilization of their existing available resources. This results in huge financial implications in expansions, and in absence of the best practices (such as adopting a lean approach), it puts increased financial pressure on the existing core business. This situation can be avoided or handled better with the lean approach where the key focus is on waste reduction and elimination bringing improvements through scientific resource management.

Lean is just a buzz word: Many of the people (most of them which are reluctant to change) give this excuse that becoming lean is something not possible in practice, and more so in an apparel manufacturing environment. They treat lean philosophy as a theoretical concept only with no or lesser practical applications. Also, due to lack of awareness and proper knowledge, people take lean as a gimmick (just limited to cleanliness, 5S posters, and few slogans of improvement) used to attract buyers and getting new business. Further, they treat lean as something very complicated and only practiced by the consultants to run their business. Such people are generally not well aware of the miracles achieved on shop-floors using lean. Here, it is imperative to understand that the key improvements are to be led by the workforce itself and no outsider (such as a consultant or lean practitioner) can bring change, till the workforce does not want to adopt it. In such scenarios, people are to be introduced on practical applications of lean (maybe by sharing success stories) and faith to be built-on toward adapting lean transformation.

Lack of know-how: It has also been observed (especially with apparel manufacturing) that sometimes people are keen to go for a lean approach but not clear on its practical execution. There has been a lot of literature, published research available that explains lean concepts, and tools and techniques, but very few of them focus on the practical execution part. Further, the step-by-step methodology (when to use which lean tool and how) seems missing and in such situations organizations are dependent on the outside consultants to lead the lean mission (Bowes, 2009). Sometimes, the consultants, being an outsider, are not aware of the holistic view of the organizations, activities, and processes. Also, the consultants have their objectives and they take lean initiatives as just one another project of their profession. This results in an incomplete journey of achieving improvement goals through lean.

1.5 Conclusion

It is imperative to understand that lean is not merely a waste reduction or process improvement approach but it is a philosophy and a lifestyle. It needs to be lived to achieve overall improvements. Human involvement from all the levels of the hierarchy is a must in such lean missions. People-centric approach, where every person involved directly or indirectly is conscious and aware of not only how to control the waste but also about the source of waste creation, is the need of the hour. It is mentioned in many literature that researchers from different parts of the world (especially from the United States) wanted to know the secret of success of Japanese manufacturing excellence and spent years of their life learning the trick but were unable to get the whole of it. The reason for the same was that it was in the blood of Japanese practitioners not in the form of a written set of rules, principles or theories, or the mathematical formulas. It is imperative to constantly challenge and push the activities and systems to achieve higher levels of excellence, through continual innovation and improvements. And there is no stop to this journey of seeking better always, this can only be achieved when you live it. Going lean is an endless journey seeking perfection in all the aspects. It took 70 long years for Toyota to become a world leader in automobile manufacturing, and they are still on the way to achieve further improvements. Hence, it may be concluded that one needs to live the lean way to get the real benefits of lean and it can only be done by traveling on a never-ending path of continuously challenging the existing solution and seeking better always.

References

Beaumont, D., Thibert, J., & Tilley, J. (2017). *Same lean song, different transformation tempo*. Retrieved from https://www.mckinsey.com/business-functions/operations/our-insights/same-lean-song-different-transformation-tempo. (Accessed 27 August 2018).

Bhasin, S., & Burcher, P. (2006). Lean viewed as a philosophy. *Journal of Manufacturing Technology Management, 17*(1), 56—72.

Bowes, P. F. (2009). Lessons from lean implementation. *Stitch World, 7*(12).

Conner, D. R. (1993). *Managing at the speed of change: How resilient managers succeed and prosper where others fail* (1st ed.). New York: Random House.

de Treville, S., Antonakis, J., & Edelson, N. (2005). Can standard operating procedures be motivating? Reconciling process variability issues and behavioral outcomes. *Total Quality Management, 16*(2), 231—241.

Ebrahim, S., Krishnakanthan, K., & Thaker, S. (2018). *Agile compendium—Enterprise agility*. McKinsey & Company.

Feld, W. M. (2001). *Lean manufacturing: Tools, techniques, and how to use them*. St. Lucie Press/APICS Series on Resource Management.

Gunarathne, G., & Kumarasiri, W. (2017). Impact of lean utilization on operational performance: A study of Sri Lankan textile and apparel industry. *Vidyodaya Journal of Management, 03*(1), 27—41.

Hammer, M., & Somers, K. (2015). *More from less: Making resources more productive*. McKinsey & Company.

Hill, A. V. (2012). *The encyclopedia of operations management.* NJ: Financial Times Press.

Jana, P., & Tiwari, M. (2018). Industrial engineering and lean manufacturing. In P. Jana, & M. Tiwari (Eds.), *Industrial engineering in apparel manufacturing—Practitioner's handbook* (pp. 267–283). New Delhi: Apparel Resources Pvt. Ltd.

JASTECA. (2020). *Japan Srilanka technical and cultural association.* [Online]. Retrieved from https://www.jasteca.org/past-winners.php. (Accessed 12 January 2020).

Lathin, D., & Mitchell, R. (2001). Lean manufacturing. *American Society for Quality Journal, 2–9.*

LEI. (2000). *Making things better through lean thinking and practice.* L.E. Institute, Producer, & Lean Enterprise Institute. Retrieved from https://www.lean.org/WhoWeAre/LeanPerson.cfm?LeanPersonId = 1. (Accessed September 2010).

Locher, D. A. (2008). *Value stream mapping for lean development: A how-to guide for streamlining time to market.* New York: Productivity Press.

Ortiz, C. A. (2006). *Kaizen assembly designing, constructing, and managing a lean assembly line.* New York: CRC Press.

Rahman, S., Laosirihongthong, T., & Sohal, A. (2010). Impact of lean strategy on operational performance: A study of Thai manufacturing companies. *Journal of Manufacturing Technology Management, 21*(7), 839–852.

Schonberger, R. (2007). Japanese production management: An evolution-with mixed success. *Journal of Operations Management, 25*(2), 403–419.

Silva, N., Perera, H. S. C., & Samarasinghe, D. (2011a). *Factors affecting successful implementation of lean manufacturing tools and techniques in the apparel industry in Sri Lanka.* Retrieved from http://ssrn.com/abstract = 1824419. (Accessed from January 2020).

Silva, S. K., Perera, H., & Samarasinghe, G. (2011b). Viability of lean manufacturing tools and techniques in the apparel industry in Sri Lanka. *Applied Mechanics and Materials, 110–116,* 4013–4022.

Sohal, A. S., & Egglestone, A. (1994). Lean production: Experience among Australian organizations. *International Journal of Operations and Production Management, 14,* 35–51.

Spear, S., & Bowen, H. K. (1999). Decoding the DNA of the Toyota production system. *Harvard Business Review, 77*(5), 96–106.

Technopak Advisors. (2011). *Lean manufacturing: The way to manufacturing excellence.* Gurgaon: Technopak Advisors Pvt. Ltd.

Wang, J. X. (2011). Introduction: Five stages of lean manufacturing. In J. X. Wang (Ed.), *Lean manufacturing: Business bottom-line based* (pp. 1–2). Boca Raton, FL: CRC Press.

Wickramasinghe, G., & Wickramasinghe, V. (2016). Effects of continuous improvement on shop-floor employees' job performance in lean production: The role of lean duration. *Research Journal of Textile and Apparel, 20*(4), 182–194.

Wilson, R., Hill, A., & Glazer, H. (2013). *Tools and tactics for operations managers (collection).* NJ: Financial Times Press Delivers.

Womack, J. P., & Daniel, T. (2003). *Lean thinking: Banish waste and create wealth in your corporation.* New York: Free Press.

Womack, J. P., Jones, D. T., & Roos, D. (2007). *The machine that changed the world: The story of lean production* (Reprint ed.). Free Press.

Lean terms in apparel manufacturing

2

Prabir Jana[1] and Manoj Tiwari[2]
[1]Department of Fashion Technology, National Institute of Fashion Technology, New Delhi, India, [2]Department of Fashion Technology, National Institute of Fashion Technology, Jodhpur, India

2.1 Introduction

As explained in Chapter 1, Lean Management in Apparel Manufacturing, lean has different approaches. Experts have explained them through different elements, principles, as well as tools. Over the years, different academicians, consultants, and industry practitioners have also combined multiple approaches and tools, and given a different name to the combined process. Today, the Lean Lexicon (fifth edition, 2014) by Lean Enterprise Institute lists 207 terms, and the popular website www.allaboutlean.com lists as many as 447 terms. According to the aforementioned website, many of the terms seem to have been invented solely to sell more products and services to industry using fancy buzzwords (all about lean, 2020). In this chapter, we have tried to list those lean terms that are popularly used or have potential application in apparel manufacturing.

We have also categorized each term to explain if it is a tool, a philosophy, a manufacturing system, or a simple measure. The nature of the tool is further categorized into diagnostic, improvement, training, or monitoring & control tool. The improvement tool is further categorized into whether it follows waste minimization or production leveling approach. Many of these terms are also found to be systemic approaches involving multiple tools, while some others are just formulae. The "Also see" section lists terms that are either synonymous or follow similar execution approaches, process steps, or outcomes. Wherever possible, the source language and originator of the term is mentioned. The terms are listed in alphabetical order for easy reference.

2.2 Lean Terms

2.2.1 A3 Report

A3 report (prepared on A3 size paper 297 mm $\times$ 420 mm or 11 in. $\times$ 17 in.) generation is a standard established practice in lean management used widely in problem-solving exercises, analysis, status, and planning exercises with graphical tools.

Category: Diagnostic tool
Also see: VSM (value stream mapping)

Lean Tools in Apparel Manufacturing. DOI: https://doi.org/10.1016/B978-0-12-819426-3.00010-2
© 2021 Elsevier Ltd. All rights reserved.

2.2.2 A-B Control

A-B control is a mechanism to regulate and synchronize the work between two work-stations or operations to avoid overproduction. The overproduction is avoided by using a balanced and synchronized way of working of the machines. The system [Machine A, Machine B, and the medium to contain and carry the work-in-progress (WIP) which is generally a conveyor] works only when there is a demand from Machine B, WIP is available (in between Machines A and B), and the Machine A is full.

Category: Waste minimization and improvement tool

Also see: Chaku-Chaku, continuous flow, just in sequence (JIS), Hanedashi, Kanban, one-piece flow

2.2.3 Affinity Diagram

Affinity diagram is a brainstorming method developed by Jiro Kawakita. It is also referred to as the KJ method in the name of its developer.

Category: Diagnostic tool

Also see: Visual factory

2.2.4 Agile Manufacturing System

Agile refers to an uncertain and complex environment with unanticipated demand patterns. This is synonymous with volatility and turmoil. Agile manufacturing systems are dynamic in responding to customer needs with faster learning and quick decision making. Such an agile system-based organization is designed for both stability and dynamism and it operates as a network of teams. Such systems are best suited for volatile, uncertain, complex, and ambiguous (collectively termed as VUCA) environments. Sometimes, these systems are also treated as the next (or advanced) version of lean manufacturing systems.

Agile manufacturing systems are also synonymous to such *quick response manufacturing* systems, which are referred to as the manufacturing systems aiming to fulfill the customer demand with a shorter lead time to fulfill the customer demands in a shorter time, such systems are designed to handle low volume high variation products.

Category: Manufacturing system

Also see: WCM (World Class Manufacturing)

2.2.5 Andon

Japanese term used for lamp. Andon is synonymous with visual control (maybe in the form of signboard) or visual feedback system for production updates/current status of achieved against the target, and sending alerts to all concerned whenever required. It also empowers workers to stop production in case there is an issue and some intervention is required to sort out the problem.

Andon line: *The term is generally used in combination to Andon (a Japanese word used for a lamp), and also called Andon Cord (maybe in the form of a line, a button, or a switch that is activated or pressed) used to indicate/communicate problems about a process.*

Category: Monitoring & control tool
Also see: Mieruka, visual factory

2.2.6 Assembly Cell

The idea of the assembly line was devised by Henry Ford in automobile manufacturing. Assembly line is a kind of manufacturing system based on the division of work when work or a piece moves from one workstation to the next and some exclusive operation or work is performed on the piece. Eventually, the output from the assembly line is a semifinished or a finished product. The iconic Model T from Ford was the first car manufactured using an assembly line.

Category: Manufacturing system
Also see: Cellular manufacturing, TPS (Toyota Production System), U-line

2.2.7 Autonomous Maintenance

A total productive maintenance (TPM) practice encouraging operators to perform basic maintenance activities by themselves.

Category: Waste minimization and improvement tool
Also see: TPM, QC (Quality Circle)

2.2.8 Baka Yoke or Poka Yoke

Ba is a Japanese term used for a place, spot, or space.

Baka yoke is another Japanese term used for idiot-proofing or to the activities that cannot be done wrongly or in an incorrect manner. It is a previous version of the term poka yoke used for mistake-proofing or error proofing.

Category: Waste minimization and improvement tool
Also see: Built-in-quality, first time through (FTT), QFD (quality function deployment), ZD (zero defects)

2.2.9 Batch-and-Queue

The term is used for keeping some inventory or buffer levels before moving forward to the next activity or the next process. Generally, such inventories are maintained for a definite time interval, where the inventories (maybe in the form of WIP) have to wait in a queue. Sometimes such inventory levels are referred to as *batch-and-push* as well.

Category: Waste minimization and improvement tool
Also see: Kanban, pull system, push system

2.2.10 Baton Touch

The term is used to organize a U-line-based manufacturing setup where operators are multiskilled and handle more than one task. The workflow is maintained in a forward direction to get the semifinished or finished output from the line.

Category: Production leveling and improvement tool

Also see: Cellular manufacturing, bucket brigade, Rabbit Chase, Shoujinka, U-line

2.2.11 Bottleneck

A bottleneck is referred to as an interruption in the flow/process where one processor workstation is unable to produce sufficient/desired output to be used as an input in the next process. Thus a bottleneck in a system affects the output where the system is unable to work at its planned capacity. Process imbalances, wrong resource allocation, sudden breakdowns, material unavailability, lack of skill or competency, over- or underproduction against targets are some of the major reasons for bottleneck creations.

Bottleneck walk: Bottleneck walk is referred to as tracing out and eliminating the bottleneck from the process. This is generally achieved by process observation, checking the inventories, and taking suitable corrective actions to control the further accumulation of WIP at the bottleneck point.

Category: Production leveling and improvement tool

Also see: Line balancing, Theory of Constraints (TOC)

2.2.12 Bucket Brigade

Bucket brigade is sometimes known as "bump-back" or "bouncing line." It is an elegant, self-organizing method to solve the problem of uneven workloads on a production line. Every worker walks with their part along the line until they meet the next worker or the end of the line. Then, the worker walks back until they meet the preceding worker (or the beginning of the line) and takes over the part. The cycle repeats. A worker with little workload will handle more processes, a worker with great workload fewer.

Category: Production leveling and improvement tool

Also see: Cellular manufacturing, baton touch, Rabbit Chase, Shoujinka, U-line

2.2.13 Buffer

Buffer is kept as a stock or WIP between two processes or workstations to avoid idleness due to run-out of work which may result in a situation of unevenness (Mura) in the system.

Category: Waste minimization and improvement tool

Also see: Pull system, push system, WIP, Kanban

2.2.14 Built-in-Quality

Built-in-quality is referred to as producing the right first time to ensure high quality to the end product or service. It focuses on not forwarding the defective parts or products to the next process; hence, the aim is to minimize the cost of poor quality.

Category: Philosophy

Also see: DQC (delivery, quality, cost), FTT, Jidoka, QFD, Six Sigma (6 Sigma), ZD

2.2.15 Bullwhip Effect

The Bullwhip Effect is also called Forrester (as the concept of Bullwhip effect was given by Jay Wright Forrester in 1961) effect and whiplash or whipsaw effect. This phenomenon of demand amplification refers to the inventory fluctuations due to a shift in customer demand. This is one of the most common problems observed in supply chains.

The products and services are produced based on the forecast, and most of the time forecasts are not accurate. This leads to fluctuations in demand. To reduce possible demand fluctuations and to avoid losses, suppliers tend to increase their inventory levels.

In the case of an increase in demand from the customer the downstream participants increase orders. However, when the demand is reduced, inventory levels are not able to be reduced accordingly. This brings inefficiencies in the supply chains and creates imbalances that lead to the generation of different wastes.

Category: Diagnostic tool

2.2.16 Cellular Manufacturing

Cellular manufacturing is one of the most popular forms of manufacturing in lean setups worldwide. A cell capable to accommodate product variations is arranged with specialized machinery along with multiskilled operators who can perform as a team. In this way a cell (set of machines with multiskilled operators) can produce products. The machine and equipment arrangements ensure a smooth, uninterrupted flow of work resulting in a seamless transition of raw materials to the semifinished or finished products.

Category: Manufacturing system

Also see: Assembly cell, TPS, U-line

2.2.17 Chaku-Chaku

Chaku-Chaku is a Japanese term used for loading pieces one by one. In the context of lean manufacturing, this a method for single-piece flow, where the output of one machine is ejected automatically and collected by an operator to be loaded to the next machine without losing any time. In such an arrangement, the workstation operates independently and ejects the piece, and the operator collects the same in a synchronized manner enabling smooth multimachine handling.

Category: Manufacturing system

Also see: A-B control, continuous flow, Hanedashi, Kanban, one-piece flow

2.2.18 Chalk Circle

Chalk Circle is a technique used in lean management for observing things on the manufacturing floor. Such exercise is also called circle exercise or standing in the circle. A circle is drawn at a designated place, and a person (maybe an engineer, lean practitioner, or a lean trainee, etc.) is asked to stand in the circle and observe things. Now and then, the coach stops by and asks what the person has seen. This exercise usually takes a few hours before the coach is satisfied with what the person has seen. This exercise and teaching method are one of the most famous techniques by Taiichi Ohno (considered to be the father of the TPS).

Category: A training tool

2.2.19 Changeover Time

Changeover is generally taken as style changeover when the last product of a particular style comes out and the first piece of the next style is loaded in the system. It includes run down (getting the line empty from the previous style), setup (modification in layout, machine setup, machine settings, etc.), and start-up (which includes learning curve for the new style).

Changeover time is referred as the time required by a production line or module or a system for switching over from one style to the next style. This is generally considered as the time between the final output of the last piece of the previous style to the output of the first good piece of the next style.

Category: Waste minimization and improvement tool

Also see: Dandorigae, Jundate, SMED (single minute exchange of die)

2.2.20 Continuous Flow

Continuous flow is referred to as producing one item or product at a time. The activities or the processes are performed sequentially and moving one item at a time (or a small and consistent batch of items) through a series of processing steps as continuously as possible, with each step making just what is requested by the next step. One-piece flow, single-piece flow, and make one, move one are some other synonymous terms to the continuous flow.

Category: Manufacturing system

Also see: Chaku-Chaku, Hanedashi, Kanban, one-piece flow, TOC

2.2.21 CONWIP (Constant Work In Process)

CONWIP (Constant Work In Process) is a type of production control similar to Kanban, but without a fixed part number assigned to the card.

Category: Monitoring & control tool

2.2.22 Customer Takt

Takt is a German term used for rhythm or beat of a drum. The Customer Takt is referred to available work or production time during a specified period divided by customer demand. This is required to determine the targeted output rate to meet customer demand.

Category: It is a measure

Also see: Cycle time, lead time, System Takt, throughput

2.2.23 Cycle Time

Cycle time is referred to as the actual observed time required to produce a part of an operation in a product or the complete product or process.

Category: It is a measure

Also see: Customer Takt, lead time, pitch time, System Takt, throughput

2.2.24 Dandorigae

This is a Japanese term used for a flexible setup or changeover the tools or dies to produce products from a machine. Some specific attachments or die are changed in a machine that can produce a large number of quantities of a specified product.

Category: Waste minimization and improvement tool

Also see: Changeover time, Jundate, SMED

2.2.25 Dashboard

The dashboard is widely used in lean manufacturing environments to summarize relevant KPI of a process or system. From the management point of view, it is a useful method to have a quick look (glance) on the relevant data using pictorial/ graphical displays of the current status of the progress.

Category: Monitoring & control tool

Also see: Andon, Mieruka, visual factory

2.2.26 DBR (Drum–Buffer–Rope)

DBR is referred to as the drum–buffer–rope concept, which plays an important role in the TOC given by Eliyahu M. Goldratt. DBR is widely applied for scheduling and production planning in business environments as a tool for flow management.

Drum: The drum is referred to as the constraint or the limiting factor in the system. The constraint is the regulatory element that controls the output rate and throughput of the process.

Buffer: The buffer is referred to as the minimum level of inventory required to ensure uninterrupted output. Buffers also indicate the amount of time in which the required work-in-process should be available to ensure smooth, uninterrupted output.

Rope: The rope is referred to as an inventory control mechanism, which is used to communicate the amount of inventory that has been consumed by the constraint. The key function of the rope is to ensure the controlled release of inventory into the process without accumulation of excess inventory.

Category: Monitoring & control tool

Also see: Chaku-Chaku, Hanedashi, Kanban, milk run, one-piece flow, TOC

2.2.27 DMAIC (Define, Measure, Analyze, Improve, and Control)

DMAIC (define, measure, analyze, improve, and control) is a data-driven (focus on analyzing the data), quality improvement strategy. DMAIC is one of the core tools of 6 Sigma philosophy and is similar to the PDCA (plan—do—check—act) cycle of lean management.

Category: Systematic approach involving multiple tools

Also see: PDCA, LAMDA (look, ask, model, discuss, and act) cycle, Yokoten

2.2.28 Downtime

Downtime is referred to the idleness in a process or stoppage of some activity or operation. Downtime is non-value-added (NVA) to the process; hence it is not a favorable situation as it is linked with the generation of various types of wastes (Muda).

Category: It is a measure

Also see: MTBF (mean time between failures), MTTF (mean time to failure), MTTR (mean time to repair)

2.2.29 DPM (Defects per Million)

DPM is referred to as defects per million. DPM is used to communicate the defect rate. At a 6 Sigma level, there are 3.4 defects in 1 million parts or products.

Category: It is a measure

Also see: DQC, QFD, Six Sigma (6 Sigma), ZD

2.2.30 DQC (Delivery-Quality-Cost)

DQC is referred to as *delivery, quality, cost.* The key focus of DQC is improving the delivery performance of the end product to the consumer. DQC is also referred to as QCD (quality, cost, delivery).

QCDF is referred as to QCD + flexibility (F)

QCDMS is referred as to QCD + morale (M) + safety (S)

Category: Philosophy

Also see: Built-in-quality, FTT, Jidoka, QFD, Six Sigma (6 Sigma), ZD

2.2.31 ECRS (Eliminate-Combine-Rearrange-Simplify)

ECRS is referred to as *eliminate, combine, rearrange, simplify*; which is an approach for process improvement and optimization.

Category: Waste minimization and improvement tool

Also see: Dandorigae, Jundate, SMED

2.2.32 EOQ (Economic Order Quantity)

Economic order quantity (EOQ) is referred to as an ideal order quantity widely used in inventory management. This approach aims to minimize inventory holding costs and ordering costs through mathematical analysis of demand or consumptions, availability or supply of products/goods, and respective lead times.

Category: It is a measure

Also see: Buffer, WIP, supermarket

2.2.33 FIFO (First In First Out)

FIFO stands for first in first out. It is a kind of material arrangement in terms of material flow (delivery sequence) widely used in pull-based systems in a lean environment.

Category: Monitoring & controlling tool

Also see: LIFO (last in first out), pull system, push system

2.2.34 First Time Through (FTT)

FTT is synonymous to the right first time or first time right (FTR). It is referred to the number of products produced correctly (at its first attempt) without any defects. FTT or FTR is a ratio of the number of correct quality pass products to the total number of products produced in a given period. FTT is generally used as one of the KPI in lean manufacturing environments.

Category: Philosophy

Also see: Built-in-quality, QFD, ZD

2.2.35 Five S (5S)

5S (seiri—sort > seiton—set in order > seiso—shine > seiketsu—standardize > seitsuke—sustain) are an important concept of lean management. This is a tool applied for workplace arrangement and standardization to improve visibility of the problems and issues and helps in quicker corrective actions.

The standardization of workplace arrangement and sustained practice to follow the set standards (in terms of habit) ensure the right items in the right place. It contributes to enhanced productivity with quick location tracing of the required materials.

In addition to the 5S, an additional S (for safety) is added as the sixth S (collectively known as 6S).

Category: Waste minimization and improvement tool

2.2.36 Flow Shop

Flow shops are generally referred to as the manufacturing setups suitable for low mix high volume productions. In such facilities, machines or processes are arranged in a predefined sequence required to manufacture products.

Category: Manufacturing system

Also see: Gemba, Genchi Genbutsu, Genjitsu

2.2.37 FMEA (Failure Modes and Effect Analysis)

FMEA refers to the *failure modes and effects analysis* that is a proven technique to estimate the probability, severity, detection likelihood, and impact of problems on products. The numerical assessment of the risk is done by determining the *risk priority number* as:

Severity × occurrence × detection.

Category: Diagnostic tool

Also see: Root cause analysis (RCA)

2.2.38 Fordism

Fordism (a philosophy developed by Henry Ford) is a mass manufacturing concept based on the assembly line which works on the division of work to optimize resource utilization. A higher level of inventories, higher throughput time, inability to handle variations in the product (due to lack of flexibility), and higher initial capital investments are some of the key shortcomings of the systems based on Fordism.

Category: Philosophy

Also see: Holonic Manufacturing System (HMS), Taylorism

2.2.39 Four M (4M)

Material, machine, man, and method are 4M in lean management. These 4M are required to produce a value of the end product as per customer's specifications.

In a manufacturing environment, material, machine, and man are resources to produce the end product, while the method is referred to as the way available resources are utilized to produce the end product.

Category: Diagnostic tool

2.2.40 Gantt Chart

Gantt charts are named after its developer Henry Gantt. These charts are a type of bar chart used to visualize the project schedule and are widely used in project management, and process planning and control activities.

Category: Monitoring & control tool

Also see: Dashboard, visual factory

2.2.41 Gemba

Gemba is a Japanese term used for the shop floor (practical environment) where actual work or process is done.

Gemba walk: Gemba walk is referred to as the visit to the shop floor (practical environment) where actual work or process is done.

Category: Diagnostic tool

Also see: Flow shop, Genchi Genbutsu, Genjitsu

2.2.42 Genchi Genbutsu

Genchi Genbutsu is a Japanese term which means to visit the actual site or shop floor and see the real happenings. This term is synonymous to Gemba walk.

Genchi: Genchi is a Japanese term used for actual site or place.

Genbutsu: Genbutsu is a Japanese term used for actual articles or objects. It focuses on collecting data based on primary observations by seeing practical or actual things.

Category: Diagnostic tool

Also see: Flow shop, Gemba, Genjitsu

2.2.43 Genjitsu

Genjitsu is a Japanese term used for being realistic, not idealistic. In lean management, it is generally referring to the real or actual situation based on the facts and conditions.

Category: Diagnostic tool

Also see: Flow shop, Gemba, Genchi Genbutsu

2.2.44 Hanedashi

Hanedashi is a Japanese term used to jump, come out, leap, or spring up. In lean manufacturing systems, it is referred to as the automatic unloading of parts or products or items from the workstation. Hanedashi (automatic unloading) is important for a Chaku-Chaku (single-piece flow, where the output from one machine and input to the next machine has to be synchronized) setup.

Category: Waste minimization and improvement tool

Also see: A-B control, Chaku-Chaku, continuous flow, Kanban, one-piece flow

2.2.45 Hansei

Hansei is a Japanese term used for taking responsibility, and commitment to improvement.

Category: A training tool

Also see: Hitozukuri, Jinbou, Jinzai Katsuyou, Jishuken, QC

2.2.46 Harada Method

Harada method is named after Takashi Harada who advocated it as a method for human resource development. In Lean management, this method is widely used for employee development. This is a method based on self-reliance to help and develop people achieving the goals.

Category: A training tool

Also see: Hitozukuri, Jinbou, Jinzai Katsuyou, Jishuken, QC

2.2.47 Heijunka

Heijunka is a Japanese term used for leveling or balancing. In lean manufacturing, it is referred to as the product leveling ensuring smooth flow of the process with minimal or no fluctuations. Heijunka focuses on flow management and aims to avoid the creation of bottleneck in the system.

Heijunka box: Heijunka box is referred to as a leveling box as well, where Kanban cards are put in a box. This is a tool to visualize the Heijunka pattern by leveling the production by distributing the Kanban cards in a production facility.

Category: Production leveling and improvement tool

Also see: Batch-and-queue, baton touch, continuous flow, Customer Takt, Kanban, DBR, line balancing

2.2.48 Hitozukuri

Hitozukuri is a centuries-old rich Japanese culture focusing on the development of people through training and coaching to become experts in a particular domain. This approach focuses on lifelong education, long-term constant training of operator to develop technical skills to achieve excellence.

Category: A training tool

Also see: Harada method, Jinbou, Jinzai Katsuyou, Jishuken, QC

2.2.49 Holonic Manufacturing System

HMS was proposed as a new manufacturing system in 1990. Such systems are self-reliant and aim to adapt to new products due to its self-adjusting capabilities.

Category: Manufacturing system

Also see: Fordism, Taylorism

2.2.50 Hoshin Kanri

Hoshin Kanri is a Japanese term used for policy deployment or strategy deployment. It focuses on setting up measurable targets (based on the resource availability), and regular assessment of the progress to achieve the same. In lean management, Hoshin Kanri is used as a strategic decision-making tool to achieve targeted results on critical initiatives of an organization.

Category: Monitoring & control tool

Also see: Kaikaku Kaizen, Nemawashi

2.2.51 Hourensou

Hourensou is a Japanese term used for frequent reporting, contacting, messaging, and touching base and discussing. It is referred to as an approach to smooth communication.

Category: Monitoring & control tool

2.2.52 Jidoka

Jidoka is a Japanese term used for automation that aims to eliminate or reduce errors. In this approach, machines and processes (generally semiautomatic) are designed in such a manner that it stops by itself whenever a problem or issue occurs. The key aim of Jidoka is not to allow any defect in a product or process to move to the next stage. The production remains at halt till the defect identified is corrected or resolved.

Category: Waste minimization and improvement tool

Also see: Baka yoke or poka yoke, built-in-quality, FTT, QFD, ZD

2.2.53 Jinbou

Jinbou is a Japanese term used for belief or trust. In lean management, Jinbou is referred to as mutual trust and belief among employees to achieve higher goals collectively.

Category: A training tool

Also see: Harada method, Hitozukuri, Jinzai Katsuyou, Jishuken, QC

2.2.54 Jinzai Katsuyou

Jinzai Katsuyou is a Japanese term used for human resource. The focus is to be on the development of available human resources and not just utilize them. In doing so, the best results may be achieved by the employees.

Category: A training tool

Also see: Harada method, Hitozukuri, Jinbou, Jishuken, QC

2.2.55 Jishuken

Jishuken is a Japanese term used for the management-driven improvement activities in an organization. In lean management, Jishuken is referred to as an intentional, workplace-focused activity as part of Gemba Kaizen. It is a kind of hands-on self-learning or learning by doing a management approach.

Category: A training tool

Also see: Harada method, Hitozukuri, Jinbou, Jinzai Katsuyou, QC

2.2.56 Jundate

Jundate is a Japanese term used for the way of setting up parts for assembly in a manufacturing setup.

Category: Waste minimization and improvement tool

Also see: Changeover time, Dandorigae, SMED

2.2.57 JIS (Just in Sequence)

JIS is referred to as the delivery or feeding of the parts or products in a predefined sequence.

Category: Waste minimization and improvement tool

Also see: A-B control, Chaku-Chaku, continuous flow, Hanedashi, Kanban, and one-piece flow

2.2.58 JIT (Just in Time)

Just in time (JIT) is a pull-based demand-driven inventory replenishment technique where parts, products, or goods are made available just at the right time in the right quantity and right quality.

Category: Waste minimization and improvement tool

Also see: Buffer, EOQ, Kanban, pull system

2.2.59 Kaikaku Kaizen

Kaikaku Kaizen is a Japanese term used for fundamental and radical changes aiming to improve productivity with lesser waste. It is also referred to as kakushin. Kaikaku Kaizen is known as a philosophy of continuous improvement and also referred to as breakthrough Kaizen.

Category: Waste minimization and improvement tool

Also see: Hoshin Kanri, WCM

2.2.60 Kaizen Blitz

Kaizen Blitz or Kaizen Event is referred to as a quick improvement achieved in a short time. This is generally a set of activities that are undertaken to fix easy-to-solve issues.

Kaizen Group: Kaizen Group is referred to as a small team focusing on improvement in a lean environment.

Category: Waste minimization and improvement tool

Also see: NVA, value-added (VA), VSM

2.2.61 Kanban

Kanban is a Japanese term used for signal or visual signs. Kanban is a pull-based popular lean management tool used for the replenishment of items and effective inventory management.

Category: Monitoring & control tool

Also see: Chaku-Chaku, continuous flow, JIS, Hanedashi, one-piece flow

2.2.62 Kitting

It is a popular term used in lean manufacturing used for feeding material to the assembly line. Kitting (making kit of different parts) is an activity of picking up a set of different parts and feeding the same to the assembly line.

Category: Production leveling and improvement tool

Also see: Kanban, pull system, supermarket

2.2.63 LAMDA Cycle

LAMDA cycle is a knowledge creation technique that is based on five key elements: look, ask, model, discuss, and act. LAMDA is used as a skill development tool and also used for the development of metaskill (the permanent skill of learning and building on new skills at a faster pace). Continuous substantive learning and deep understanding are the key points of a LAMDA cycle.

Look: Practical exposure and hands-on experience on the shop floor through primary observations.

Ask: Critical questioning to find out the root causes.

Model: Developing models to predict the performance using data analysis, simulation, and prototype development.

Discuss: Brainstorming and discussion on the models or solutions developed.

Act: Testing and validation of the models or solutions.

Category: Systematic approach involving multiple tools

Also see: DMAIC, PDCA

2.2.64 Lead Time

Lead time is referred to as the total time of the process right from the beginning of the input in the process to the time when the piece or product is out at the end. Lead time is also taken as the minimum time required in manufacturing the product. There are different lead times in practice depending on the situation and references, such as order lead time, production lead time, process lead, and time.

Category: It is a measure

Also see: Cycle time, Customer Takt, pitch time, System Takt, throughput

2.2.65 LIFO (Last In First Out)

LIFO is a kind of material arrangement in terms of material flow (delivery sequence) where the part or piece, which has arrived at the last, comes out at the first.

Category: Monitoring & controlling tool

Also see: FIFO, pull system, push system

2.2.66 Line Balancing

Line balancing is a technique widely used for balancing the machines (in a line) or processes (in a sequence) based on the work content respective to the activity. The objective is to achieve higher machine or process productivity with optimized output with lesser chances of a bottleneck in the system.

Category: Production leveling and improvement tool

Also see: A-B control, batch-and-queue, baton touch, bottleneck, Chaku-Chaku, continuous flow, DBR, Customer Takt, Heijunka, Kanban

2.2.67 Little's Law

Little's law is a method to determine the lead time of a process or an activity. This is based on the WIP, lead time, and the average time of the process for the activity.

Little's law equation can be written as

Several items in the queue = Arrival rate × average time spent in the queue.

Category: Formula

Also see: Buffer, DBR, TOC, WIP

2.2.68 MTO (Make to Order)

Make to order (MTO) is a production approach where the products are produced only after specific demand or order is raised. This is a kind of pull (demand-driven) production approach with the manufacturing of products with low quantities with more product variations.

MTO is also referred to as build to stock.

Setsuban Kanri is a Japanese term used for MTO-based production management concepts.

Category: Philosophy

Also see: Make to stock (MTS), pull system, push system

2.2.69 MTS (Make to Stock)

MTS is a production approach where the products are produced for an inventory (stock) in anticipation of future demand. This is a kind of push (supply-driven production approach) with the manufacturing of products with high quantities with lesser product variations.

Category: Philosophy
Also see: MTO, pull system, push system

2.2.70 Mass Customization

Mass customization is catering to individual product requirements (but at a larger scale and competitive price).

Category: Philosophy
Also see: Mass Production

2.2.71 Mass Production

Mass production is producing in large quantities. The key advantage of mass production is that it involves higher resource utilization with achieving benefits of the Economies of Scale.

Category: Philosophy
Also see: Mass customization

2.2.72 Mendomi

Mendomi is a Japanese term used for taking care of the workers where they are treated as a family member. This is a human resource development approach where the organization takes care of the workforce and entrust the sense of loyalty and commitment.

Category: A training tool
Also see: Harada method, Hitozukuri, Jinbou, Jinzai Katsuyou, Jishuken, QC

2.2.73 Mieruka

Mieruka is a Japanese term used for effective visualization or making it visible. The key aim of Mieruka is to make the process simple, and easy to understand. This is a concept developed and practiced in TPSs and widely used lean manufacturing systems.

Category: Monitoring & control tool
Also see: Andon, dashboard, visual factory

2.2.74 Milk Run

Milk run is a technique for providing materials to different stations in sequence, according to the feeding schedule. Milk run aims to improve the flow of materials between facilities by routing and it minimizes the waiting time significantly. The other advantages of milk run include reduced inventory levels and faster response times.

Category: Production leveling and improvement tool

Also see: Chaku-Chaku, DBR, Hanedashi, Kanban, one-piece flow, TOC

2.2.75 MTBF (Mean Time Between Failures)

MTBF is the average time between two subsequent breakdowns of a machine or a process. Higher MTBF is an indicator of increased reliability of the system or machine. MTBF is generally used for repairable systems.

Category: It is a measure

Also see: Downtime, MTTF, MTTR

2.2.76 MTTF (Mean Time To Failure)

MTTF is the time that an item, machine, or system is expected to work until it fails. Higher MTTF is an indicator of increased reliability of the system or machine. MTTF is generally used for nonrepairable systems.

Category: It is a measure

Also see: Downtime, MTBF, MTTR

2.2.77 MTTR (Mean Time To Repair)

MTTR is the average time required to repair a machine or a system after a breakdown, and getting it ready to work again. Lesser MTTR indicates as a higher maintainability of the system which means the system can be repaired quickly if there is any breakdown.

Category: It is a measure

Also see: Downtime, MTBF, MTTF

2.2.78 Nagara

Nagara is a Japanese term used for doing something. Generally, it is used in the context when someone is involved in more than one thing at a time.

Nagara Fixture: Term used for a fixture used to carry parts or tools in an assembly line; such fixture travels along while the worker is working.

Nagara-Switch: Term used for a switch or a flexible rod activated (either by pressing or by swiping) for faster actuation.

Category: Monitoring & control tool

Also see: SMED

2.2.79 Nagare

Nagare is a Japanese term used for the flow of people, things, fluids, or gases. In the context of lean manufacturing the same is used for smooth production flow in a manufacturing setup.

Category: Monitoring & control tool
Also see: Changeover time, DBR, Dandorigae, Jundate

2.2.80 Nemawashi

Nemawashi is a Japanese term used for the process of acquiring acceptance and preapproval for a proposal by evaluating first the idea and then the plan with management and stakeholders to get input, anticipate resistance, and align the proposed change with other perspectives and priorities in the organization.

Category: Philosophy
Also see: Hoshin Kanri

2.2.81 NVA (Non-value-added)

NVA are the activities that don't add or contribute value to the end product but consume resources.

Necessary-NVA (NNVA): NNVA are the type of activities that don't add any value to the end product but is essential to perform, as without them the process of work cannot be completed.

Category: It is a measure
Also see: Seven waste (7 waste), value-added (VA)

2.2.82 OEE (Overall Equipment Effectiveness)

OEE (overall equipment effectiveness) is a popular TPM tool used to measure the overall effectiveness of a system or a process or a machine by considering availability, performance, and quality of the output. OEE is widely used as a framework to measure productivity loss in a system.

$$OEE = Availability \times performance \times quality$$

Category: It is a measure
Also see: Autonomous maintenance, TPM

2.2.83 One-piece Flow

One-piece flow or single-piece flow is a popular practice in a lean manufacturing system that aims for a faster throughput with a lesser level of WIP. As the WIPs are lesser, such systems also promise quick changeover (QDC) and faster learning curve, however, to get the best results, the line balancing and synchronization has to be precise.

2.2.84 Pacemaker Process

A pacemaker is referred to any process or activity (in a value stream) that sets the pace for the entire process or stream.

The pacemaker activity or process is generally towards the consumer's end of the value stream. For example, the apparel manufacturing stitching process may be considered as the pacemaker process, as this is the process where the raw material is getting converted into semifinished or finished garments. Usually as a common practice, in apparel manufacture, production planning is done concerning the stitching process as the key process.

Category: Production leveling and improvement tool

Also see: Bottleneck

2.2.85 Pareto Principle

Pareto principle is popularly known as 80:20 rule. It is named after its developer Vilfredo Pareto. According to this rule, 80% of the outcomes are controlled or decided by 20% of the activities or factors. For example, 80% of the total profit is generated by 20% of the product categories, or 80% of the maintenance expenses are incurred by 20% of the machines. This rule is very much followed in most of the situations or conditions; hence this is a widely used principle in many of the areas including inventory management, finance and project planning, etc.

Category: Diagnostic tool as well as monitoring & control tool

Also see: Buffer, EOQ, vendor managed inventory (VMI)

2.2.86 PDCA (Plan–Do–Check–Act)

The cycle PDCA, also referred to as PDSA (plan–do–study–act), is one of the key tools of lean management. It is also known as the Deming circle or Shewhart cycle. PDCA cycle provides a much-required opportunity for regular assessment of progress against targeted goals. Also, it helps in the provision of necessary fine-tuning or modification (if required) and taking corrective action or informed decision to ensure achieving the goals.

Plan (establish a plan and anticipated results)

Do (implement the change)

Check (evaluate progress done so far against the targeted goal)

*Act*s (standardize and stabilize the process through appropriate reviewing and assessment; restart the cycle)

Category: Systematic approach involving multiple tools

Also See: DMAIC, LAMDA cycle

2.2.87 Pitch Time

Pitch time is referred to the average time taken by an operator to finish the task or the activity assigned. For example, if the total time of a process is 10 minutes, and there are five operators in that system work in a sequence, then pitch time shall be 2 minutes. Pitch time is a useful term in line balancing or resource balancing, to ensure a synchronized balanced flow. To achieve a balanced workflow it is recommended that the individual operator's work content is equal/near to the pitch time.

Category: It is a measure

Also see: Cycle time, Customer Takt, lead time, System Takt, throughput

2.2.88 Pull System

Pull system is a demand-driven approach where things, products, or services are only produced against the demand. This approach is a lean-based approach backed up by JIT ensuring lesser levels of WIP to control or minimize the possibilities of waste (Muda) generation. Pull-based systems are recommended in lean management.

Category: Philosophy

Also see: Chaku-Chaku, DBR, Hanedashi, Kanban, milk run, one-piece flow, push system

2.2.89 Push System

Push system is an opposite concept of pull, where more quantities of products are manufactured in anticipation of future demand by the consumer or by the next process. This system aims to ensure the availability of work (in the form of WIP) to avoid chances of process idleness or run-out due to the nonavailability of material or parts. Many a time, this results in an accumulation of excess inventories and creation of bottlenecks which is a waste (Muda in lean management).

Category: Philosophy

Also see: Chaku-Chaku, DBR, Hanedashi, Kanban, milk run, one-piece flow, pull system

2.2.90 QC (Quality Circle)

QC is a famous problem-solving technique used in lean management. The key characteristics of this approach are the voluntary involvement of the workforce to address the issues and recommend solutions (QC members) meet at regular intervals, discuss and understand the issues, and suggest appropriate solutions after thorough brainstorming. This is also referred to as quality control circle.

Category: Philosophy

Also see: Harada method, Hitozukuri, Jinbou, Jinzai Katsuyou, Jishuken, QC

2.2.91 QFD (Quality Function Deployment)

QFD is a quality management approach aiming to provide the end product or services precisely as per the customer requirements. Keeping the key focus on specific customer requirements, the process is designed accordingly to add value and avoid/minimize the waste due to overprocessing.

Category: Philosophy

Also see: Built-in-quality, DQC, FTT, Jidoka, ZD

2.2.92 Rabbit Chase

Rabbit Chase is a way of working in lean manufacturing setups (especially in U-line setups) where one operator moves to the next workstation (or operation) as soon as they finish the one work. By doing so, the operators chase each other in the forward direction of the material flow. Such systems are most suited for cellular or modular systems where multiskilled operators are employed to handle more than one workstations.

Category: Production leveling and improvement tool

Also see: Baton touch, cellular manufacturing, Shoujinka, U-line

2.2.93 RCA (Root cause analysis)

RCA is a problem-solving approach that is similar to 5 Why analysis, Fishbone diagram or Ishikawa diagram, or cause and effect diagram. The RCA approach is widely used to find out the possible root causes of an effect on an issue. Kadai Souzouryoku is a Japanese term used to identify core problems or issues in the process.

5 Why technique is used to identify the root cause of the problem. Repeated questioning in a series (starting with "why" for seeking the reason) results in the determination of the real reason for the problem or issue which further helps in taking an appropriate corrective course of action.

Fishbone diagram is also known as the Ishikawa diagram (as this technique was developed by Kaoru Ishikawa) or cause and effect diagram. These diagrams are widely used to find out the possible root causes of an effect on an issue.

Category: Diagnostic tool

Also see: FMEA

2.2.94 Seven Waste (7 Waste)

Waste is termed as Muda in the Japanese language. This is an important term widely used in lean management, referred to all the activities or processes which do not add any value to the end product but consume resources. Seven wastes in lean

Lean terms in apparel manufacturing 39

management are motion, transportation, waiting, defect, overprocessing, inventory, and overproduction.

8 Waste: In addition to this, excess unused human resource/manpower is also considered as a waste (Muda) and treated as the eighth waste.

Category: Philosophy

Also see: Three M (3M)

2.2.95   Shoujinka

Shoujinka (also spelled as Shojinka) is a Japanese term used for manpower reduction or producing with lesser people. In lean management, it is referred to as the flexible manufacturing systems (cellular or modular systems) where a team of few operators works according to the production requirements.

Category: Waste minimization and improvement tool

Also see: Cellular manufacturing, Rabbit Chase, U-line

2.2.96   Six Sigma (6 Sigma)

6 Sigma (6σ or 6 Standard Deviation) technique was developed and first practiced by Motorola to achieve operational excellence. 6 Sigma aims to achieve a high level of accuracy by just producing a maximum of 3.4 errors in 1 million or 3.4 parts per million.

Category: Philosophy

Also see: Built-in-quality, DQC, FTT, Jidoka, QFD, ZD

2.2.97   SMED (Single Minute Exchange of Die)

SMED is an important technique widely used worldwide for QDC to achieve a faster setup. It is aimed to keep the setup time in a single-digit minute, that is, less than 10 minutes.

Category: Waste minimization and improvement tool

Also see: Changeover time, Dandorigae, Jundate, TPM

2.2.98   Spaghetti Diagram

Spaghetti diagram is a visual representation of flow using a continuous line on a chart. Such charts are used for optimizing the walking distances (reducing Muda of motion and transportation) through layout improvement. The movement of the worker, material, or machine is represented by a line that looks like spaghetti.

Category: Diagnostic tool as well as monitoring & control tool

Also see: Visual factory

2.2.99 Standard Work

Standard work is the defined way of executing an activity or operation in a standardized manner. The work sequence and design of the workstation according to specific product requirements are well specified, and the same to be followed by the respective workforce.

Hyōjun Sagyō is a Japanese term used for standardized work which is treated as one of the best practices in lean management.

Category: Waste minimization and improvement tool

2.2.100 Supermarket

In a lean manufacturing, setup supermarket is a kind of storage area where different materials are stored and kept separated according to its specification and flow requirements (as per the supply/feeding sequence to the next process). To maintain the smooth functioning and controlling the inventories, upper limit and lower limit levels are followed in the supermarkets.

Category: Production leveling and improvement tool

Also see: Buffer, EOQ, Kanban, JIT

2.2.101 System Takt

Takt is a German term used for rhythm or beat of a drum. It is used to measure the demand in time per quantity of the product. The System Takt or Customer Takt is referred to as to the available work or production time during a specified period divided by customer demand. This is required to determine the targeted output rate to meet customer demand.

Category: It is a measure

Also see: Cycle time, Customer Takt, lead time, pitch time, System Takt, throughput

2.2.102 Taylorism

Taylorism is a philosophy of Scientific Work Management given by Frederick Winslow Taylor. This is an important concept of Industrial Engineering based on scientific analysis of the work to enhance productivity and economic efficiency. In 1911 the famous book *The Principles of Scientific Management* was written by Taylor, which provides deeper details of Taylorism.

Category: Philosophy

Also see: Fordism, HMS

2.2.103 TOC (Theory of Constraints)

TOC is a management philosophy introduced by Dr. Eliyahu Goldratt as introduced in the book "The Goal" in 1984.

Lean terms in apparel manufacturing

TOC is a scientific approach for overall improvement which encourages continuous identification and systematic elimination or control of the limiting factor (constraint or a bottleneck). According to TOC, the strength (or performance) of a system is determined by constraint activity or process which is also referred to as the weakest link of the chain. To improve the strength (performance) of the systems, it is essential to improve upon this limiting factor. Further, the next limiting factor should be identified and eliminated to ensure continuous improvement.

In TOC, DBR is one important technique used for production synchronization with optimum inventory level and work-in-process.

Category: Philosophy

Also see: Chaku-Chaku, DBR, Hanedashi, Kanban, milk run, one-piece flow

2.2.104 Three M (3M)

Muda, Mura, and Muri are the popular 3M in lean management used to indicate wasteful (NVA) activities in the process.

Muda: A Japanese term used for waste. According to lean, waste is anything that consumes resources without contributing any value to the final product.

Mura: A Japanese term used for unevenness. The major reason for Mura is imbalanced resource planning due to lack of proper level scheduling and varied pace of work.

Muri: A Japanese term used for overburdening or overloading. Muri is when the available resources are pushed to produce more output (maybe at a higher pace for a longer duration) than their capacity.

Category: Philosophy

Also see: Seven waste (7 waste)

2.2.105 Throughput

Throughput is the rate at which goods are produced. Throughput time is the time required to get the first piece out from the line or process or system. This is generally the difference of time when the same piece was input in the line to the time of output.

The relation between throughput, material flow time, and inventory in a supply chain is explained by

Little's law as inventory = Throughput × material flow time

Category: It is a measure

Also see: Cycle time, Customer Takt, lead time, pitch time, System Takt

2.2.106 TPM (Total Productive Maintenance)

TPM is a maintenance philosophy incorporating the skills of all employees and focusing on improving the overall effectiveness of the facility with a total participation of the workforce. The TPM philosophy focuses on proactive and preventative maintenance to enhance the overall effectiveness of the system.

Category: Philosophy

Also see: Autonomous maintenance, SMED

2.2.107 TPS (Toyota Production System)

TPS is the core of lean management. This is a philosophy and culture adopted by the employees at Toyota Company to achieve manufacturing excellence. There are several tools and techniques followed to minimize the waste in the process and maximize value addition in the end product. The TPS aims to achieve easier, better, faster, and cheaper improvements.

Seisan Hoshiki is a Japanese term used for the production system. TPS is called Toyota Seisan Hoshiki.

Category: Manufacturing system

Also see: Cellular manufacturing, Rabbit Chase, Shoujinka, U-line, WCM

2.2.108 Turtle Diagram

Turtle diagram is a schematic visual representation tool used for process improvement. Turtle diagram represents the resources required (in terms of input) to produce the desired results (in terms of output), along with other key characteristics of the process such as process metrics and information required for the process.

A turtle diagram consists of two lanes: primary lane and secondary. The primary lane includes details related to the inputs, process, and output. The secondary lane includes details related to the key resources required to achieve the desired results. Such key resources include equipment & materials, personnel & competencies, documentation & methods, and objectives & targets.

Category: Monitoring & control tool

Also see: Dashboard, Mieruka, visual factory

2.2.109 U-Line

This is a manufacturing setup where workstations are arranged in U shape. This kind of setup is generally practiced in cellular or modular manufacturing. It makes multiskilled operators handle more than one machine with minimal movement. Other key advantages of such arrangements include ease of supervision, flexibility to change the capacity (by adding or removing workers), QDC, and higher manpower utilization.

Category: Manufacturing system

Also see: Cellular manufacturing, Rabbit Chase, Shoujinka, TPS

2.2.110 Value-added (VA)

Value-added (VA) actions are the activities that add value to the end product from the consumer. The key objective of lean management is to enhance or maximize value addition in the product or services.

VA ratio (VAR) is the ratio of time spent on value-added activities (adding value to the end product or service in the context of the consumer) to the total time consumed from the receipt of the order to the delivery of the product or service. Increasing the VAR is one of the key objectives of any lean initiative.

Category: It is a measure

Also see: NVA, VSM

2.2.111 VMI (Vendor Managed Inventory)

VMI is a pull-based supply chain management concept where the inventory is managed by the vendor or supplier even at the customer's manufacturing site. This is an important aspect of quick response supply chains ensuring effective inventory management.

Category: Waste minimization and improvement tool

Also see: Buffer, EOQ, Kanban, JIT

Visual factory: *The visual* factory is the setup where visual information throughout the workplace is encouraged using visual indicators, displays, and controls.

Visual management: Visual management is referred to as the manufacturing environment that can be observed with symbols, visuals, markings. The targets, performances, achieved goals, instructions, and other relevant information are visually depicted in a visual management setup.

Category: Monitoring & control tool

Also see: Dashboard, Mieruka

2.2.112 VSM (Value Stream Mapping)

A value stream is the set of all the activities right from the value creation a production till the delivery. It is also referred to as a process of increasing the value of the product or services by value addition to it.

VSM is also called material and information flow mapping. It is an important tool of lean management used to identify the VA in a process. VSM is a management tool to analyze the current state and design the future state of a value stream of a specific product. A set of standard symbols is used in VSM which makes it a powerful visual tool to display critical steps in a value stream. A VSM involves two keymaps named as current state value stream map and future state value stream map. A current state map is the visual representation of the process and its activities as per the existing or current situation, while a future state map depicts the improved (recommended for future) process to achieve a higher level of performance.

VSD (value stream design): VSD is referred to as the structured diagram which indicates the expected material and information of respective value stream for a particular product.

VSL (value stream loop): VSL is referred to as the segments or parts of a value stream, where the boundaries of such segments are marked by supermarkets. From the management point of view, having such VSLs makes the future state implementation easy to manage.

Category: Diagnostic tool

Also see: Kaikaku Kaizen, NVA, value-added (VA)

2.2.113   WCM (World Class Manufacturing)

WCM is referred to as the set of practices, concepts, techniques (many of such techniques have opted from lean manufacturing or Toyota Manufacturing System), focus on producing the best products, and aiming at achieving global competitiveness.

Category: Philosophy

Also see: Agile manufacturing system, Lean, TPS

2.2.114   WIP (Work-in-Progress)

WIP is referred to as the unfinished or semifinished work or the inventory inside a process or a system. Keeping higher levels of WIP than required is a waste (Muda) in lean management, and lean philosophy advocates for keeping the WIP as minimum as possible to maintain a smooth workflow.

Category: Waste minimization and improvement tool

Also see: Buffer, EOQ, pull system, push system

2.2.115   Yamazumi Chart

Yamazumi is a Japanese term used for pile or stack. Yamazumi charts are a method of visual display in the form of a stacked bar chart. Such charts are colorful representations of a process in the form of stacked up bars or lad charts used for effective communication.

Category: Diagnostic tool as well as monitoring & control tool

Also see: Dashboard, Mieruka, visual factory

2.2.116   Yokoten

Yokoten is a Japanese term used for sharing learning laterally across the organization. This is related to horizontal deployment or sideways expansion in the context of copying and practicing lean concepts to achieve further improvement.

Category: A training tool

Also see: DMAIC, PDCA, LAMDA cycle

2.2.117 ZD (Zero Defects)

ZD is an ideal condition or expectation in practical manufacturing environments. Knowing that it only exists as an idea, some organizations that are seeking for manufacturing excellence, wish to achieve the same. ZD is also referred to as producing no errors or no defective items, which is only possible when there is zero waste in the process. The lean management with a key focus of improving value addition aims to achieve ZD. Concepts such as 6 Sigma are also some landmarks on the path to achieve ZD.

Category: Philosophy

Also see: Built-in-quality, DQC, FTT, Jidoka, QFD, Six Sigma (6 Sigma)

Fundamentals of lean journey

Manoj Tiwari
Department of Fashion Technology, National Institute of Fashion Technology, Jodhpur, India

> *It is not the strongest of the species that survive, nor the most intelligent, but the one most responsive to change.* — Charles Darwin

3.1 Introduction

The advancements of technology and global competitiveness have opened up new opportunities to the consumers where they have more apparel product options readily available at a competitive price. This has resulted in increased pressure on apparel manufacturers to ensure product delivery at the earliest with the best price to the consumer. This has also resulted in dynamic demand patterns, where product cycles are getting shorter (Mourtzis & Doukas, 2014). The era of long orders with no or minimal variations is no more relevant, and the same has been replaced with orders with smaller quantities with varied product specifications. Brands are becoming more and more conscious about meeting their consumer expectations and failing to which is a matter of brand image and eventually a question on survival in this cutthroat competitive environment. This reminds of "rule of tens" (refer Fig. 3.1), where every defect occurred as the level of product development costs in multiple times if it is identified by the consumer. Initially, the cost of an error or a defect may cost a little amount (say one dollar) in getting it corrected on the spot, but if the same defect is passed to the next process, then the cost of defect increases in multifold. Eventually, if the consumer receives a defective product and discovers this defect, then the impact of the same is huge in terms of value as well as the brand image (Locher, 2008).

Traditionally, apparel manufacturing is a human-driven sector where automation is relatively lesser in comparison to other industries such as automobiles. Several factors such as the nature of raw materials, the design complexity of the product, frequent changes in the product specifications, and tight time lines make apparel manufacturing vulnerable toward optimum utilization of available resources. Lean manufacturing aims to achieve the competitive advantages and a 90% reduction in lead time, 90% reduction in inventory levels, 90% reduction in cost, and 50% improvement in labor productivity can be expected by a traditional mass manufacturer if lean manufacturing is applied

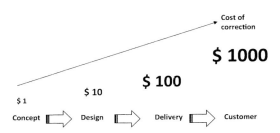

Figure 3.1 Rule of Tens (Locher, 2008).

(Bhasin & Burcher, 2006; Lathin & Mitchell, 2001). Lean philosophy has been successfully addressing these issues in several industries (especially automobile manufacturing) since the last 70 years but apparel manufacturing is yet to be benefitted. Lean manufacturing tools and techniques are applied worldwide and it has global superiority in cost, quality, productivity, flexibility, and quick response (Schonberger, 2007). The application of lean manufacturing helps in the successful operating of manufacturing organizations in a challenging and competitive environment (Bashar & Hasin, 2018). In the last few years, people have realized the relevance of adopting lean way in apparel manufacturing, but it is still at a very nascent stage (Jana & Tiwari, 2018).

Very few lean tools and techniques are applied in apparel manufacturing, and that too with a limited scope. Not meeting delivery deadlines, delays in receiving raw materials, quality-related issues, lack of latest manufacturing progress data, higher levels of unplanned downtimes, and poor resource utilization are some of the frequently observed situations in apparel manufacturing (Deshpande, 2016; Mazumder, 2015). This chapter discusses the challenges in adapting lean way, an approach to start with the lean initiatives, and different tools used as lean problem-solving in the context of apparel manufacturing.

3.2 Value addition

Fundamentally, the traditional approach of mass production aims to achieve a limited goal that is "good enough." It results in an acceptable level of defects, some predefined level of inventories, and a few standardized products. This is in contrast to the lean thinking. Lean production aims to achieve "better always." And to achieve the same, it focuses on zero tolerance with defect (to achieve zero defect), a minimum level of inventories, and increased product variations. To achieve more from less, the key focus of the lean approach is the identification, reduction, and if possible elimination of the wastes from the processes. The term "Muda" (a Japanese word used for waste) is synonymous to waste and popularly used among lean practitioners. As lean is regarding getting maximum output from minimum possible inputs, anything that is not adding value to the final product is a waste and should be removed/controlled to achieve higher productivity.

To initiate any improvement in the process, it is very important to identify the waste, and that can be seen from the value addition point of view. To initiate

Fundamentals of lean journey

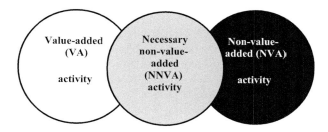

Figure 3.2 VA, NVA and NNVA activities. *VA*, Value-added; *NVA*, non-value-added; *NNVA*, necessary but non-value added.

identifying the wastage in the process, the activities should be categorized as value-added (VA), non-value-added (NVA), and necessary but non-value added (NNVA). According to the research conducted by the Lean Enterprise Research Center (LERC), United Kingdom, most of the physical environments (manufacturing or logistics flow) have value addition of just 5%, where remaining part (95%) is covered by NVA (60%) and NNVA activities (35%) (Hines, Found, Griffiths, & Harrison, 2011). This indicates that there is always a huge potential for improvements in increasing the VA ratio.

VA activities: The activities may be referred to as which are essential to perform (refer Fig. 3.2) and without which it is not possible to get the final product, as required by the customer. Such activities are adding value to the final product which may be in terms of quality, functionality, or any other feature or characteristics as per customer's specific requirement.

NVA activities: The activities may be referred to as which just consume money and time and not contribute (or add value) in achieving the final product, as required by the customer. These activities are the waste in a process and need to be eliminated or reduced (if elimination is not possible).

NNVA activities: These activities fall between value-added and NVA activities that need to be performed necessarily despite there is no value-addition to the final product by performing such activities. In pursuit of process improvement, efforts should be done to eliminate or reduce such activities.

3.2.1 Criteria for value addition

Learning to recognize the waste is a challenge. Until unless we are not sure whether the activity is adding value to the final product or not, we cannot realize the course of action for process improvement. Three-point criteria (Beels, 2019; Tiwari, 2010) to easily judge the value addition and categorize the activities accordingly are suggested as follows.

3.2.1.1 Significant transformation

The activity performed should have a significant transformation which may be in the form, physical shape, characteristics, or properties. The step or activity perform should transform the work toward achieving the final product. This condition is valid for tangible as well as intangible products. While developing the software, applying logical programming may be required for adequate functioning of the software solution, or using an algorithm to get some output is a nontangible phase change. Bulk cutting of fabric rolls after spreading to manufacture apparels or stitching of cut panels to get semi-finished or finished products is a physical transformation.

3.2.1.2 Achieving the quality standards as required by the buyer

The activity or the process must be done to ensure the desired quality of the final product. Here it is essential that the work done is right the first time. Parameter settings in a machine to get desired output, conformance to the quality standards by performing some operations, such as attaching buttons in shirts according to buyer specifications may be considered as examples to these criteria.

3.2.1.3 Payment to the activity

The customer or the buyer should be ready to pay for the activity or the work performed. This means that the activity is considered while the cost of the product and customer is ready to pay for the same as it is part of the specifications of the final product.

To qualify for value addition, one activity must fulfill all three criteria at the same time. In the case of noncompliance, such activity should be considered as *Non Value Adding activity*. After the segregation of value-adding and non-value-adding, further screening should be done for the non-value-adding activities. It is recommended to check, if the activity or work is essential or if there is a possibility to eliminate it. If the activity or the work cannot be eliminated, it should be classified as a NNVA activity. Further efforts should be put toward improving the NNVA activities maybe by applying improved method, reengineering, or maybe by using some alternate ways. Excerpt from a data collection format for classifying activities as VA/NVA/NNVA is shown in Fig. 3.3. We can see the duration of every microactivity listed in the "time taken" column (refer Fig. 3.3), the respective process symbols are indicated against each microprocess, and the waste categories for every microprocess are also indicated.

Another popular approach for classifying any activity as VA or NVA is based on the five types of activities: operation, inspection, transportation, delay, and storage. Operation type is considered as VA, whereas the other four types are commonly categorized as NVA. Each NVA is further questioned whether the activity is necessary? However, there lies the challenge. The decision regarding necessity is often become subjective, without objective guidelines in place.

Let us take some examples; inspection of cut parts after cutting, when we ask whether this activity is necessary? The answer is yes, because during spreading the fabric defects were allowed to be there (of certain percentage as per quality standard), with an anticipation that some defects will fall in a dead zone of marker and

Fundamentals of lean journey

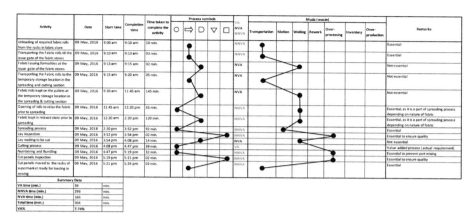

Figure 3.3 Time consumed in VA, NVA and NNVA activities. *VA*, Value-added; *NVA*, non-value-added; *NNVA*, necessary but non-value added.

rest will be identified through cut parts inspection and replaced by fresh-cut parts. Therefore the cut parts inspection becomes a necessary activity for the system to function (even the customer does not pay for it). Hence it can be classified as NNVA.

In case of the random in-line inspection of garments, when we ask whether this activity is necessary? The answer is yes from the quality management point of view. When we further ask whether we can eliminate random in-line inspection, then the answer would also be yes; when the sewing operators start to produce the right first time and zero-defect products, the random in-line inspection will become unnecessary. Hence it is an NVA.

So, we have seen the same inspection activity can be NNVA or NVA depends on the scenario. Similarly, transportations of fabric from the mill to apparel manufacturer or transporting fabric roll from the fabric store to the cutting department can be NVA or NNVA depending on the business situation, manufacturing system, etc.

Delay and storage are the other two categories where no apparel VA happens. Often, we have seen knit fabric is spread but cutting is delayed by hours. This delay is intentional and allowing the fabric to shrink before cutting. Often misinterpreted, this delay can be classified as NNVA. This process of natural residual shrinkage is ensuring quality in the product.

Another deliberate delay between ironing and polybagging activity is for natural evaporation of residual moisture from garment after ironing/pressing operation. This delay is also VA activity, which is often wrongly classified as NVA. Storage of cut parts between cutting and sewing department is NVA from the customer point of view, but from the manufacturing system point of view, it is necessary to maintain the buffer stock, therefore can be classified as NNVA.

From the above examples, it is clear that although value-added categorization may be easy, the categorization of NVA and NNVA is difficult, complicated, and subject to subjective interpretation. One suggested approach is the process of categorization through differentiating the customer; external and internal (Ladia & Jana, 2014). External Customer is the buyer, that is, the person buying the finished product at the retail store. Internal customer is within your company who uses your products or services to manufacture its product.

Activity which adds value to the product for the External Customer is considered to be VA.

Activity which does not add value for the External Customer but adds value to or is necessary for the next process (Internal Customer) or adds value or is necessary for the previous process (Internal Customer) is considered to be NNVA.

Activity that does not add value to the product for either the External Customer or the Internal Customer is considered to be NVA.

If we look at certain examples, the concept will be clear. In a sewing operation (with progressive bundle unit system), when the operator ties or unties the bundle, the activity is NVA for an external customer, as the ultimate product quality or specification is not affected by such activity. However, the activity may be necessary for the next operation to perform satisfactorily. Although tying and untying may be NVA for an external customer, it is still value-added for the internal customer.

The nonstandardization of classification of VA, NVA, and NNVA is obvious from the fact that some of the explorations in apparel manufacturing reported very high value-added time of 30.2% (Ladia & Jana, 2014) in presewing activities to 32.2%−33.3% in sewing activities (Singh & Singh, 2014). The same study (Ladia & Jana, 2014) also found the average NVA time to be 53.65% ranging from min. 39.98% to max. 61.57% and the average NNVA time was found to be 16.15% ranging from min. 10.07% to max. 27.30%.

A supply chain diagnostic study to devise a road map for lean transformation at a leading textile manufacturer in the world (with a manufacturing base in India) was conducted. The project aimed at diagnosing the internal supply chain using Value stream mapping (VSM) for a major product family. The VA at the current state VSM was observed to be 0.006% and 0.014% for open-end spinning and ring spinning, respectively. In future state VSM, the VA was proposed to be increased to 0.025% and 0.034% for open-end spinning and ring spinning, respectively (Priya, Saini, & Tiwari, 2011).

Here it is important to note that the context of calculating VA plays a vital role. The VA figures may vary drastically as it is purely dependent on the criteria followed while deciding VA, NVA, and NNVA. It can be observed from research done by Ladia and Jana (2014) that VA percentage was around 30%, which seems very high in comparison to other established research studies such as the findings by the LERC, United Kingdom, where it was observed that the most of the manufacturing or logistics flow environments have VA of just 5% (Hines et al., 2011), whereas the VA figures observed by Priya et al. (2011) seem relative toward

Fundamentals of lean journey

the lower side. Such situations can be understood further with an example; let us assume that the manufacturing lead time of an apparel manufacturer is 30 days (8:00 hrs shift) which garments standard minute value of 20 minutes including all the operations such as spreading and cutting (15 minutes), stitching (2 minutes), and finishing and packing (3 minutes). Now there may be different scenarios of VA for the same product given as follows:

Scenario 1, VA at individual sewing operator level: Assuming the cycle time for an operation is 1 minute, and the only needle working (when stitching is happening) is considered as VA. The average needle time observed for that operation is 15 seconds, in such scenario VA for that particular operation will be 25% [(15/60) × 100].

Scenario 2, VA at sewing line level: Assuming that the cut panels of a particular garment are loaded in the line at 8:00 am on day X, and the same garment was out from the line at 11:00 am on day X + 2, this makes 19.00 hrs spent by the garment in the stitching line. As the standard time for sewing the garment was 15.0 minutes, the VA maybe 1.316% {[15/(19 × 60)] × 100}.

Scenario 3, VA at apparel manufacturing plant level: Assuming that the order quantity for that particular style was 10,000 units, and the production lead time (from receive of the fabric % trims to the final QA and loading of the pieces for final delivery from the manufacturing site) is 30 days. In such case, VA shall be 1 04% {[(15 × 10,000)/ (60 × 30 × 8)] × 100}.

In such different practical scenerios, the criteria of deciding an activity as VA, NVA and NNVA becomes critical, and one needs to be careful while doing such categorization.

3.3   3M of lean manufacturing

Muda (waste), Muri (overburden), and Mura (unevenness) are popularly known as 3M of lean. These are the key lean terms that are interconnected and often used together. All three terms are from the Japanese language and were coined in the Toyota Production System (TPS). 3Ms are an indication of inefficiencies or the hurdles that work as barriers to achieve the desired success (Liker, 2004). In other sense, these are nothing but NVA activities in the process. This means these are consuming resources but not adding any value to the end product. To achieve excellence through elimination or control of these 3Ms is the ultimate goal of any lean initiative. Muda refers to the waste in a process, which is categorized into seven categories according to TPS (types of wastes discussed in detail in the following pages).

Muri is referred to as overburden in the process. An overburden may be due to any reason such as pressure to produce more from the available resources. This may affect the efficient functioning of the available infrastructure and man power. An overburden may be due to poor planning or not able to foresee the demand or may be due to a fluctuating demand pattern. Lack of standard operating procedures

(SOP) may also be the reason for Muri in a system. Sudden breakdowns of machines and absenteeism are some of the outcomes of Muri, which eventually ended up in producing Muda (wastes) (Hines et al., 2011).

Mura is referred to as unevenness in the process. Mura may be closely linked with Muda and Muri, as one of the main cause of unevenness in a system is process imbalance, and process imbalance (poor resource allocation) is also an indication of waste in the process and overburden of resources. The unevenness in a system may be due to bottlenecks in a chain, where the flow is interrupted creating pressure on other machines/subprocesses. This again may cause overburdening or Muri. The unevenness in the process may also result in quality-related issues, fluctuations in the output rate. To sum up, it may be concluded that each of the M is responsible for creating the other two Ms, and all generally present parallelly in a system.

3.4 Types of wastes

In lean, waste is categorized into seven types. All the waste categories are interlinked to each other if observed closely (refer Fig. 3.4 and Table 3.1), and also it may be realized that the waste, irrespective of its type, is the prime source of productivity loss that brings inefficiencies in the system.

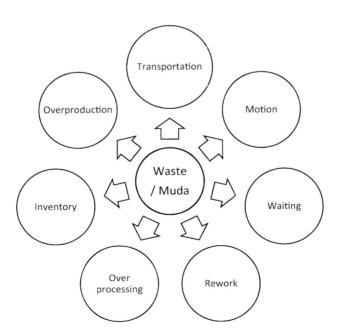

Figure 3.4 Types of wastes.

Fundamentals of lean journey

Table 3.1 Examples of different types of Muda (waste).

Type of Muda (Waste)	Activity example	Reasoning
Transportation	1. Movement of fabric rolls from the fabric store to spreading and cutting section. 2. Movement of semi-finished or finished products from one process to the next process.	The physical movement of material
Motion	1. Movement of personnel from one department to another. 2. Movement of supervisors while overseeing the activities in the respective department. 3. A machine operator moving away from his/her workstation to call a mechanic in case of breakdowns.	The physical movement of personnel
Waiting	1. Fabric rolls waiting to be issued for spreading and cutting. 2. Sewing machines waiting for cut panels and sewing trims to commence the stitching process. 3. Merchandisers waiting for approvals from buyers to proceed on the operations.	Material waiting Machine waiting Personnel waiting
Rework	1. Recutting of the fabric as a panel replacement activity due to faults in the fabric or damage while sewing. 2. Correcting on the garments defects (alterations). 3. Rewashing of garments to match the shade. 4. Opening up and correcting the packed garments after a final inspection failure.	Right quality not achieved for the first time
Overprocessing	1. Applying excess pressure and temperature hoping for better attach (sticking) for interlining, which resulting in stains on the fabric.	Lack of process standardization

(*Continued*)

Table 3.1 (Continued)

Type of Muda (Waste)	Activity example	Reasoning
	2. Multiple runs of wash cycles to achieve the right shade. 3. Excess sharpening of the straight knife cutter by the operator while cutting of plies. 4. Application of extra stitches and seams (multiple strokes) in hope of achieving better seam strength. 5. Quality inspections conducted in presewing, sewing, and postsewing processes. The best scenario is the right first time, where no inspection/quality check is required.	
Inventory	1. Getting the fabric and trims in advance (general practice in apparel manufacturing is keeping fabric inventory of 2 weeks) 2. Keeping buffers [in terms of work in progress (WIP)] in the sewing lines (general practice in apparel manufacturing is keeping WIP of 2 days in assembly lines) 3. Keeping in the inventory of expensive machines and machine parts without understanding the consumption (demand) patterns.	Dead material stock staying in the process without any significant progress Inability to understand the concept of uninterrupted flow in the internal and external supply chain
Overproduction	1. Excess production (maybe in cutting, sewing, or any other department). 2. Producing more than desired or orders quantity anticipating quality defects and rejections in future (we generally cut and stitch 3%−5% extra quantity than the order quantity). 3. Producing more than the demand to avoid shortage.	Inability to understand the demand patterns Belief in the Push approach as a prime tool for performance enhancement

3.4.1 Muda of transportation

This is related to any movement of material and/or machine that does not add any value to the final product. Such losses are observed vary commonly in the organizations. Unnecessary and excessive movement of machines and goods results in wastage of time and effort. Poor plant set ups where the arrangement of the facilities accordingly to the process flow is not done may also be responsible for such losses. The other sources of such waste may be a wrong method selection in performing the task. This results in an excess man power requirement to undertake the same activity, the possibility of imbalance in the process of creating bottlenecks. One of the biggest side effects of Muda of transportation is the requirement of excess space.

In the context of supply chain and distribution, Muda of transportation results in excess expenditures in fuel, energy, time, as well as delayed availability of products to the consumers with added cost.

3.4.2 Muda of motion

Unnecessary physical movements of the human fall under this category. Poorly designed workstations and methods are one of the prime sources of such wastes. Traveling a longer distance (maybe within the setup or outside) results in excess time, efforts, and energy with excess physical stress. Muda of transportation and Muda of motion are very much interlinked as the source of both types of waste is same or similar, which include the poor design of layout (where the facilities are located in such a way, one has to move/walk/travel more), poor ergonomic work environments causing excess body motion to the worker that affects the physical health of the work results into reduced efficiency and increased time to complete the task.

3.4.3 Muda of waiting

Unavailability of product, services, or work at a desired or proposed time is waiting. This is one of the most common types of waste observed across all the industries. This waste may be considered as the result of the first two types of wastes that are Muda of transportation and Muda of motion, as increased work in progress (WIP), process imbalances, poor methods, and long distances traveled are responsible for wastage of time in waiting. We observe such wastage of time (Muda of waiting) in day-to-day apparel manufacturing, few examples to cite are waiting for sampling approvals, production delays due to nonavailability of necessary raw materials, poor process balancing resulting into higher WIP, and bottlenecks, waiting of unfinished, semi-finished, or finished products to be processed further.

3.4.4 Muda of rework

This category of wastage is related to the defects in the process results in deterioration in the quality of the output. Cost of quality (or poor quality) is one of the prime concerns in any manufacturing process. In the absence of achieving the right first time, and subsequently, activities involved in rework (to correct the faulty or defective work) is a huge cost, bring all inefficiencies including excess work, excess material (required to make the corrections as well as wastage incurred in making the product the first time), excess time, process imbalance (creating WIP and bottlenecks), excess man power consumption (in terms of manufacturing as well as in terms of supervision and inspection, etc.), delayed delivery and eventually loss of profitability to the organization and loss of reputation which is vital for survival in the competitive business environment.

3.4.5 Muda of overprocessing

This type of waste is referred to as the extra processing or inappropriate processing is done (knowingly or unknowingly) on the product. Many times this is done to improve the quality or performance of the product but NVA as the process or activity does not meet the criteria of VA (please refer Section 3.2.1 discussed in this chapter). In simple words, overprocess may be understood easily with this example, suppose one applies fertilizers in the agriculture field to achieve the maximum yield, however, the excess quantity of the fertilizer may destroy the crop as well. This is nothing but a matter of having the right or a balanced mix of ingredients (due to lack of process standardization) which is most of the time ignored, hence it results in the Muda of overprocessing.

It is a commonly observed scenario that in the absence of the SOP people do hit and trial to achieve the results and end up wasting the material, incurring defects, rework, wastage of time, and other resources. There are several examples in apparel manufacturing that may be cited under this category. Few of such examples are applying extra/additional stitches, bar tacking, reinforcement using rivets, etc., to improve the strength of the seam than prescribed in the technical specification of the product. Consuming excess chemicals (in the absence of standard recipe of the washing treatment), overrubbing of surface (hand-sand, whiskers, destruction, etc.) than appropriate quantity to get wash effect. Excess exposure of the material to heat treatment (heat-sealing) under the impression that excess heat may yield better results. In the context of day-to-day office work, examples of such waste may be data duplication where the same data entry is done by people at different points or departments, and excessive report generation which results in errors, wastage of time, and wastage of man power.

3.4.6 Muda of inventory

Keeping excess WIP is one commonly observed scenario, it is generally referred to as to keep a higher level of stocks or inventories than required in the process. Many a time, excess WIP is kept to avoid the situation of idleness in the process. Higher levels of dead inventories result in several operational issues and challenges which are difficult to handle and control. Despite shorter fashion cycles, typical apparel manufacturing is known for its higher process inventories. Excess inventory is nothing but a blockage of capital resulting in dead assets. It also leads to unwanted

Fundamentals of lean journey 59

expenditure on storage and security with the possibility of higher defects as well as the possibility of obsolescence. To understand the losses due to high inventories, an example of the year 2004 of the steel industry may be taken when a significant rise in steel prices was witnessed globally. Following the same pattern, in August 2004, the steel prices were at an all-time high. It was anticipated that the same trend shall continue and that results in keeping safe buffer stocks of steel to avoid further losses. After some time, suddenly the steel prices started declining and people were left with huge inventories of unused steel which was purchased at higher prices. This resulted in big financial losses and difficulty in selling the goods at a competitive or lower price with a huge shrinkage in the profit margins. In the context of apparel manufacturing, larger quantities of raw materials require bigger space as well as protection from theft, sunlight, and rain, etc. This is purely nonvalue adding activity, with a possibility of obsolescence of material in situations of reduced demand in the future. Excess inventories created due to wrong processing may result in bigger losses as well, for example, cutting inventory based on wrong fabric or faulty pattern or excess stitching WIP with measurement or fit related issues may result in shipment failures quality grounds. It may be observed here that the magnitude of an error can be defined by a controlled inventory level.

3.4.7 Muda of overproduction

This category of wastage is considered as the "Mother of all the Mudas." The reason for the same is that in most of the cases all the other kinds of wastes are developed from the excess production only. Hence it results in various types of inefficiencies and losses. Lean thinking believes in producing lesser just to meet specific demands, while overproduction is just opposite to this thought. This is nothing but the traditional push-based thinking where excess production of goods and services asks for a higher level of inventories in terms of raw material, extra space to store the excess production and inventories. This also leads to the chances of error or quality-related issues. Sometimes the errors are due to an unwanted pressure to produce more, where SOP are skipped. The situation of overproduction also leads to process imbalances creating bottlenecks causing Muda of waiting or idleness in the process.

Apart from all the seven above-listed wastes, a different type of waste which is Muda of excess man power is also witnesses, though it is not considered among the typical waste categories as per lean thinking. The excess man power in a system that leads to an unused human resource is still not recognized completely, but surely it creates losses to the system. Reduced levels of man power utilization, higher man against a specified number of machines, and reduced productivity levels per person are some key indicators for Muda of excess man power. It also results in increased liabilities and additional expenses, which are completely undesirable in the competitive business environment of current times.

Poor work allocation, where human competencies are not fine-tuned with the job requirements, is one commonly observed scenario. Skill assessment of individuals must be done very carefully, and a competency mapping should be done religiously following a scientific approach. A mismatch between job allotted and the skill sets of personnel results in financial losses due to inappropriate wage distribution, possibilities of increased operational errors, and quality issues due to lack of desired skills to perform the task. More importantly, this may affect the morale of the person and his/her motivation toward work.

In apparel manufacturing preproduction processes, motion was found to be the biggest contributor accounting for 48.36% of all waste on an average. Defect and Reworks was 16.21%, Inappropriate Processing was 13.75%, Transport was 11%, and Waiting Time was least at 10.66% (Ladia & Jana, 2014). The systematic elimination of waste has resulted in Sri Lankan mass garment manufacturers reducing average stock holding days from 85 to 65 days to an average between 20 and 40 days (Financial Times, 2011).

In comparison to manufacturing, a typical office setting deals with paperwork, phone calls, and meetings, etc. Such an office setting is very common in preproduction functions in any apparel manufacturing organization. Transportation wastes are uncommon in an office setting but often have a lot of information related to wastes. Unnecessary paperwork, missed phone calls, data translation errors, and poorly run meetings with too many participants are all common forms of office wastes. It is possible to categorize office waste using a manufacturing waste categorization scheme, but other schemes designed with the office setting in mind yield more meaningful findings. Lean consultants from Kaufman Global have created one such waste categorization scheme designed for the office setting. As per this scheme, wastes are grouped under four main categories: people's energy waste, process waste, information waste, and people's work waste. People's energy waste results from a failure to harness an employee's potential. The possible causes of people's energy wastes include poor focus, poor structure, no ownership, excessive control, tampering, assignment issues, and improper goal alignment. Process wastes are the wastes from an inefficient structure, interaction, and execution of complex business processes. The possible causes of process wastes are checking, work-around, nonstandard, unbalanced flow, and suboptimization. Information wastes arise from inefficient data flow between activities and across connections. Causes of information wastes are due to poor handoffs, translation, missing information, irrelevant information, and inaccuracy. People's work wastes are similar to the waiting, motion, and overprocessing counterparts found on a factory floor. The causes of people's work wastes are waiting, motion, and processing. This categorization scheme for waste suits knowledge workers in an office setting as they can relate to these categories more often than Ohno's manufacturing types of waste (Kaufman Global, 2007). Although no implementation or experimentation of this classification is found in the apparel manufacturing industry, this holds tremendous potential for application in the apparel preproduction setting.

3.5 The lean way for improvement and its challenges

Lean is all about producing more with lesser resources. In the cutthroat competitive environment, where it is not only a matter of doing business and survive but also to strive continuously for achieving operational excellence. In such situations, optimum utilization of resources (man power, material, machines, information, space, etc.) to get desired on-time quality output is the only way, and being lean is the most appropriate approach. Lean philosophy is very much capable of solving out the issues and challenges in a practical environment and that has been witnessed again and again in several industries. According to Locher (2008), successful adaption to lean practices results in a 40%–90% reduction in lean times, 30%–50% reduction in process times, and 30%–70% enhancement in the quality performance (Locher, 2008). Despite such promising improvement, the adapting lean approach and its applications are not truly observed in the apparel manufacturing where lean initiatives are yet to be accepted whole-heartedly (Silva, Perera, & Samarasinghe, 2011b).

Although it is difficult to attribute any particular tool to any organizational improvement, organizations that have implemented lean manufacturing systematically have been able to gain many benefits out of that implementation. The most achieved benefits are lower defects (14%), improved employee satisfaction (14%), lower inventory (13%), and lower lead time (12%) (Silva, Perera, & Samarasinghe, 2011a). Through lean implementation in Bangladesh, line balancing and labor productivity increased by 11% and 24.5%, respectively, whereas WIP, alteration, rejection, and spotting were reduced by 85.4%, 10.67%, 33.34%, and 75%, respectively (Ahmed, 2013). Another study shows that efficiency has been improved by 44%, lead time has been reduced up to 45%, reduction in absenteeism reduced by 57%, labor turnover reduced by 50%, besides improved results in quality performance such as Right First Time (RFT) (Chowdhury, Alam, & Rupaul, 2019). The same trend of the appreciated result was observed in the reduction of man power (22%), increase in productivity (14%), increase in line efficiency (32%), increase in capacity per hour (19%), and increase in SMV target fulfillment (23%) by applying lean techniques such as 5S, value stream mapping, Kanban, Kaizen, cellular layout, just in time, etc. in a sewing line (Jahan, Debnath, & Biswas, 2018).

In apparel manufacturing, there are several myths and misconceptions related to adopting the lean way and that works as challenges in successful lean implementations. Most of the apparel manufacturing plants still operate with the traditional approach and rigid to change (Raju & Modekurti-Mahato, 2014). The traditional approach in apparel manufacturing involves cutting down the costs by firing people, downsizing, and working with higher WIP are still considered as key tricks of being economical. Apart from it, typical apparel manufacturing setups struggle with additional issues such as poor quality, lower productivity levels, poor space utilization, longer assembly lines with higher levels of WIP, higher machine downtimes, and poor work allocation due to lack of work standardization. This results in longer lead times and higher operating costs.

Some of the majorly observed challenges to lean implementation in apparel manufacturing may be cited as misconceptions including lean is for standardized work, unique nature of apparel manufacturing, lean implementations are not sustainable, lean should not be implemented by the organizations already doing well, lean should be carried out by somebody, lean is just a buzz word, and lack of know-how of the lean implementations. The challenges of lean implementations in apparel manufacturing have been discussed in Chapter 1, Lean Management in Apparel Manufacturing.

3.6 Lean way of manufacturing

Being lean is a never-ending process aiming for continuous improvement and achieving newer heights of operational excellence. This requires a great deal of constantly reviewing and assessing the existing situation, identifying the possible areas of improvement. This way of relentless reflection and commitment to continuous improvement are the key traits of a learning organization. Gaining knowledge and experience through regular analysis and critical review, planning for further improvements, and taking necessary steps to attain the set targets in a phased manner are some of the essential milestones in a lean journey (Technopak Advisors, 2011).

Lean is not just about applying its tools and techniques but much more than this. Becoming lean starts much before than its practical implementation. Smeds (1994) considers lean interventions as an innovation, which is related to the generation of a new idea and its implementation into a new product, process, or service to create wealth (Smeds, 1994). Further, based on the nature of change an innovation brings in the organization, innovations may be classified into two categories: (1) incremental innovation, and (2) radical innovation. The incremental innovation improves the old process or product within the existing structure, whereas the radical innovation results in new business opportunities, new strategies, and new structures (Burgelman, 1983, 1988; Urabe, 1988). The radical innovations are not necessarily to be the big ones or big jumps in an organization but are the collective result of smaller changes occurring at consort and bringing significant advantages (maybe in form of decreased costs or higher quality) to the organization. The lean intervention should be considered as radical innovations, as the lean improvements in the business process can create radical impact by converting an organization into a lean enterprise (Smeds, 1994).

The very first point is the cultural change by making people aware of the existing situation and establishing an urge for improvement. Here human element becomes vital, especially in apparel manufacturing which is primarily a human-driven industry. People are the biggest and the most valuable resources, and any improvement cannot be achieved without active and value-added involvement of the workforce. There are several pieces of evidence through proven research that there is no significant difference in implementing the hard practices at the lean plants, but this is the extensive implementation of soft practices which make lean transformation success at such lean plants (Bhasin, 2012; Bortolotti, Boscari, & Danese, 2015; Shah & Ward, 2003).

Creating a conducive work environment for any change (such as lean initiative) is a challenge. Any change is difficult and painful, and it is not something which can be done overnight. Changing the social environment of work takes time to receive acceptance from the workforce (Karlsson & Åhlström, 1996; Sparrow & Otaye-Ebede, 2014). Continuous improvement can only be done with contribution from all of the employees, where they continually seek and eliminate the sources of process imperfections (Ni & Sun, 2009; Wynder, 2008). Continuous improvement in an organization requires a lot of efforts for a longer duration of time by each of the member (Malik & YeZhuang, 2006). Given the same, employees' views on the impact of new working methods on the shop floor becomes very important. It is critical to understand employees' task performance and the new work environment (Losonci, Demeter, & Jenei, 2011). It takes time in bringing all the stakeholders on board. To convince people for change and to engage them as active stakeholders in the journey of improvement, it is imperative to work on some core questions. Such questions may be like why the change is required? What will be the role of people in carrying out change? What are the expectations (long-term and short-term) from the change? What will be the expectations from different teams while change process? What will be the mechanism for measuring the improvements or the improvement assessment mechanism? As any change aiming for improvement involves cost, time, and efforts, it has to be planned and implemented carefully. The road map for change should be designed in a planned and logical manner to ensure progress in the right direction. A constant assessment and monitoring mechanism of the transformation should be devised and practiced religiously. The start point of improvement: to become competitive and world-class, it is very important to produce correct for the very first time at each of the manufacturing processes. In the context of apparel manufacturing, production is the area that has relatively high human intervention and it is often seen as the hub of the problems as well. Given the same, the recommended area to initiate change should be the production floor. The other reason to initiate the improvements from the production floor or sewing floor because a sewing machine is a point where the raw material is getting transformed into a semi-finished or a finished product. Hence, a sewing machine can also be treated as a point of revenue (value) creating. An organization with a poorly managed, inefficient product facilities cannot be expected to deliver the right product at the right time at the optimum price. It is important to inject autonomy in the work, and the shop floor workers should be involved in the diagnose and control of the problems. This may be an effective way to start with the lean intervention. Shop floor employees enjoy being involved in the decision-making process to identify and rectify the issues (Forza, 1996; Wall, Jackson, & Davids, 1992). This helps in recognizing their contribution to continuous improvements and encourages them to learn further for improved task-oriented behaviors (Forza, 1996).

3.7 A framework to initiate lean interventions

It is very important to devise a long-term strategy before initiative lean interventions. Thoughtful insights and groundwork result in a solid foundation for the sustainable improvement process. Smeds (1994) proposes a generic framework to manage innovative change toward a lean enterprise. The management of change frameworks starts with devising the strategic vision and an umbrella strategy to carry out the improvements (Smeds, 1994), it includes establishing the need for change. Most of the time such an umbrella strategy is referred to as the mission of the organization, which is nothing but a way on how to achieve the vision. Hence, the terms of vision and mission are generally used together. As indicated in Fig. 3.5, once the vision and mission are finalized, the assessment of the present state, identification of challenges in the existing system, and opportunities are discussed in a participative manner. This includes activities such as value chain analysis, assessment against lean practices benchmarks, and development of the model for process redesign. The next step is experimentation and selection of the future state, which include simulation exercises (with an assessment of different future state models), debriefing workshops, and finally deciding on the best model of the future state. Practical implementation of change to achieve the future state

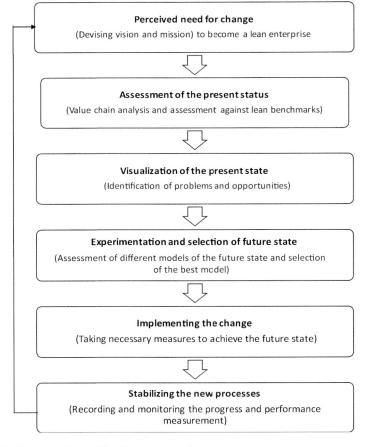

Figure 3.5 Framework to initiate lean interventions.

Fundamentals of lean journey

model is one of the most important activities. It includes adapting to the new processes, its assessment, and visual communication of the processor developments during the implementation process.

For a sustainable change and to achieve continuous improvements, it is essential to stabilize the new process. It requires documentation of the new process, constant measurement of the changes, and assessment of progress while continuous improvement. Once the process is stabilized and has achieved the targeted results (improvements), the need for new major reorganization should be established, and the same process should be repeated.

A recommended step-by-step methodology (refer Fig. 3.8) to initiate the improvement as a part of lean initiatives is discussed in the next paragraphs.

3.7.1 How to start

Any kind of improvement starts with a critical review of the existing process or activities and identifying the areas of potential improvements. It is essential to work out the plan for identifying the areas of improvement. Once the vision, or strategy, or policy comes to mind in lean implementation, the Hoshin Kanri appropriately fits the purpose. It is a management technique that was first developed by Professor Yoji Akao in Japan in the 1950s. The Japanese word Hoshin means "direction" or "compass needle." Kanri means "control" or "management." This reflects the intention of the technique to let the strategic goals of the organization guide every decision and action (Jacobson, 2017). The term roughly embraces four key elements of business management namely: Vision, Policy Development, Policy Deployment, and Policy Control (Hutchins, 2008). While Hoshin Kanri is seen as a glorified cross between a to-do list and a PDCA (Plan, Do, Check, Act) (Roser, 2020), Hutchins mentioned Hoshin Kanri and Japanese-style total quality management (TQM) are intrinsically related to each other (Hutchins, 2008). The four key elements of Hoshin Kanri are presented in the form of Hoshin Kanri X-Matrices. The central Hoshin Kanri box has four parts: long-term (breakthrough) objectives start at the bottom or south and move clockwise. The annual objectives that are mentioned at left-hand or West direction in the Hoshin Kanri box) are followed by the top-level priorities (how?) (as mentioned at the top or North direction in the Hoshin Kanri Box). These top-level priorities are followed by the final targets to improve (as mentioned at the right-hand or East direction in the Hoshin Kanri Box). These X-Matrices are also referred to as a table of A3 sheets (Kesterson, 2015).

Although not popularly used in apparel manufacturing, Hoshin Kanri may be a useful tool tracking the strategy deployment linking the cascading structure from breakthrough objectives to the annual objectives down to top-level improvement priorities ensuring that there are tangible targets and that the efforts are resourced. Fig. 3.6 shows a Hoshin Kanri box of an apparel manufacturing organization, showing slightly different orientation where the start point is left triangle (West), that decides what is to be done (general policies). The second step is the bottom triangle (South) that shows how much of what is to be done. In the third step (the top triangle, North), the strategies or projects are decided (to achieve the global effects).

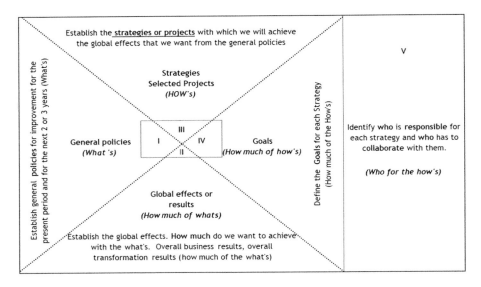

Figure 3.6 Hoshin Kanri box.
Source: Courtesy Corrigo Consultancy Pvt. Ltd, India.

The final step (the right triangle at East) lists how the microtargets and goals will be achieved.

Fig. 3.7 shows the Hoshin Kanri X-Matrix. In the first step, the general policies are listed (serial no. 1 to 5) in the columns; like on-time delivery, improvement on RFT quality, and so on.

In the second step, each Hoshin (what) is further elaborated with numerical targets (how much) and their linkage with each Hoshin are shown by an "X" in the south-west corner intersections grid. For example "on-time delivery" is linked with the following seven activities (refer to Table 3.2) and "Improvement on RFT Quality" is linked with three activities and so on.

In the third step, strategies (sub-Hoshins or micro-Hoshins) for achieving each of five Hoshins (listed in the first step) are listed above the north triangle. For example, the serial no. for "Improvement on RFT quality" was 2; the three strategies linked to this are serial numbered as 2.1, 2.2, and 2.3. Similarly, the improvement of targets or goals are substrategies and numbered as 1.1.1, 1.2.1, and so on. The relationship between two adjacent sets of Hoshins and sub-Hoshins is shown by "X" in four respective corner intersections grids and there may be one Hoshin linked to more than one sub-Hoshins and one sub-Hoshin may be linked to multiple Hoshin and so on.

Finally, the resource persons are also linked with the goals clearly showing who coordinates and who participates. Although we see the items listed under global effects are based on a target (e.g., reduce downtime by 50%), Roser argued that these items should be based on a process, not on a target. In the original Hoshin Kanri, there is more focus on having a "process rather than having numeric targets".

Fundamentals of lean journey

This helps in implementing better countermeasures against repeated problems and developing the quality skills of the people (Roser, 2020).

3.7.2 Continuous improvements

The lean journey is all about continuous improvements, which is also called as Kaizen (a Japanese term used for continuous improvements). Hence the lean

ABC Limited [Improvement Plan 2018-2019]

Improvement items (How's):

No.	Description
5.4	Review and improve all Policies
5.3	Review and improve recruitment & training
5.2	Multiskill Operators
5.1	Establish Kaizen Drop Boxes & conduct events
4.4	Create Standard Work SOP
4.3	Process Analysis to eliminate waste
4.2	Work on BEC on all machines
4.1	Work on QCO in Modules & Cutting
3.4	Introduce Supplier Kanban
3.3	Improve the Sourcing Methods & Policy
3.2	Review and improve order management method
3.1	Improve the IT Support System
2.3	Introduce Andon Signals
2.2	Apply Poka Yoke process to control Quality
2.1	Implement Source Inspection
1.10	Create Cut Kit Supermarket
1.9	Improve the Planning Methods
1.8	Implement Lean Metrics
1.7	Implement Visual Management System
1.6	Implement 6S
1.5	Establish Kanban to drive pull production
1.4	Reduce Bundle Size to one-piece-flow
1.3	Train Supervisors on Line Balancing Chart
1.2	Create Cut to Box modules
1.1	Training on Lean Manufacturing System

Central X-Matrix:

- (I) Objectives (What's)
- (II) how much of the what's
- (III) How's
- (IV) How much of the how's

Objectives (What's), left axis:
- 5 — Motivated Workforce
- 4 — Reduction in Manufacturing Cost
- 3 — Materials Inhouse Ontime
- 2 — Improvement on RFT Quality
- 1 — Ontime Delivery

Targets (how much of the what's):

No.	Description
1	Improvement in SDP from x to 97%
2	Reduction in Downtime from x to y
3	Reduction in Absenteeism from x to y
4	Reduction in LTO from x to y
5	50% Reduction in WIP inventory
6	Reduction in MLT from x to y
7	The serial number for
8	Reduction in Rework % from x to y
9	Reduction in Off Grades from x to y
10	Reduction in Inventory Holding from x to y
11	Reduction in OT from x to y
19	
20	

How much of the how's (detailed targets):

No.	Description
1.1.1	100% of the Executives, Staff, Sups to be trained by end Jul 07
1.2.1	100% of the modules to be cut to box by end Aug 07
1.3.1	100% of Supervisors to be trained on Line Balancing by end Aug 07
1.4.1	100% of Cut to Box Modules to have 1 piece flow by end Aug 07
1.5.1	FOW to RMW Process to work on Kanban Signals by End Oct
1.6.1	75% to be scored on 6S audit by end Sep 07
1.7.1	Signs of VMS to be seen by End Aug 07
1.8.1	100% of Employees to be aware of Lean Measures by End Aug
1.9.1	Have an automted & sophisticated Module by End May07 with a plan to actual
1.10	Cut Kit Supermarket with Kanban to be made available by May07
2.1.1	RFT to increase from x to y by end Aug 07
2.2.1	Off Grades reduction from x to y by end Oct 07
2.3.1	100% of the modules to have and use andons to stop for quality issues
3.1.1	Have a IT system to support from Order Management to Delivery by End Aug 07

Figure 3.7 Hoshin Kanri X-Matrix.
Source: Courtesy Corrigo Consultancy Pvt. Ltd, India.

initiative programs are also called Kaizen programs. A clear vision needs to be devised by the top management as the very first step and accordingly, in the alignment of the vision, a long-term strategy (mission) for lean movement (kaizen program) should be worked out. Any Kaizen program is generally involved in a kaizen event, which can also be the start point of the kaizen program or kaizen mission (Ortiz, 2006). This starts with the formation of a small group of 8–10 personnel from different areas who work together to address a particular area identified. The team members should be made aware of the vision of the organization and the

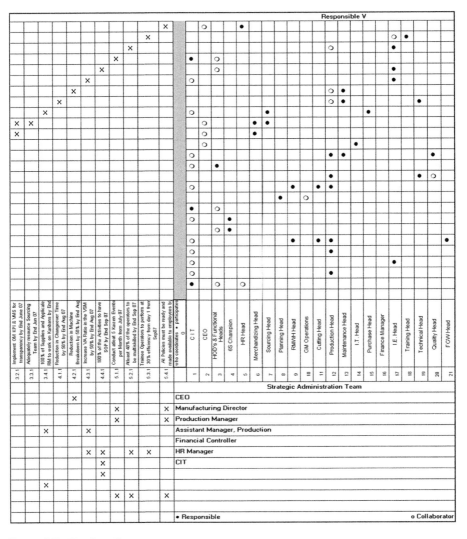

Figure 3.7 (Continued)

Table 3.2 Linkage of Hoshin objectives with targets.

"Ontime delivery" is linked with following seven	"Improvement on RFT Quality" is linked with following three
• Improvement in supplier delivery performance (SDP) from x to 97% • Reduction in downtime from x to y • Reduction in absenteeism from x to y • Reduction in labour turn over (LTO) from x to y • 50% Reduction in work in progress inventory • Reduction in manufacturing lead time (MLT) from x to y • Increase in productivity by 15%	• Increase in productivity by 15% • Reduction in rework % from x to y • Reduction in off grades from x to y

long-term strategy should be clearly explained. It is very important to have clear guidelines or plans (Wilson, 2010) to keep this exclusive group of employees engaged in the task allotted and make them accountable. The kaizen even needs to be supervised closely, as this helps in executing the kaizen event in a more organized, directional, and controlled manner and prevents dilution of efforts while achieving the targeted results eventually working toward making the vision a reality. A step-by-step recommended methodology to initiate a kaizen program is shown in Fig. 3.8. As indicated in Fig. 3.8, top management plays a vital role in any kaizen program, as they are instrumental in working out vision and mission and it is communicating to the employees (Worley & Doolen, 2006). It also confirms the commitment of the top management to carry out a successful kaizen program or lean mission.

The appointment of kaizen champion is one very critical activity for any kaizen program. A kaizen champion is a person fully responsible to carry out the kaizen movement at the organization. A kaizen champion should be an experienced lean practitioner possessing skills in project management and planning. The kaizen champion should have the necessary skills and hands-on experience of executing 5S, standard work, waste reduction, visual management, and project management (Ortiz, 2006). Considering the importance of the kaizen program, it is recommended that the kaizen champion should be exclusively appointed to the kaizen work. A kaizen governing committee should also be formed by the top management. Such a committee is responsible for monitoring the progress of the kaizen events. Guiding and helping out kaizen teams to carry out the improvements is one key responsibility of such a committee. A kaizen governing committee should have members from all the concerned departments. For example, in the case of apparel manufacturing, a kaizen governing committee may have Store Manager, Merchandising Manager, Quality Manager, Production Manager. Process planning and control manager, Industrial engineer, Finishing and packing manager, Human resource manager, and kaizen champion as the members.

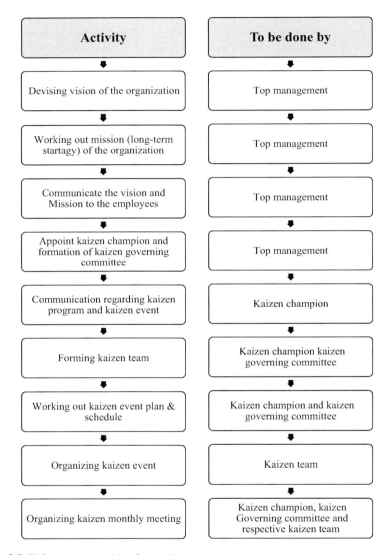

Figure 3.8 Kaizen program (step-by-step).

Lean philosophy considers human resource as the most vital and driving force toward achieving the transformations. The involvement of human resources from all the levels of the organization is always encouraged in a kaizen program. Kaizen teams are a critical link in the journey of continuous improvement. Formation of kaizen teams is done to identify the areas of improvements and activities to be undertaken. Kaizen team is the one who is going to execute the improvement on the shop floor. The kaizen team should consist of at least one person from the concerned area or department depending on the event. For example, a kaizen event for

Fundamentals of lean journey

cycle-time improvement of waist-band attach of a five-pocket jeans (refer Fig. 3.9 for kaizen communication event board), kaizen team members should be the industrial engineer, line supervisor of the respective line, waist-band attach operator, quality supervisor, and the maintenance person. The most experienced person from these team members should be appointed as the team leader who is responsible for leading and guiding while kaizen event. Apart from these persons, one person from top management should also be a member of the kaizen event team. This shall further boost the commitment of top management to carry out the change and also encourage people to form middle and lower management to execute the kaizen event. It is also recommended that the members of the kaizen governing committee actively participate in the kaizen event. This helps in keeping the improvement process in the right direction and maintain the focus of the event.

A kaizen communication system should be developed by the kaizen champion, where a clear and detailed announcement of the kaizen program and the kaizen events is deone. Such a kaizen communication helps in improving the awareness and seriousness of the lean initiatives. Addressing the common queries of the employees and developing a holistic approach toward the imprcvements is another key objective of such communications. Please refer to Fig. 3.9 for a sample of a kaizen communication board. Some details such as the plan of the kaizen event, area of intervention, expectations of the kaizen event, etc. are to be communicated. For an effective and wider circulation of information, notice boards may be placed at the locations where more and more people can see such boards.

The day-to-day activities and improvements under the kaizen events (with all necessary relevant details of the activity, team, problem statement, challenges, and the outcome in terms of improvements, etc.) should also need to be communicated to all the levels of hierarchy in the organization. A newsletter with all such details and updates should be published at a regular interval. Such newspapers may be circulated on a weekly or monthly basis depending on the quantum of improvements. The kaizen event calendar should also be mentioned in such newsletters so that people are aware and updated for the upcoming happenings, and concerned people are well prepared for the event.

Proper detailed documentation of the issues, improvements, status of achieved goals with all supporting data is a must for every kaizen event. Each of the kaizen events must be well recorded as well as well reported in an organized manner. A kaizen tracking system should be developed, where all the relevant information is

Kaizen Communication Event Board			
Kaizen Event	Line No.	Event Schedule	Team Members
Waist-band attach to 5-Pocket Jeans	12	24 January 2020 4:00 pm	1......... (Team Leader) 2............................... 3............................... 4...............................
Objective		To reduce the cycle time of waist-band attach operation	

Figure 3.9 Kaizen event communication board.

Kaizen Event Tracking Sheet												
Kaize n Event Name	Team members	Duratio n	Objectives	Expectatio ns	Results achieved	Pre-planning	Estimat ed cost (INR)	Actual expens es (INR)	Next course of actions	Schedu le for the next activiti es	Responsible person	Status
Waist -band attach to 5 Pocke t Jeans	1........ (Team leader) 2......... .... 3......................... 4.........................	24 January 2020 4:00 pm	1. To reduce the cycle time of waist-band attach operation. 2. To improve the operator utilization 3. Enhanceme nt of Machine productivit y.	1. Cycle time reduction by 5% 2. Operator utilization by 8% 3. Enhanceme nt in machine productivit y by 6%	1. Cycle time reduction by 3% 2. Operator utilization by 5% 3. Enhanceme nt in machine productivit y by 2%	1. Method study 2. Time study 3. Waistband material arrangeme nt	5000	3500	1. Further method improveme nt 2. Modificati on in the Work-station layout	11/02/2 0 16:00	1........ (Team leader) 2......................... .... 3......................... .... 4.........................	Not complet ed

Figure 3.10 Kaizen event tracking sheet.

recording to track the progress of each of the kaizen event. The kaizen event tracking format (refer Fig. 3.10) should essentially have key details of the kaizen event, kaizen team member (who carried out the kaizen event on the floor), duration of the event, the key objectives (s) and the expectations of the kaizen event, preplanning done for the event, financial implications (estimated and actual), results achieved, next course of actions required for further improvements with its schedule (time line), task or responsibility allocation to each of the team member, and the status of the progress against the targeted results. A kaizen event tracking system must be updated constantly and should be discussed by all concerned at a regular interval in the monthly kaizen meetings. A kaizen meeting discussion should be well structured (discussion on current status, activities undertook report of the last meeting discussion, and next course of action) and focused on the agenda only, and it should not be allowed to deviate. Each of the kaizen events should be quickly but completely discussed covering all the aspects of the respective kaizen event.

3.8 Step-by-step implementation of lean tools

It takes time in lean implementation, as it starts with the cultural change, the change of social environment (Karlsson & Åhlström, 1996; Sparrow & Otaye-Ebede, 2014). Further, it is important to understand that lean is not for the short-term but it is a journey of long-term continuous improvements (Womack & Jones, 1996). Further, it is to be mentioned that the implementation order of lean tools is more or less same in different industries, however, the organizational culture must be carefully studied and analyzed before any lean intervention (Silva et al., 2011b). It requires fundamental rethinking and cultural change, change in routines, and actions by the management as well as the shop floor personnel (Wickramasinghe & Wickramasinghe, 2017). Lean implementation is considered as the diffusion of innovative technology (as a function of time), which involves several innovative steps toward adapting new routines, new and innovative methods, changed behavior, and actions by the workforce (Powell, 1995; Wickramasinghe & Wickramasinghe, 2016). Here time becomes a key determining factor for a successful lean adaptation.

Given the same, lean is a process of appropriate and committed application of the innovative technology (Klein & Sorra, 1996). According to Silva et al. (2011b), the application of lean tools depends on several factors such as the nature of the industry, plant size, and the technological capabilities of the country. 5S and visual management tools are one of the most applied tools in Indian and Sri Lankan industries. Single Minute Exchange of Die (SMED) is a very popular and widely applied lean tool in the Sri Lankan apparel industry, which is contradictory to SMED applications worldwide. Six sigma is heavily used by US industries. Application of VSM is observed at a larger scale in US and Canadian industries compared to small scale Indian industries. The Australian industry focuses more on TQM over kaizen, kanban, and group technology (Silva et al., 2011b).

Though lean transformation is a long journey, some tools can be implemented in the beginning phase. The research was conducted to confirm tools such as 5S and visual management. These tools are capable of making shop floor changes visible and help in quickly improving the financial aspects of the organization. The next focus should be on operation stability and flow management which may be achieved by the application of takt time, one-piece flow, cellular manufacturing, and SMED (Silva et al., 2011b).

Different work environments have different issues, and accordingly the objectives for improvement. Every single organization is unique in terms of its recourses, capabilities, and limitations. A common solution that suits all the organizations may not be possible, hence organizations used to adopt customized solutions with selected tools and techniques. As discussed earlier, lean initiatives should be taken in a logical phased manner. It may not be necessary to apply multiple tools at the same time. Pavnaskar, Gershenson, and Jambekar (2003) discuss the misapplications of lean tools and techniques and classify these misapplications as: (1) applying the wrong lean tool for problem-solving, (2) applying single lean tool as a panacea, and (3) applying all the lean tools as a solution to each problem (Pavnaskar et al., 2003). Loayza, Olave, Perez, Rojas, and Raymundo (2020) propose a lean thinking-based model for apparel companies and suggested to implement VSM to visualize the value streams in the organization. Implementation of VSM helped in determining the current situation (through the identification of waste in the process along with cycle time and lean time in the production process) of the process. The VSM exercise guided in setting up objectives and production management and control indicators. Other lean tools including 5S and work standardization were applied to maintain work areas and to create process stability (Loayza et al., 2020). Andrade, Cardenas, Viacava, Raymundo, and Domingue (2020) propose a lean tool-based methodology to reduce the lead times in the textile industry through a reduction in production time and reduction in quality-related issues. In the proposed model, VSM was applied as a diagnostic tool to identify the problems (wastes) in the production process. Lean tools 5S and Kanban were integrated to achieve the correct information flow, planning of work method, and reduction in the manufacturing times. Further, the Deming cycle (Plan–Do–Check–Act) was applied to ensure continuous implementation and improvements. This resulted into a significant impact on the improving productivity (increased by 41%) and reducing quality issues (defective products reduced by 25%) and led to higher revenue generation (increased company income by 84%) and increased sales (nonfulfilled orders reduced by 26%) (Andrade et al., 2020). Ferdousi and Ahmed (2009) investigate the manufacturing performance improvement in Bangladeshi garment firms and observed that Kanban, just-in-time (JIT), 5S, Pull production, total productive maintenance (TPM), and Kaizen were the key lean tools applied for improvement. The study also revealed significant improvements in reduction of production unit costs (minimum reduction of USD 0.03 to maximum reduction USD 2.00 per unit cost), total productivity improvement (minimum 10% to maximum 60% productivity improvement), reduction in lead time (minimum 3 days to maximum 30 days reduction in lead time), quality improvement (minimum 10% to maximum 80% in quality improvement), and reduction in manufacturing cycle times (from 5 to 20 minutes, indicating time savings of 12.5%–33.33%) (Ferdousi & Ahmed, 2009). According to Nawanir, Teong, and Othman (2013), flexible resources (such as human involvement and empowerment, multifunction teams, and multifunctional resources), cellular layouts, pull system/kanban, small-lot production, quick setup (such as quick changeover and SMED), uniform production level (such as uniform work allocation and production scheduling), quality at the source (such as process quality control and quality circles), total productive maintenance (such as autonomous maintenance and workplace arrangement with 5S), and supplier networks (such as JIT delivery by supplier and supplier development) are some of the common practices of lean manufacturing (Nawanir et al., 2013). According to Marudhamuthu, Krishnaswamy, and Pillai (2011), based on their research on the Indian garment export industry, lean implementation results in increased productivity at a lower cost and producing the best product and service quality. This results in an improved process environment with reasonable improvement (Marudhamuthu et al., 2011). Saleeshya, Raghuram, and Vamsi (2012) state that lean implementations in the Indian textile industry include activities such as process analysis, rapid setup and quick changeover, visual management (using color coding for volume-mix identification), applications of kaizen, and quality circles (Saleeshya et al., 2012). According to Bashar and Hasin (2018), lean implementation results in several benefits including a reduction in inventory levels, quality improvement, productivity enhancement, reduction in delivery time (lead time), improvement in on-time delivery, cost reduction, and reduction in cycle time. VSM and 5S used for waste identification, whereas kaizen, one-piece flow, standardized work, production leveling, cellular manufacturing, visual management, SMED, Kanban and JIT, TPM, and TQM are used as waste elimination tools (Bashar & Hasin, 2018).

From the literature cited, it may be concluded that it is good to start the lean journey by applying the tools which can create an awareness in the employees and can trigger the sense of change. VSM (as a diagnostic tool to identify waste in the process), 5S, and visual management tools are quite capable for the same (refer Fig. 3.11; the lean tools mentioned in the figure are just indicative,

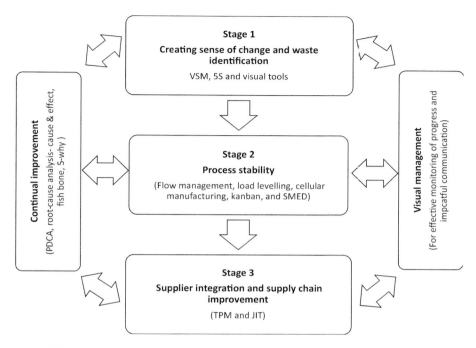

Figure 3.11 Lean implementation.

there may be other additional tools as well depending on the nature of the organization and specific requirement). In the next stage, the focus may be on process stability that involves operational standardization. Operational standardization may be achieved through flow management which may involve the application of tools such as takt time-based load leveling (Heijunka), cellular manufacturing, kanban, and SMED (as a quick changeover tool). Simultaneously, visual management tools may be parallelly applied for effective monitoring of progress and impactful communication to all levels of organizational structure. Other lean tools such as PDCA, root-cause analysis (maybe with cause and effect, fishbone diagram, and five-why analysis), and standardized work may be applied regularly for continuously investigating the potential areas for sustainable improvement. Once the organization has achieved significant stability in lean initiatives, tools such as TPM and JIT may be applied as measures for supplier integration and improving the effectiveness of the supply chain.

For the sustainability of lean improvement (to become a lean enterprise), it is imperative to understand that each of the lean tools is an integral part of the organization culture, and any lean intervention aims for continual improvement. Given the same, once a lean tool or technique is applied it has to be practiced throughout without having a feeling that as we are done with a particular lean tool and achieved the desired results, there is no need to continue with the same.

3.9 Conclusion

Lean philosophy has tremendous proven potential for continuous improvements (also known as Kaizen), and lean tools and techniques are widely used in different industries worldwide. Organizations have witnessed significant success through becoming a lean enterprise. Many of the apparel manufacturing organizations in a different part of the world are practicing lean approach and trying to achieve operational excellence; however, the nature and structure of lean interventions may not be same as it is in the other industries (such as automobile manufacturing which has achieved greater success through lean initiatives). It has been observed that that apparel manufacturing organization tweaks the lean tools making lean interventions customized as per their specific requirements without compromising the sole objective, which is achieving continuous and sustainable improvements. It is rightly advised by Womack and Jones (1996) that achieving lean manufacturing takes time, and it should not be treated as the panacea to solve short-term problems (Karlsson & Åhlström, 1996). Accordingly, models and approaches to lean implantations change from organization to organization. It seems reasonably justified to apply the lean tools as per specific requirements, as the key essence of lean philosophy is not the lean tools but the cultural and social change of the organization where zeal for continuous improvement is injected into each of member participating in the lean transformation.

References

Ahmed, R. (2013). An analysis of the change and volatility in the apparel industry of Bangladesh after MFA era. *Journal of Textile and Apparel Technology and Management, 8*(1).

Andrade, Y., Cardenas, L., Viacava, G., Raymundo, C., & Domingue, F. (2020). Lean manufacturing model for the reduction of production times and reduction of the returns of defective items in textile industry. In G. Di Bucchianico (Ed.), *Advances in design for inclusion. AHFE 2019. Advances in intelligent systems and computing* (954, pp. 387–398). Switzerland: Springer Nature.

Bashar, A., & Hasin, M. A. (2018). Lean manufacturing awareness and its implementation status in the apparel industry in Bangladesh. *International Journal of Lean Enterprise Research, 2*(3), 202–217.

Beels, M. (2019). *What is value added vs. non-value added work?* (Michigan Manufacturing Technology Center) Retrieved August 2019, from www.the-center.org: <https://www.the-center.org/Blog/February-2019/What-is-Value-Added-vs-Non-Value-Added-Work>.

Bhasin, S. (2012). Prominent obstacles to lean. *International Journal of Production and Performance Management, 61*(4), 403–425.

Bhasin, S., & Burcher, P. (2006). Lean viewed as a philosophy. *Journal of Manufacturing Technology Management, 17*(1), 56–72.

Bortolotti, T., Boscari, S., & Danese, P. (2015). Successful lean implementation: Organizational culture and soft lean practices. *International Journal of Production Economics, 160*(2), 182–201.

Burgelman, R. A. (1983). Corporate entrepreneurship and strategic management: Insights from a process study. *Management Science, 29*(12), 1349–1364.

Burgelman, R. A. (1988). An evolutionary perspective on strategy making. In K. Urabe, J. Child, & T. Kagono (Eds.), *Innovation and management: International comparisons* (pp. 63–80). Berlin, Germany: De Gruyter & Company.

Chowdhury, A., Alam, Z., & Rupaul, A. (2019). *Lean manufacturing: Its effect in supply chain performance in Bangladesh readymade garments: Case study review*. Dhaka: www.Fashionpress24.com. Retrieved 22 March, 2020, from <http://www.fashion-press24.com/lean-manufacturing-its-effect-in-supply-chain-performances-in-bangladesh-readymade-garments-a-case-study-review>.

Deshpande, A. (2016). Shop floor control in apparel production. *StitchWorld*, 34–40.

Ferdousi, F., & Ahmed, A. (2009). An investigation of manufacturing performance improvement through lean production: A study on Bangladeshi garment firms. *International Journal of Business and Management, 4*(9), 106–116.

Financial Times. (2011). *Impact on lean implementation at Bodyline*. Colombo, Srilanka. Retrieved 02 March, 2020, from <http://www.ft.lk/article/50964/impact-on-lean-implementation-at-bodyline>.

Forza, C. (1996). Work organization in lean production and traditional plants: What are the differences? *International Journal of Operations & Production Management, 16*(2), 42–62.

Hines, P., Found, P., Griffiths, G., & Harrison, R. (2011). *Staying lean: Thriving not just surviving*. New York: Productivity Press.

Hutchins, D. (2008). *Hoshin Kanri: The strategic approach to continuous improvement*. Burlington: Gower Publishing Limited.

Jacobson, G. (2017). *KaiNexus*. Retrieved 22 May, 2020, from <https://blog.kainexus.com/improvement-disciplines/hoshin-kanri/hoshin-kanri-what-why-and-how>.

Jahan, I., Debnath, D., & Biswas, S. (2018). Improvement of productivity through implementation of lean manufacturing tools in RMG industry. *International Journal of New Techniques in Science and Engineering, 5*(4).

Jana, P., & Tiwari, M. (2018). *Industrial engineering in apparel manufacturing*. New Delhi: Apparel Resources Pvt. Ltd.

Karlsson, C., & Åhlström, P. (1996). Assessing changes towards lean production. *International Journal of Operations and Production Management, 16*(2), 24–41.

Kaufman Global. (2007). *Office kaizen: Making lean work in service environments*. Kaufman Global.

Kesterson, R. (2015). *The basics of Hoshin Kanri*. Florida: CRC Press.

Klein, K., & Sorra, J. (1996). The challenge of innovation implementation. *Academy of Management Review, 21*(4), 1055–1080.

Ladia, U., & Jana, P. (2014). *Value stream mapping of pre-sewing processes in ABC Company*. New Delhi: National Institute of Fashion Technology.

Lathin, D., & Mitchell, R. (2001). Lean manufacturing. *American Society for Quality Journal*, 2–9.

Liker, J. K. (2004). *The Toyota Way: 14 management principles from the world greatest manufacturer*. New York: Mc-Graw Hill.

Loayza, L., Olave, S., Perez, M., Rojas, J., & Raymundo, C. (2020). Management model based on the lean thinking method for medium-sized peruvian companies in the apparel sector. In A. T., T. R., C. S., & C. A. (Ed.), *International Conference on Human Interaction and Emerging Technologies (IHIET) 2019: Human Interaction and Emerging Technologies. 1018*, pp. 896–902. Switzerland: Springer Nature.

Locher, D. A. (2008). *Value stream mapping for lean development: A how-to guide for streamlining time to market*. New York: Productivity Press.

Losonci, D., Demeter, K., & Jenei, I. (2011). Factors influencing employee perceptions in lean transformations. *International Journal of Production Economics, 131*, 30−43.

Malik, S. A., & YeZhuang, T. (2006). Execution of continuous improvement practices in Spanish and Pakistani industry: A comparative analysis. In *IEEE International Conference on Management of Innovation and Technology*, 2, pp. 761−765. Singapore.

Marudhamuthu, R., Krishnaswamy, M., & Pillai, D. (2011). The development and implementation of lean manufacturing techniques in Indian garment industry. *Jordan Journal of Mechanical and Industrial Engineering, 5*(6), 527−532.

Mazumder, S. (2015). Lean wastes and its consequences for readymade garments manufacturing. *Global Journal of Researches in Engineering: G Industrial Engineering, 15*(1), 14−20.

Mourtzis, D., & Doukas, M. (2014). The evolution of manufacturing systems: From craftsmanship to the era of customisation. In V. Modrák, & P. Semančo (Eds.), *Handbook of research on design and management of lean production systems* (pp. 1−29). Hershey: IGI Global.

Nawanir, G., Teong, L. K., & Othman, S. N. (2013). Impact of lean practices on operations performance and business performance. *Journal of Manufacturing Technology Management, 24*(7), 1019−1050.

Ni, W., & Sun, H. (2009). The relationship among organisational learning, continuous improvement and performance improvement: An evolutionary perspective. *Total Quality Management, 20*(10), 1041−1054.

Ortiz, C. A. (2006). *Kaizen assembly designing, constructing, and managing a lean assembly line*. New York: CRC Press.

Pavnaskar, S. J., Gershenson, J. K., & Jambekar, A. B. (2003). Classification scheme for lean manufacturing tools. *International Journal of Production Research, 41*(13), 3075−3090.

Powell, T. (1995). Total quality management as competitive advantage: A review and empirical study. *Strategic Management Journal, 16*(1). 15−37.

Priya, S., Saini, H., & Tiwari, M. (2011). *Supply chain diagnostics devising a roadmap for lean transformation using value stream mapping*. Gandhinagar: National Institute of Fashion Technology, National Institute of Fashion Technology, Department of Fashion Technology.

Raju, P. G., & Modekurti-Mahato, M. (2014). Impact of longer usage of lean manufacturing system (Toyotism) on employment outcomes − A study in garment manufacturing industries in India. *International Journal of Services and Operations Management, 18*(3), 305−320.

Roser, C. (2020). AllAboutLean.com. Retrieved 13 July, 2020, from <https://www.allaboutlean.com/dfma-1/>.

Saleeshya, P., Raghuram, P., & Vamsi, N. (2012). Lean manufacturing practices in textile industries − A case study. *International Journal of Services and Operations Management, 3*(1), 18−37.

Schonberger, R. (2007). Japanese production management: An evolution-with mixed success. *Journal of Operations Management, 25*(2), 403−419.

Shah, R., & Ward, P. T. (2003). Lean manufacturing: Context, practice bundles, and performance. *Journal of Operations Management, 21*(2), 129−149.

Silva, S., Perera, H., & Samarasinghe, G. (2011a). Retrieved from http://ssrn.com/abstract=1824419: http://ssrn.com/abstract=1824419.

Silva, S., Perera, H., & Samarasinghe, G. (2011b). Viability of lean manufacturing tools and techniques in the apparel industry in Sri Lanka. *Applied Mechanics and Materials, 110-116*, 4013−4022.

Singh, S., & Singh, J. (2014). *Value Stream Mapping of trouser sewing floor.* Jodhpur: National Institute of Fashion Technology.

Smeds, R. (1994). Managing change towards lean enterprises. *International Journal of Operations & Production Management, 14*(3), 66−82.

Sparrow, P., & Otaye-Ebede, L. (2014). Lean management and HR function capability: The role of HR architecture and the location of intellectual capital. *The International Journal of Human Resource Management, 25*(21), 2892−2910.

Technopak Advisors. (2011). *Lean manufacturing − The way to manufacturing excellence.* Gurgaon, NCR: Technopak Advisors Pvt. Ltd.

Tiwari, M. (2010). *Look at your business through LEAN Lens.* (Fibre2Fashion) Retrieved January 2019, from www.fibre2fashion.com: <https://www.fibre2fashion.com/industry-article/5074/look-at-your-business-through-lean-lens>.

Urabe, K. (1988). Innovation and the Japanese management system. In K. Urabe, J. Child, & T. Kagono (Eds.), *Innovation and management: International comparisons* (pp. 3−25). Berlin, Germany: De Gruyter & Company.

Wall, T., Jackson, P., & Davids, K. (1992). Operator work design and robotics system performance: A serendipitous field study. *Journal of Applied Psychology, 77*(3), 353−362.

Wickramasinghe, G., & Wickramasinghe, V. (2016). Effects of continuous improvement on shop-floor employees' job performance in Lean production: The role of Lean duration. *Research Journal of Textile and Apparel, 20*(4), 182−194.

Wickramasinghe, V., & Wickramasinghe, D. (2017). Autonomy support, need fulfilment and job performance in lean implemented textile and apparel firms. *Research Journal of Textile and Apparel, 21*(4), 323−341.

Wilson, L. (2010). *How to implement lean manufacturing.* USA: McGraw-Hill Companies, Inc.

Womack, J., & Jones, D. (1996). Beyond Toyota: How to root out waste and pursue perfection. *Harvard Business Review, 74*(9), 149−151.

Worley, J., & Doolen, T. (2006). The role of communication and management support in a lean manufacturing Implementation. *Management Decision, 44*(2), 228−245.

Wynder, M. (2008). Employee participation in continuous improvement programs: the interaction effects of accounting information and control. *Australian Journal of Management, 33*(2), 355−374.

Lean problem-solving

4

Manoj Tiwari and Yuvraj Garg
Department of Fashion Technology, National Institute of Fashion Technology, Jodhpur, India

4.1 Introduction

In lean thinking, waste (Japanese term for waste is Muda) is referred to anything (activity, process, etc.) that doesn't add any value to the end product but consumes resources. Any kind of waste in the process results in inefficiencies; hence, these are non−value added (NVA) activities. The activities, based on its value addition to the end product, could be classified into three categories: VA, NVA, and necessary but NVA (Jana & Tiwari, 2018). Different types of waste and criteria of value addition are been discussed in detail in Chapter 3, Fundamentals of Lean Journey, kindly refer to for further details in this regard.

Identification of waste is an essential requirement to start any kind of improvement (Stewart, Charles, & David, 2001). It helps in unrevealing the issues and problems in the existing process and opens up the path for improvement. This makes waste identification as the very first and critical step in the never-ending journey of lean with continuous improvement. There are several tools used for waste identification and its elimination (problem-solving) in the lean philosophy; however, a few of them are applied in the apparel manufacturing (Development Commissioner Micro, Small & Medium Enterprises, 2010). Some of the most common tools, such as value stream mapping (VSM), Standardized work, PDCA (plan−do−check−act) cycle, root-cause analysis (RCA), and 5-Why analysis, have been discussed in the following sections.

4.2 Kaizen

Kaizen is made up of two Japanese ideograms—Kai means thinking and Zen means good. It is a process of continuous improvement. The first real kaizen events were recorded as "quality circles." Quality circle is a cross-functional team that focuses on analyzing a problem, finding the cause of a problem, and implanting the solution (Mika, 2006).

Kaizen is a continuous activity, while Kaizen Event or "Blitz" is different from the Kaizen. Kaizen Event or "Blitz" is a less expensive, and short-term problem-solving tool (Farris, Aken, Doolen, & Worley, 2008). It is also referred to as a rapid improvement event. Kaizen Event or "Blitz" is a focused and structured improvement project, using a dedicated cross-functional team to improve a targeted work area, with specific goals, in an accelerated timeframe (Letens, Cross, & Aken, 2006).

Lean Tools in Apparel Manufacturing. DOI: https://doi.org/10.1016/B978-0-12-819426-3.00008-4
© 2021 Elsevier Ltd. All rights reserved.

Kaizen improvement may be linked to the strategic purpose of the company, such as productivity improvement, defect reduction, downtime reduction, floor area utilization, inventory reduction. The Kaizen event is implemented by Kaizen Team. The members of the Kaizen Team are generally related to the area or section from where the problem mainly exists and are from those people who perform the task on the work floor. A typical Kaizen Team may consist of the production or process engineer, the quality control checker, the line operator, and a member from the top management (Ortiz, 2006). The key activities in the Kaizen Event are as follows:

- team selection and training,
- current state mapping,
- finding the opportunity for improvements,
- improvement of choice and implementation,
- result presentation, and
- documentation for follow-up action (Aken, Farris, Glover, & Letens, 2010).

Usually, the companies nominate a Kaizen Champion for the successful implementation of kaizen events. Kaizen Champion is a person in the organization who is fully dedicated to the planning and execution of kaizen events. For example, the lean manufacturing engineer can be a Kaizen Champion. Another important aspect is the formation of the Kaizen Governing Committee which helps to schedule and monitor the kaizen events. In a typical manufacturing unit, the member of this committee might be from different departments of the company: the Kaizen Champion, the production manager, the Human resource manager, the quality manager, and the engineering manager (Ortiz, 2006).

The activities of the Kaizen Event are being tracked through a Kaizen Event Tracking Sheet. This sheet is periodically reviewed by Kaizen Governing Committee. This sheet summarizes the details, such as the tile of Kaizen Event, Kaizen team leader, Kaizen team member, Date and length of the event, Strategic purpose, Estimated cost, Event spending, Actual results, Due date of action items (Ortiz, 2006).

Kaizen is being practiced widely in apparel manufacturing setups in India, there are several success stories, one such is at Matrix Clothing (a leading apparel export house in India). Kaizen was implemented as an improvement tool to bring the cost-effectiveness in operations. The reduced work-in-progress (WIP) is one key improvement where an average WIP of 5–5.5 units was maintained between two workstations in the assembly line. Changeover time was reduced by 60%, while sewing efficiency was improved from 57% to 65%. The impact of Kaizen was witnessed on quality as well where in-line and end-line defects were reduced from 20% to 12% and 14% to 8%, respectively. Garment rejections came from a 3% level to 1.5%. Another significant improvement observed was in fabric storage time, which reduced from 30 to 17.75 days (against the target of 12 days). Other improvements were witnessed in reducing the marker turnaround time (from average 5 to 2 days) and reducing the trims issue time (time required in reaching trims from stores to the floor) by 13.2% (Saran, 2010).

Madura Garments which is a group company of Aditya Birla group (now known as ABFRL: Aditya Birla Fashion & Retail Limited) initiated the Kaizen program to all its apparel manufacturing units with a target of 2 Kaizens per employee per year. As a result, the efforts culminated in 6458 Kaizens (Stitch World, 2010). A few of the examples of Kaizen accomplished are listed here.

Kaizen: Problem of feed dog adjustment in a Feed of the Arm machine.

Solution: The Allen screw is usually accessible if the cover was removed by unscrewing 12 screws in the machine. This usually takes around 12–15 minutes. A hole was drilled (refer Fig. 4.1) on the cover through which an Allen key can be inserted for adjusting the feed dog without removing the cover resulting in time-saving (Stitch World, 2010).

The Allen key was inserted for adjusting the feed dog (Stitch World, 2010).

Kaizen: Changed the conventional method of holding the embroidery frame which is by fixing pins around it.

Solution: The operation was unsafe as the pins sometimes hurt the operator. Also, there were incidents where the fabric was damaged by pins. The new design used the wooden frame (refer Figs. 4.2 and 4.3) which held the embroidery frame more securely and the operator could do his job without any fear of injury (Stitch World, 2010).

Before: The conventional method of holding the frame by fixing pins around it (Stitch World, 2010).

Figure 4.1 Kaizen 1: feed dog adjustment in a Feed of the Arm machine.

Figure 4.2 Kaizen 2: holding the embroidery frame (before).

Figure 4.3 Kaizen 2: holding the embroidery frame (after).

After: The new safer design holds the embroidery frame without pins (Stitch World, 2010).

Kaizen: Developed Automatic Waistband Lining making machine.

Solution: Waistband is constructed with five to six tapes by attaching in four to five steps based on the style in a manual process. To improve the productivity, folders were designed which feed the tapes based on different styles. By attaching folders to the machine the number of operations was reduced to one. All bias-cut tapes are loaded to a roll stand which continuously feeds tapes to the machine. Moreover, to eliminate the operator who controls the machine a new Electro-Pneumatic System (refer Fig. 4.4) was also developed which can be operated with a remote controller. The machine eliminates the operator intervention but for loading/replenishing material and quality checking. The machine works with minimum manual intervention. Behind the gripper, there is a stretching and winding device for the clean finish and rolling of the lining. To prevent the machine from running out of tape, an optical sensor is attached to the machine which will stop immediately when the tape is empty (Stitch World, 2010).

Automatic waistband and lining making machine requires no operator and the lining is disposed of in a roll (Stitch World, 2010).

Kaizen: Carrying of different kinds of tools by a service engineer.

Solution: A mobile tool and component were devised to save the time wasted in searching and bringing the required tools to the breakdown/problematic spot. This trolley (refer Figs. 4.5 and 4.6) was very compact and easy to move to any corner of the factory. It houses an array of tools, including double end spanners, ring spanners, screw spanners, ranch spanner, ratchet spanner, screwdrivers, Allen keys, jewelry file, flat file, half-round

Figure 4.4 Kaizen 3: Automatic Waistband Lining making machine.

Figure 4.5 Kaizen 4: carrying of different kinds of tools (before).

Lean problem-solving

Figure 4.6 Kaizen 4: carrying of different kinds of tools (after).

Figure 4.7 Lean problem-solving.
Source: Courtesy Corrigo Consultancy Pvt. Ltd, India.

file, hammer, and drilling machines. Also, there is a panel board with three-phase connection, testing, and soldering facilities. Even grinding and drilling machines can be carried along. Hence, a lot of time is saved while doing preventive or breakdown maintenance (Stitch World, 2010).

Tools were earlier difficult to carry individually.

A mobile and compact trolley was devised to save time while searching and bringing the tools to the breakdown spot.

The Kaizen is treated as an umbrella under which several tools and techniques are applied. Please refer to Figs. 4.7–4.9 for different kaizen applications in the apparel manufacturing setups.

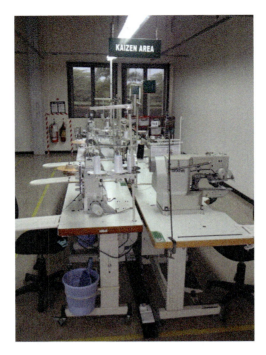

Figure 4.8 Kaizen area.
Source: Courtesy Corrigo Consultancy Pvt. Ltd, India.

Figure 4.9 Kaizen display board.
Source: Courtesy Corrigo Consultancy Pvt. Ltd, India.

4.3 Value stream mapping

The only thing that matters to a customer is the value. The value is something that consumer wants in the end product and that can be summed up as receiving the desired product as per the specifications, at the right time, and the competitive cost.

Hence, the key goal of any manufacturing is to achieve OTIFEF (on-time in-full and error-free). A nonconformance to OTIFEF indicates inefficiencies in the process leading to the generation of wastage which eventually eats up the profit margins of the organizations. Given the same, it is imperative to identify the wastes in the value stream. The waste identification in the value stream can be done by mapping the value stream, and the lean technique used for the same is known as VSM. According to Jones and Womack (2002), the VSM is a process of observing information and materials flow as it happens; visual humanization, and then working out the future state with improved performance.

VSM is a visual display technique used to analyze the materials and information flow in a production process (Wang, 2011). A set of tools is used while applying VSM (ILO, 2017). VSM is a pencil and paper tool used to see and understand the flow of material and information as a product makes its way through the value stream (Jana & Tiwari, 2018). VSM is a visual representation of every process in the material and information flow following a product's production path from customer to supplier, usually referred to as SIPOC (supplier−input−process−output−customer) (Locher & Keyte, 2016). A value stream perspective means working on the big picture, not just individual processes, to improve the whole, and not just optimizing the parts. One requires to follow the value streaming of the product across many firms and even more facilities. A value stream refers to all the actions (both VA and NVA) which are currently required to bring a product through the main flow essential for every product: (1) the production flow, from raw material to the arms of the customer, and (2) the design flow, from concept to launch of the product (Damelio, 2012; Rother & Shook, 1999).

VSM is a strong tool that is capable to depict the complete fact-based information of the entire stream of activities in a measurable way. It also communicates the VA and NVA activities. Usage of a common language (in terms of standard symbols) and common measures of lead time, process time, and cycle time make it easy to communicate and understand.

A step-by-step procedure to implement the VSM in a 5-pocket denim garment manufacturing factory is presented in the following subsections.

4.3.1 Preparation for value stream mapping

Reduction in lead times and waiting times results in faster delivery to the consumer (Feld, 2000). It also results in reducing operational costs. To work in this direction, it is imperative to understand and analyze the entire value stream step-by-step. VSM may also be considered as the study of the flow of material and information.

The interruptions in the value stream, such as the material flow (maybe due to disturbed/imbalance process where a bottleneck is created and WIP getting accumulated), or interruptions in the information flow (where the right information is not able to reach at right time to the right persons) result into longer waiting times. Because of the same, the prime objective of any VSM becomes studying and analyzing the flow and then suggesting corrective measures to improve this flow of the value stream. VSM needs to be applied in an organized manner to achieve the best results. The recommended steps while performing VSM are mentioned in Fig. 4.10. As indicated in Fig. 4.10, the first step is planning & preparation for VSM, which involves the formation of a cross-functional team with a VSM manager, who is going to undertake the VSM activity, and selection of product family of the process to study. The VSM methodology is also worked out by the VSM as part of the preparation process.

Preparation for the VSM starts with a critical examination of the existing process with some key questioning on what type of product, process, or service to be mapped; what is to be mapped; the start and end point of the mapping; and the scope of the VSM exercise to be undertaken. It is also very important to set the business objectives and measures of the improvements. The VSM team members are required to be introduced to the rules and guidelines of the VSM exercise at the preparation stage.

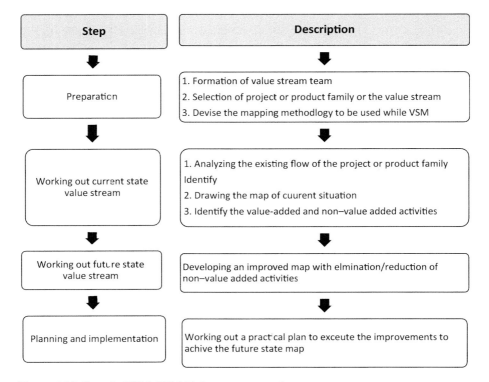

Figure 4.10 Steps in VSM. *VSM*, Value stream mapping.

Identifying the family for VSM: *The selection of a product or process family is one important task before conducting VSM*. A family is a group of products or services sharing similar characteristics and processing steps (Cudney, Furterer, & Dietrich, 2014; Locher, 2008).

In the context of apparel manufacturing, generally, there are several styles (with design variations) of the same product category that are manufactured at a time. Such different product styles may have different process steps, for example, some of the styles may have different sampling steps while others may have different steps, while the product is getting manufactured.

An analysis based on criteria such as steps followed in the manufacturing process, dry process and washing, and the order quantities may be carried out. There is a possibility that styles with the complicated process may be of higher value (in terms of revenue generation) such orders may have smaller order quantities, while on the other hand, there may be styles with the relatively simpler process but with larger order quantities of not that higher value. Here, one needs to see the styles that are creating most of the business for the organization, and accordingly, product family or process may be selected for the VSM study. A product or service matrix may be created for easy analysis and selection of the product family for VSM (McLean, 2015). Table 4.1 represents a sample product matrix in the context of apparel manufacturing.

The product matrix for three different styles has been shown in Table 4.1, as indicated in different styles follow different steps in the manufacturing process, also it varies with

Table 4.1 Product matrix.

Process	Activity	Product		
		A	**B**	**C**
Preproduction activities	Prototype sample	X	X	X
	Fit sample	X	X	X
	Size-set sample	X	X	X
	Salesman/photoshoot sample		X	
	Preproduction sample	X	X	X
	TOP/sealer sample		X	X
	Shipment sample	X	X	
Fabric management	Shade blanket	X	X	X
	Shrinkage	X		
Spreading & cutting	Spreading & cutting	X	X	X
Embroidery	Embroidery	X		
Sewing process	Standard Process as 5-Pocket Jeans	X		
	Moderate process (up to three additional steps in the Standard Process)		X	
	Difficult process (more than three additional steps to the Standard Process)			X

(Continued)

Table 4.1 (Continued)

Process	Activity	Product		
		A	**B**	**C**
Dry process	Hand-sand (global/local)	Local	Global	Local
	Whiskers/chevrons	Whiskers		Chevrons
	PP spray/resin		X	X
	Laser effect			X
	Destruction		X	
	Other dry process			
Washing process	Rinse wash	X		
	Softener wash			
	Light (bleach) wash with wash cycle of 3.5−4.0 h		X	
	Medium wash with wash cycle of 2.5−3.5 h			X
	Dark wash with wash cycle of 1.5−2.5 h			
	Other washes			
Finishing & packing process	Standard Process as 5-Pocket Jeans	X		
	Moderate process (up to three additional steps in the Standard Process)		X	
	Difficult process (more than three additional steps to the Standard Process)			X
Order quantity	Small (up to 5000 units)			X
	Medium (between 5001 and 8000 units)		X	
	Large (between 8001 and above)	X		
Value generation	High (unit price above 15 USD)	X		
	Medium (unit price between 10 and 15 USD)		X	
	Low (unit price below 10 USD)			X

TOP, Top of the production.

the order quantity and the value generation (in terms of price per garment). A style for VSM may be selected after a comparative analysis of different styles. After the identification of the product family a project needs to be selected. For example, in the case of apparel manufacturing, it may be a recently completed order (maybe referred to as PO—purchase order) from the chosen product family. Once the project is selected, the complete information related to the project right from suppliers to customers should be summarized in a single sheet. The standard term referred to as such sheet is SIPOC (supplier−input−process−output−customer). Please refer to Fig. 4.11 for a sample SIPOC sheet. The key process activities that are carried out in the production process are identified, beginning with the "supplier" and ending with the "customer" (Locher, 2008).

The information on the SIPOC sheet starts from the left-hand side with information related to suppliers (fabric and trim suppliers), and the customer (buyer) details are mentioned toward the right-hand side sheet. The main process (apparel manufacturing)

Lean problem-solving 91

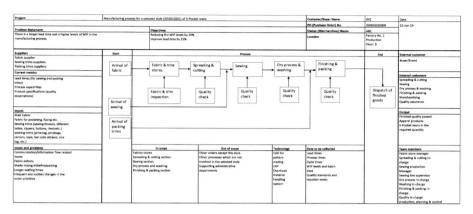

Figure 4.11 SIPOC (supplier−input−process−output−customer) sheet.

steps are indicated in the center of the sheet. One of the biggest advantages of a SIPOC sheet is the identification of clear boundaries where a VSM is to be performed. This helps in defining the scope of the VSM. Also, it helps in identifying the relevant people or departments in the VSM exercise for the selected product family and project.

4.3.2 Conducting value stream mapping in a practical environment

After the homework is done with working on some key tasks such as the selection of product family, project group, and the SIPOC sheet, it is the time to carry out the VSM exercise in the practical environment. As mentioned earlier as well, VSM is a graphical tool that represents the selected process and related activities in a single sheet; it requires some standard symbols (King & King, 2015). The application of these symbols makes it easy to draw and understand the process and activities. Refer Fig. 4.12 for steps to undertake current state VSM.

It is important to follow standard symbols while conducting VSM, as it helps to make an easy and effective communication of the activities and processes performed throughout the value stream. In a manufacturing setup the goods or products are manufactured through production flow. This is possible with a visible flow in the form of material flow, but apart from it, there is an information flow that also exists (maybe in invisible mode) which regulates or controls the material flow. Both material and information flow are equally important in any lean initiative. Based on this, there are separate categories of VSM symbols to represent material flow as well as information flow. Apart from it, there are exclusive symbols to show inventory, process metrics, and to show the timelines. The most frequently used standard symbols in VSM are mentioned in Table 4.2.

Step 1: Identifying the customer requirements

Customer focus is key to any lean initiative. All the activities performed are aimed to add value to the end product; hence, meeting customer requirements is a prime concern. Wang (2011) defined the voice of customers (VOCs) as a deep "truth" about the customer based on their behavior, experiences, beliefs, needs, or desires, which is relevant to the task or issue and "rings bells" with target people.

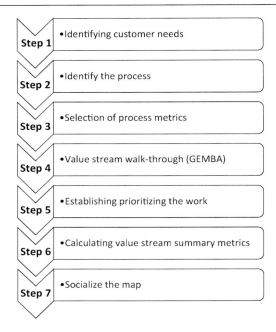

Figure 4.12 Current state VSM steps. *VSM*, Value stream mapping.

The VOC concept puts the customer, not the producer, first in a quest to deliver world-class manufactured products (Wang, 2011).

The information that is related to customer requirements, such as what is the product, daily demand, order size, lead time, and quality expectations (maybe in terms of acceptable tolerance/defect limits) should be mentioned in the data box (refer Fig. 4.13). Other details such as first-time pass percentage, cut-to-ship ratio, and on-time shipment percentage about a particular buyer may also be mentioned.

Step 2: Identify the process

Process plays an important role in manufacturing, and VSM studies as well as it is the place where the actual raw materials are getting transformed into products. A process may have several subprocesses that are required to manufacture a product. For example, the apparel manufacturing process primarily involves subprocess as spreading & cutting, sewing, dry process & washing, and finishing & packing. It is also important to understand the flow channels of information required to manufacture the good and then how this information is utilized at different processes. It is recommended to have a separate process box for every process. Generally, a process starts and the end is considered when the material (maybe in the form of raw material, cut panels, semifinished, or finished goods) is handed over from one process to the next (refer Fig. 4.14). The processes should be considered based on their impact on the total lead time and the total processing time. Some basic information such as lead time and the processing time for a given process should essentially be recorded to get some useful inferences. It is recommended not to have too many details in the process box, as it may confuse the team, and may not have a significant impact on the VSM exercise (King & King, 2015).

Table 4.2 Value stream mapping (VSM) symbols.

Symbol	Name	Description
	Supplier or customer	This is usually the start point or the end point of the process in the VSM sheet. Suppliers are marked at the top left corner, while the customers are represented at the top right corner of the VSM sheet.
	Process flow	It represents a process where an activity is done to manufacture the product or service. It can be a process or operation, machine or workstation, department, section where the material flows.
	Shared process	It represents a process where an activity is shared by more than one value stream to manufacture the product or service. It can be a shared process or operation, machine or workstation, department, section where the material flows.
	Data box	Data box is used to mention useful and relevant information pertaining to specific process. Data box are generally positioned just below the process.
	Work cell	It is used to indicate a section where multiple processes are integrated in a manufacturing cell.
	Inventory	To represent inventory between two workstations or processes.
	Shipment/ delivery	To represent the movement of raw material from suppliers to the manufacturing site and delivery of the finished goods from the manufacturing site to the customers/buyers.
	Material movement (push)	Material movement from one process to the next process.
	Supermarket	To represent inventory supermarket.
	Pull	Physical pulling of material from the supermarket by the downstream process.
$-FIFO\rightarrow$	FIFO movement	Used to represent FIFO inventory movement.
	Safety stock	To represent safety stock, which is a buffer inventory kept to avoid process idleness due to run out of the work-in-progress.
	External shipment	To represent external shipment, while receiving raw materials from suppliers or delivering finished goods to the customers/buyers through external transportation.

(Continued)

Table 4.2 (Continued)

Symbol	Name	Description
Process Planning	Production control	To represent centralized production control (in the context of apparel manufacturing it may be used for PPC) widely used for planning and scheduling of orders and processes.
→	Information flow (manual)	To represent the information flow manually done. Manual flow of information may be through personal discussion, memos, circulars, orders, physical work-orders, entries in registers, etc.
↝	Information flow (electronic)	To represent the information flow using some electronic medium. Electronic flow of information may be through ERP solutions, mobile text messages, fax, email, Intranet, Internet, LAN, WAN, etc.
▭	Production Kanban	To represent production Kanban as a medium/signal to start or initiate the production depending on the demand from the next process/customer.
▭	Withdrawal Kanban	To represent withdrawal Kanban as a medium/signal to transfer parts from a supermarket to the receiving process.
▽	Signal Kanban	To represent the minimum point of the inventory between two processes. This happens when in-hand inventory levels are reduced to a minimum level.
Ұ	Kanban post	Kanban post is required to keep the Kanban signals for pick-up.
◎	Sequenced pull	This is used to provide necessary signals to produce a predetermined product in required quantity.
OXOX	Load leveling	To represent level the production volume and mix to maintain a balanced production flow.
▢	MRP/ERP	To represent process scheduling using systems like MRP or ERP.
👓	Go see	To represent scheduling fine-tunings based on actually seeing the inventories on floor.
⌇	Kaizen burst	To represent improvement in the process.
Ⓠ	Operator	To represent operator/workforce in the process
⌐_	Timeline	To represent lead time and cycle time on a timeline indicating VA and NVA time. Cycle times (VA) are indicated on top of the timeline, while lead times (NVA) are mentioned at the bottom of the timeline.

ERP, Entreprise resource planning; *FIFO*, first-in−first-out; *MRP*, material requirements planning; *NVA*, non−value added; *PPC*, production planning & control; *VA*, value added.

Lean problem-solving 95

```
Product = 5 Pocket Denim Jeans
Daily demand = 2500 garments
Order size= 40,000–60,000
Garments
Lead time =90/75 days
Defect-% = 1%
```

Figure 4.13 Data box for VOC. *VOC*, Voice of customer.

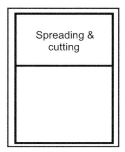

Figure 4.14 Process box for spreading and cutting.

Step 3: Selection of process metrics

Process metrics are the indicators of the process performance. There are different process metrics used in different industries. In the area of apparel manufacturing, processing time, lead time, change-over-time, number of people, batch size or bundle size, defect level, etc. are one of the most frequently used metrics. It is not necessary to use all the metrics in a process box (refer Fig. 4.15) but only the most relevant and useful metrics which help achieve the objective should be recorded. There may be some other different metrics depending on the specific requirements. These metrics are required to visualize the specific value stream, and the VSM team may even introduce using some new metrics as well (Locher & Keyte, 2016).

Step 4: Value stream walk-through

Value stream walk-through is a critical step to the VSM as it provides an essential opportunity to practically observe the activities on the shop floor. Gemba is a synonymous term used in Japanese for going to the floor to observe the work how it is performed. A value stream walk should be done thoroughly and sufficient time should be devoted to study and observe the process in detail. Each of the process (which is in the scope of the VSM) should be covered in this exercise. The focus should be on understanding the challenges and issues about the process and on the contribution of this step-in adding value to the end product. The findings and observations of the values stream walk-through should be well recorded after agreement or consensus as a result of an in-depth discussion on each of the points (King & King, 2015).

Step 5: Establishing prioritizing the work

To understand the criticality of the information to perform and prioritize a process, it is important to know how people set the work priorities, or how they learn which work

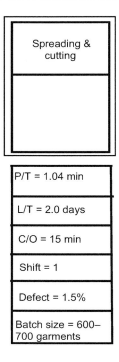

Figure 4.15 Process matrix.

to do when. Many times, it is being observed that the information is not "complete and accurate," which leads to confusion among the workers. In such situations of unclear instructions, people do work as per their understanding, which may not be in the line of the actual work priority. Such situations are very often observed in apparel manufacturing, where several work-orders are processed simultaneously, and we see a delay in shipments due to wrong scheduling and inappropriate prioritization at different processes.

"Complete and accurate," an important measure of information quality, is the percentage of time that all necessary information is received, and weather information is accurate (Wang, 2011). As indicated in Fig. 4.16, if the information is not complete and accurate, it is recorded in the data box as "iterations" (Locher & Keyte, 2016).

Step 6: Calculating value stream summary metrics

This step is related to the assessment of the value stream performance based on the data collected in the previous steps. The lead time and process time for the individual process are mentioned on a timeline at the bottom of the sheet. As a standard practice, lead times are shown on the upper side of the timeline, while the process times are mentioned at the bottom of the timeline (refer Fig. 4.17). The values may be indicated in terms of a range of time as well; however, we generally prefer using average values for each of understanding the calculation for determining value addition (King & King, 2015).

Step 7: Socialize the map

This is an important step to get the VSM work validated by others and make it acceptable by the larger set of people. The VSM exercise is initially done by the

Lean problem-solving 97

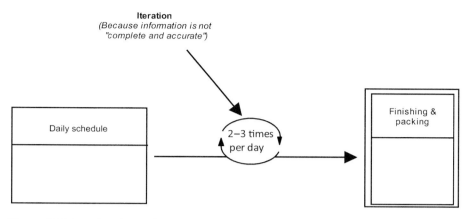

Figure 4.16 Iterations in a value stream.

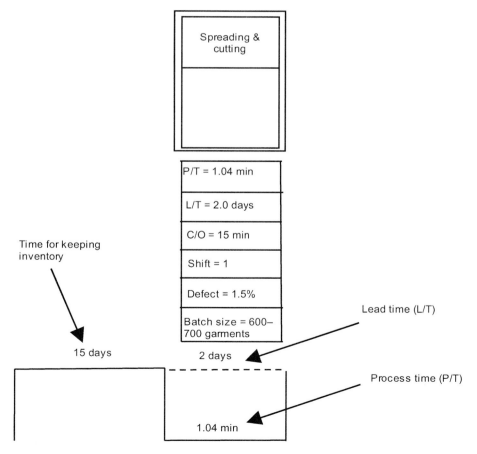

Figure 4.17 Value stream summary matrix.

VSM team only, and the inferences or results indicated in the chart are based on their observations. Sometimes, the observations mentioned in the chart don't reflect the true picture as some other points (which can only be known by the people who work on the floor and spent significant time practicing). Given the same, it is important to share the initial or draft VSM with others to get more inputs and perspectives for further improvement. Sharing the draft VSM results in acceptance of the efforts of the VSM team by the other members of the organization.

4.3.2.1 Current state value stream mapping

A current state VSM is a graphical representation of the existing processes and designed to capture a gap in the current working performance, to lead to an understanding of the major operational processes as they are today, and to identify root causes about why performance is required and what it is.

Please refer to Fig. 4.18 for a sample current state VSM for a 5-Pocket Jeans manufacturing process of a particular style and a work-order. As indicated in Fig. 4.18, the VSM starts from the suppliers (indicated at the top left corner) of fabric and trims (sewing trims and packing trims), and it ends at the buyer or customer (indicated at the top right corner) with the delivery of finished garments from the apparel manufacturer.

In the current state VSM sheet, the material and information flow is represented in an organized manner using the standard VSM symbols. The upper part of the sheet (refer Fig. 4.18) indicates the information flow, where the electronic exchange of information between the buyer and the apparel manufacturer to place the order (or maybe issuing PO) is shown. On the other hand, the electronic exchange of information between the apparel manufacturer and the fabric & trims suppliers is mentioned for sourcing of raw materials. The central section of the sheet indicates the planning [production planning & control (PPC)] of the apparel manufacturer, which is responsible for planning and scheduling of processes and activities at the apparel manufacturing site. Generally, in typical apparel manufacturing plant, plans (based on the priority depending on the delivery date, and raw material arrival) are prepared and issued to the concerned departments or processes with necessary instructions on a monthly, weekly, and/or daily basis. Such information may be circulated manually or electronically; in this case (as in Fig. 4.18) the information flow is manual. The lower part of the sheet indicates material flow from one process or department to another. The material (as raw materials in stores) moves further in batches to upcoming processes (such as spreading & cutting, stitching, washing, and finishing & packing) and transforms into finished garments at the end. At the bottom of the sheet, cycle times and lead times are indicated for respective processes. The relevant information about a specific process has been mentioned in the data boxes.

The daily demand for the order mentioned in the current state VSM sheet is 2500 units of 5-Pocket Jeans. The order size for the given buyer and style varies from 8000 to 12,000 units (average order size is 10,000 units). The lead time from placing the order to receive the finished goods is ranging from 75 to 90 days. The fabric lead time, which is placing the order for fabric purchase to receive of fabric by the apparel manufacturer, is between 45 and 60 days, while the trims lead time is between 30 and 45 days. It takes an

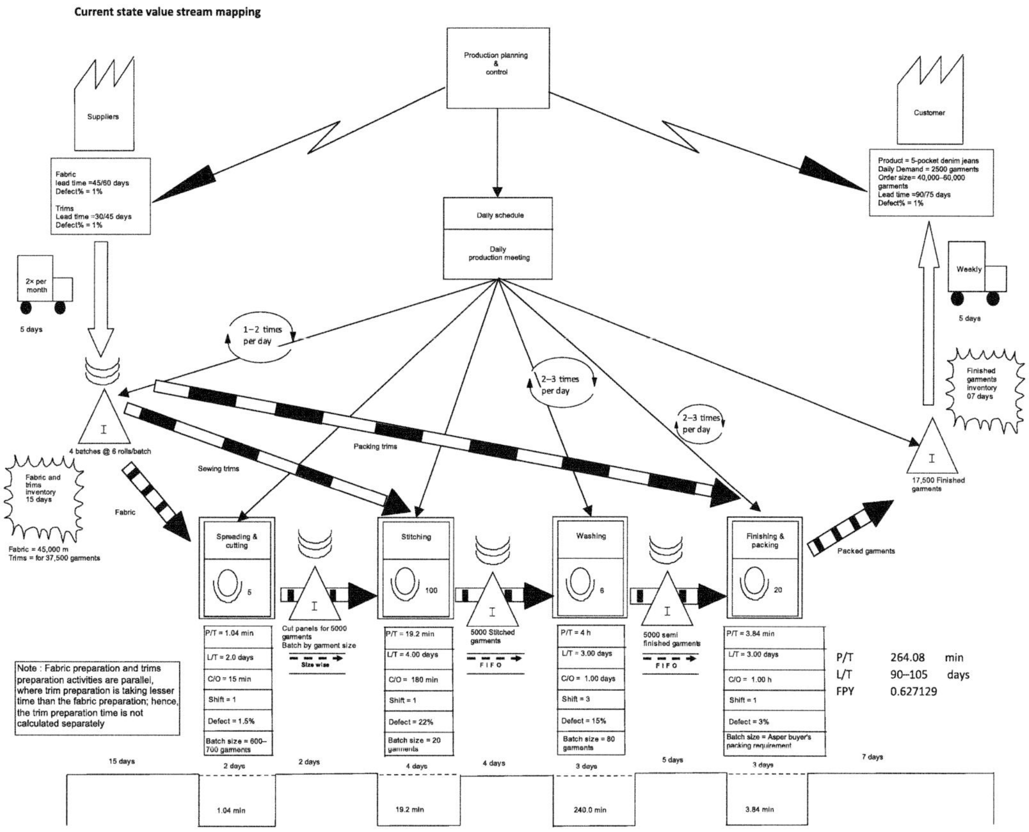

Current state value stream mapping

Figure 4.18 Current state value stream mapping.

average of 5 days in transit for transporting the raw materials from the fabric & trims manufacturers to the apparel manufacturer, and for transporting finished apparel goods from the apparel manufacturer to the buyer using external transport.

The fabric and trims were arranged at least 15 days before the planned cut date, which means an average of 15 days' inventory of raw materials was maintained in the fabric and trims store. Considering average fabric consumption 1.20 m per garment (fabric width 60 in.), and the daily demand of 2500 units, about 45,000 m of fabric is required in the fabric stores. Similarly, an inventory of sewing and packing trims for 37,500 garments (daily demand × 15 days' inventory) is maintained. An excess of a 3% measure of fabric and trims is procured considering the possibility of rejections and other quality-related issues. The fabric comes into roll form with a meterage of approximately 120-m fabric per roll. The fabric preparation starts with the fabric inspection conducted at the fabric stores. The shade & shrinkage testing and shade blanket are prepared for the development of shade band and internal shade-wise segregation of the rolls. The recipe used in developing a shade blanket is the same after the washing requirements of the garments. Finally, the rolls are arranged in shade-wise order for bulk cutting (Hasan, 2016). This entire exercise of shade-shrinkage testing and development of shade blanket takes around a week.

Following the plan and schedule, as received from the PPC, the required fabric rolls are sent to spreading & cutting. On average, a lay (a group of fabric layers to be cut at a time, also referred to as cut in apparel manufacturing) has 600−700 garments that consume approximately six fabric rolls. This makes daily feeding of fabric to spreading & cutting in form of four batches per day containing approximately six fabric rolls per batch. The cutting of fabric is done shade wise following the internal shade sequence; however, feeding of cut panels to sewing is done size-wise. This is done to avoid mixing panels and to have better control of material movement. The relevant current state process data for the process steps are mentioned in Table 4.3. On an average 2 days, WIP is kept between the spreading & cutting and stitching, and the subsequent processes. Means, for the demand of 2500 units per day, an inventory of about 5000 units (approx.) is kept as a buffer between two subsequent processes. The sewing trims are supplied to the stitching section as per the instructions from the merchandiser, and it was issued at once for the entire order quantity. The subsequent processes (stitching, washing, and finishing & packing) work as per the work-orders and instructions issued from the PPC

Table 4.3 Current state process data.

Process step	Process time (P/T) (min)	Lead time (L/T) (days)	Changeover time (C/O) (min)	Shift	No. of workers per shift	Batch size	Defect %
Spreading & cutting	1.04	2	15	1	5	600−700	1.5
Stitching	19.2	4	180	1	100	20	22.0
Washing	240	3	480	3	2	80	15.0
Finishing & packing	3.84	3	60	1	20	As per buyer's packing requirement	3.0

Lean problem-solving 101

daily. Stitching is done using an assembly line with a progressive bundle system with a bundle size of 20 panels per bundle. The material movement from the stitching process onward is as per FIFO (first-in—first-out) basic. The finished garments are made ready for shipment a week before the shipment date. This is done to ensure last-minute troubles in case there is any quality-related issue observed in the final quality inspection as conducted by the buyer QA. This results in an average 17,500 units of packed garments kept as inventory in the finished goods warehouse, which is ready to be shipped.

4.3.2.2 Future state value stream mapping

The common wastes observed in the existing process (through current state VSM) and development of future state VSM

The defects and iterations are typically the most evident waste from the review of current state mapping, and the key focus of the future state. We can summarize the defects by calculating another measure, named "first-time pass yield" (Locher, 2008). As mentioned in Table 4.4, "first-time pass yield" is measured by multiplying the decimal equivalent of the defects.

Another crucial defect in denim jeans production is related to shade mismatch during denim washing or wet processing process. Huge stock of inventory of washed garments is commonly seen between the washing process and finishing & packing process. The variations in fabric characteristics of raw denim such as within roll variation and between roll variation are commonly cited reason for the after-wash shade mismatch.

The overprocessing or NVA processing related to the development and optimization of the washing parameters at various stages is being observed from the current stage of VSM. The rewashing or rework attempts are made while adjusting and optimizing the washing parameters or recipe, namely, cycle time, temperature, and concentration of washing chemicals, at various stages of the production process, such as proto sample development, fit sample, preproduction (PP) samples, bulk fabric in-house, and bulk washing of stitched garments.

At the proto sample development stage, the wash samples in the form of leg panels are developed and submitted to the buyer for approval. Optimization of washing parameters (recipe) is again performed for the PP samples and fit samples to achieve the correct and the buyer approved shade, and look & hand feel. Further, once the bulk fabric is in housed, a shade blanket is prepared by stitching the swatches, usually of size $12'' \times 12''$,

Table 4.4 First-time pass yield [current state value stream mapping (VSM)].

Process	First-time pass yield (current state VSM)
Suppliers	0.99
Spreading & cutting	0.985
Stitching	0.78
Washing	0.85
Finishing & packing	0.97
First-time pass yield (FPY%)	0.63 or 63% ($=0.99 \times 0.985 \times 0.78 \times 0.85 \times 0.97$)

from each fabric roll. The shade blanket is washed and is used for shade segregation of the fabric rolls based on after-wash shade. Further, cut order planning and fabric spreading are done based on these shade segregation or shade lots (Tyler, 2008). Although the spreading and cutting of fabric rolls are done based on after-wash shade categories or lots, *yet the feeding of the cut panel from cutting to sewing section is done based on size.*

The people engaged in the development of washing parameters or recipe spend an unwarranted time and effort to reinvent the wheel. Why would the merchandiser, washing department, R&D section, and the wash development team reinvent the wheel?

Another common waste within the current state is excessive waiting and inventory among the various departments. For instance, higher waiting for size and color-specific packing trims in the finishing and packing department is usually observed. The store department issues the trim to packing departments based on verbal instruction and pushes system. This type of trim issuing system is almost completely based on a forecast, rather than real-time demand that changes frequently. A large quantity of trims is issued to various departments, such as sewing, washing, cutting, finishing, and packing. Usually, these departments are situated at different locations. The traditional system of inventory control is associated with higher process time due to a high level of inventory, low product availability, and higher hidden defects.

Another prominent waste observed in the current state is the sewing room defects and rework. Generally, sewing defects become more prominent and are often created during the garment washing process. For an instance, the damages to chain stitch in both the inside seam and the waistband attach may lead to the opening of seam during washing, typically because of the requirement of rigorous washing process to match the shade, look, and the hand feel to meet the buyer requirements. Future state planning is summarized in Table 4.5.

Creating future state and implementation strategies

The creation of a future state is a really powerful tool. The future stage VSM incorporates a cross-functional working from various departments and develops the plan for the future that should be achievable within a reasonable amount of time, such as less than 1 year.

The particular improvement efforts decided by the VSM team are mentioned as "kaizen burst icon" within the future state (refer Fig. 4.19). This kaizen burst means an activity that must be commenced to achieve the future state and several kaizen bursts are used (Wang, 2011). Generally, the team will not have 100% certainty about achieving all the ideas generated in the future state. The team leaders may achieve a 70%−80% success rate in the improvements. The team may develop creative ideas to improve the production process. The current beliefs and practices will be challenged.

Table 4.5 Future state mapping planning sheet.

Parameter	Current state	Future state
Process time (min)	264.08	264.08
Lead time (L/T) (days)	90−105	75−90
First-time pass yield (%)	63	83

Lean problem-solving 103

Figure 4.19 Kaizen burst.

The basis of the future state is to recognize the need for both internal and external customers, to judge the level of service required by customers or voice of the customer, to check current performance against customer need, to identify the wastes, to improve flow and reduce interruption, and to improve processes. The need of customers is perceived in term of the "level of service" required by customers. The three commonly perceived services are lead time, quality level, and cost. Second, an important question is: How often we check our performance against customer needs (Locher, 2008)?

Identifying the customer actual needs

The late deliveries or shipments are one of the typical reasons for customer's discontent (Abernathy, Dunlop, Hammond, & Weil, 1999, p. 108). The current state mapping specifies that the customer demands a lead time of 75−90 days. However, the actual lead time is 90−105 days. Therefore there is a need to reduce the lead time by 15 days. The VSM team aims to achieve a reduction in lead time as per buyer requirements, which is 15 days. In addition, the number of iterations, interruptions to flow, and process inventory need to be reduced to achieve the desired lead time.

From quality aspect the first-time pass yield (FPY%) is 63% within the current state. The VSM team has targeted an improvement in the first-time pass yield to 83%. The VSM team believes that this target may be achieved within the next 12 months. The team decides to address the needs of internal customers in following future state issues.

Working on increasing the first-time pass yield in washed garment program

The VSM team proposes to improve information quality related to washed shade, to incorporate shade-wise product feeding from cutting to sewing department & from sewing to washing department, and to create a library for the recipe of washing parameters vis-à-vis washing standard required by buyer (Locher, 2008).

First, the VSM team proposes to frequently check the process against the wash standards of garments as required by the buyer. The buyer provides the comments and feedback on the washing developments at various stages at different times, such as proto leg panel submission, fit sample, PP sample submission, and production sample submission. Although washing standards are shared with all stakeholders in the PP meeting, yet the detailed comments and feedback are not properly communicated. Also, the size or dimension of the buyer approved washing standard is not of enough size. Generally, the approved sample standard is manhandled at various stages and is prone to change in shade due to time lag between shade approval from buyer to bulk garment production and washing.

In the current state the planning department does the cut order planning on the basis of an after-wash shade blanket which has $12'' \times 12''$ swatch from each roll. The rolls of similar after-wash shade are classified, spread, and cut together. Shade lot labels are issued to the sewing department for differentiating the shade lot.

The VSM team proposes a kaizen burst (refer Fig. 4.20 for future state VSM) to prepare a checklist for shade-based cut order planning vis-à-vis actual bulk product flow across the various departments such as fabric store, spreading, cutting, stitching, and washing. The team proposes the development of visual shade boards (refer Fig. 4.20 for future state VSM) with an aim to implement buyer requirements, feedback, and comments across various stages to improve communication.

Finally, the wash development team may use the concept of lean product development (Womack, Jones, & Roos, 2007) and develop a shade-wise library of washing parameters. Generally, the R&D department does two types of work: one is to develop new shade or look from scratch and the second is to reuse existing washing parameters or recipe of already developed shade, look, or had-feel. The VSM team recognizes that most of the washing recipe—related activities involve the reuse of exiting parameters instead of developing new parameters. Often, the people involved spending a lot of time reinventing the wheel. The washing department may be encouraged to reuse existing washing parameters or recipe instead of developing a new recipe from scratch every time and the *library of washing parameters* will help in streamline this process.

Working on increasing the first-time pass yield in the stitching section

Generally, one of the prevalent reasons for sewing room defects is linked to the garment washing program. The inherent sewing defects become more prominent and are also caused by the garment washing process. For an instance, the damages to chain stitch both in-side seam and waistband attachment may lead to the opening of seam during washing, typically because of a rigorous washing process needs to match or correct the shade, look, and hand feel as required by the buyer. This leads to higher iterations or rework. The VSM team plan to implement a kaizen burst (refer Fig. 4.20 for future state VSM) to improve the sewing quality during stitching and to reduce stitch defects that are created during the washing process by optimizing the rigorous washing parameters or recipes such as cycle time, temperature, and concentration.

Implementing a shade-wise and size-wise feed of cut panels to the sewing section

As mentioned earlier, the spreading and cutting of fabric rolls are done based on after-wash shade lots, *yet the feeding of the cut panel from cutting to sewing section is done based on size and not on basis of shade*. This is another reason for the shade mismatch. Second, the feeding from the sewing section to the washing section is also based on FIFO. FIFO means that the first unit brought is first to be issued (Das & Patnaik, 2015). The VSM team proposes to implement the *shade-wise as well as size-wise feeding to the sewing* section and to implement shade-wise feeding from stitching to the washing section (refer Fig. 4.20 for future state VSM). The shade-wise feed will help to limit iteration or rework in the washing and sewing section (Karrer, 2012).

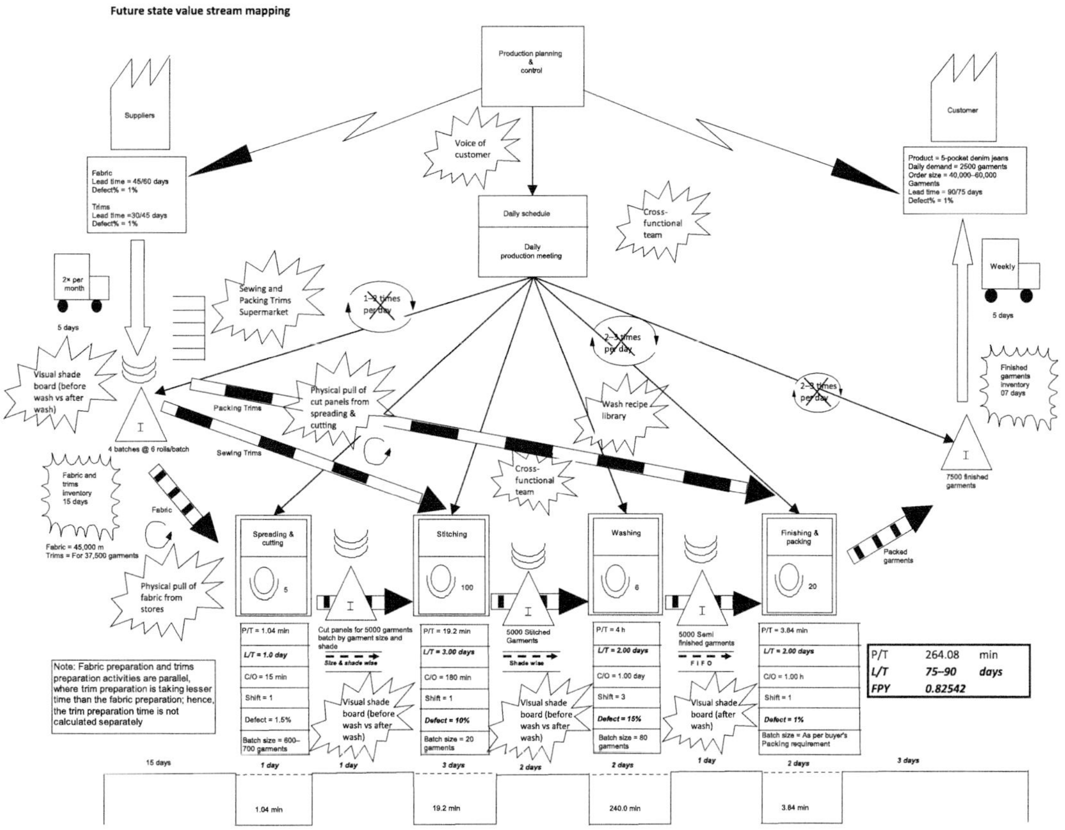

Figure 4.20 Future state VSM. *VSM,* Value stream mapping.

Implementing a pull system for fabric and trims in process inventory

Another aspect proposed in the future stage is a pull system that is based on signals (Kanban) to initiate the replenishment of inventory (fabric, sewing trims, cut panels, and packing trims) at different points (refer Fig. 4.20 for future state VSM). A demand-driven pull-based inventory replenishment system will be incorporated (Taylor & Vatalaro, 2005). A physical pull is proposed between fabric store and spreading & cutting, to receive the fabric lot-wise specific to the work-order or cut. This will ensure shade-wise issues and cutting of panels, and eliminated chances of mixing of panels while stitching. Similarly, the physical pull of cut panels from spreading & cutting to sewing will be established to ensure shade-wise and size-wise feeding in sewing according to the work-order. A supermarket is created for sewing and packing trims, where all necessary sewing and packing trims respectively to orders were arranged and the same may be collected by the person from the next step (i.e., sewing or packing) using a Kanban card. The Kanban card used has some key information: name of the item required, item type, work-order number, style number, quantity needed, and date when supply ended. This will ensure no release of excess sewing & packing trims, and eliminate the chances of loss and mixing of trims (refer Table 4.6 for sample Kanban card) (Hirano, 2009; Shopfloor Series, 2002).

This method regulates the flow of material by replenishing whatever has been consumed and work is based on consumption. The future state incorporates both production Kanban and withdrawal Kanban. This system will result in improved product accessibility, lower inventory, and improved visibility which in turn will reduce defect and process time.

The targeted improvements after lean interventions at the different points of the manufacturing process have been highlighted in bold & italics in the future state VSM (refer Fig. 4.20 for future state VSM). It can be observed from Fig. 4.20, that at many of the process, the lead time (L/T) is estimated to be reduced. Further, the first-time pass yield has forecasted to be increased to 82% due to proposed interventions to reduce or eliminate the iterations (for replacement, alterations, and rejections) at different points (refer Table 4.7) in the manufacturing process. This will collectively result in the forecast for future lead time (L/T) of 75−90 days against the existing lead time of 90−105 days.

Table 4.6 Sample Kanban card.

Kanban card for sewing & packing trims replenishment						
Item name	Item type	Work-order number	Style number	Quantity	Date	Check

Table 4.7 First-time pass yield [future state value stream mapping (VSM)].

Process	First-time pass yield (future state VSM)
Suppliers	0.99
Spreading & cutting	0.985
Stitching	0.90
Washing	0.95
Finishing & packing	0.99
First-time pass yield (FPY%)	0.82 or 82% (=0.99 × 0.985 × 0.90 × 0.95 × 0.99)

4.4 A3 problem-solving

Originally A3 was developed by the Toyota Motor Corporation in the 1960s, as a tool to solve problems and continuous improvements. It was made on A3 paper (297×420 mm) and is based on the A3 thinking approach. According to Roser (2020), the idea of using a single sheet to show all the relevant information was given by quality Guru Joseph Juran (1904–2008). Subsequently, the A3 report technique was developed at the Toyota Motor Corporation under the guidance of Juran (Roser, 2020).

A3 reports describe how a consensus to a complex decision can be made efficiently (Bassuk & Washington, 2013). An A3 sheet is a one-page plan that visualizes, documents, and communicates a story or project (as a structured problem-solving). Generally, it consists of several textboxes and graphs to emphasize the issues, describe the current state, details the future state, and the resources required for achieving the future state (McLean, 2015; Stenholm, Mathiesen, & Bergsjo, 2015). According to Shook (2008), A3 reports are not just problem-solving techniques or communication tools, but these are powerful tools leading to effective and fact-based countermeasures. The A3 reports are successfully used in decision-making, planning, developing proposals, and, of course, in problem-solving. Further Shook referred A3 as a management process as well, where A3 is applied to standardize the methodology for innovation, planning, and problem-solving. It also helps in developing foundational structures for sharing a broader and deeper form of thinking (Shook, 2008).

Background of the A3 tool is the "A3 thinking" which is founded upon scientific or logical mindset toward problem-solving. The "A3 thinking" emphasizes on the hunt for the root cause by tracing the source or origin of every shop floor problem, and the focus is on the logical methods to solve the situation that is in one's circle of influence or control. Second, an important feature of the "A3 thinking" is objectivity: any shop floor or other problem should be measurable, quantified, detailed, and specific. The objective approach emphasis is to see the process clearly and not merely falling in the trick of blaming some other group, individual, or department. For instance, the root cause may be determined with the help of the Five Whys tool. Third, the "A3 thinking" strongly emphasizes on both the results and process. For example, performance objective might be to aim for the highest results, such as the highest production, profits, and quality, and achieving the lowest price, lead time for the product or service. However, the results should not come at the expense of the process and people, else the result will be short term. Further, it is very important to focus on synthesis, alignment, and coherency viewpoint. A3 reports are concise and brief single-page reports that elaborate on the status of the problem area. For example, the emphasis is on visualization, sketch, or graphs, rather than a word to communicate the message clearly and efficiently. Also, the A3 thinking place high importance on consensus from all the parties. Alignment in A3 thinking refers to a written communication or consensus horizontally as well as vertically across the organization (Bassuk & Washington, 2013; Sobek & Smalley, 2008).

A3 is a flexible tool and can be applied in several problem-solving situations. The overall flow of the report incorporates the PDCA. The typical flow of the A3 sheet is described in Fig. 4.21. A3 sheet consists of seven sections, in addition to the title, which are explained in the following.

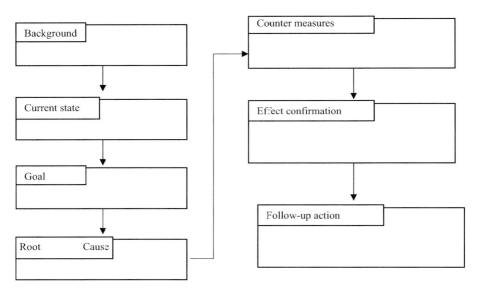

Figure 4.21 Typical flow of the A3 sheet.

Title: The report begins with a title that introduces the content of the A3 report to the audience, for example, improving the productivity of the line, reducing the sampling time, or reducing the sewing defects.

1. *Background:* The purpose of background information is to describe the overall context of the situation. It may identify product, service, serial number, software used, and customer specifications that are under consideration. It may suggest how the issue to be worked upon in the A3 sheet is related to the perspective of the overall goals of the organization or company.
2. *Current condition and problem statement:* It depicts the current state of the process and highlights the key factors in the current state and summarizes the key information related to project.
3. *Goal statement:* This section depicts the clear target for the situation while clarifying the measure of performance.
4. *RCA:* Suitable tools, such as RCA, Five Whys, or Pareto charts, may be used for identifying and summarizing the causes of the problem.
5. *Countermeasures:* This step deals with allocating the action items to root causes which are identified in the previous steps. Also, it may specify the concerned persons who will implement the countermeasures along with the location and due date by which the improvement is to be implemented.
6. *Check/confirmation* of effect.
7. *Follow-up action* (Bassuk & Washington, 2013; Saad et al., 2013; Sobek & Smalley, 2008).

The key objective of the A3 approach is to get all the necessary data on a single paper (A3 size sheet) to get the glance of the entire story in a visual manner. Here, it is imperative to get the data directly from the shop floor (*Go to Gemba*), as the shop floor is the only place where activities are happening, and we have the opportunity to witness the real situation.

An A3 sheet and a pencil are two essential items for A3 problem-solving. It is recommended to use a pencil as it gives freedom of corrections/changes in the content (by simply using an eraser) as and when required. This makes an easier representation of the actual situation (as observed on the shop floor) and a workable solution to the problem/issue observed with multiple iterations. However, considering the limitations of legibility in handwriting (while manually creating A3 sheet using paper and pencil) advance technological tools (such as tablets and interactive smart boards available), the author is in opinion to use such smart tools (that are equally user-friendly and flexible to any changes or modifications) as an alternative to conventional tools. Here, it is important to understand that there is no standard format for the A3 report, it's all about making the tool fit the problem, and it is up to the observer choosing the most appropriate tool (that is fitting the best) in a specific situation/problem. An A3 can be considered as a visual representation of PDCA in a single sheet. The "Plan" covers points such as background, current state, problem statement, target, problem analysis, and comparing and proposing the solution. Corrective actions, implementation plan and schedule, and people or process responsible are part of "Do." The "Check" is about monitoring the solutions implemented and measuring the progress, while "Act" is about follow-ups and making the solutions sustainable, and standardizing the process.

Let's understand the A3 problem-solving with an example of an apparel manufacturing environment. An export-oriented 5-Pocket Jeans manufacturing plant was facing a serious issue of high alteration levels. The average monthly alteration levels were around 22%. Most of the time, jeans manufacturing involves a rigorous dry process and washing process, which result in damages in the stitched garments leading to high alteration rates of the washed garments. Several defects including raw edge, broken stitches, damaged leather patch, raveling, a pleat at bottom Hem, roping at bottom Hem, distortion in the destruction of dry process, pocket bag opening, raveling buttonhole, label damage, zipper damage, etc. were visible after washing process only. Raw edge (with seams open out at different parts of the garment) was one of the major defects observed responsible for higher alteration levels and garment rejections. Here, it is important to note that due to the heavy dry process and washing treatments, the alterations in the washed 5-Pocket Jeans are difficult to correct. Correcting a defective piece (such as seam open after washing) leads to stitch marks and exposing of the unwashed surface of the garment (which was inside the seam earlier). The corrections done in such garments are visible and results in rejection of garments in the final quality check. Eventually, such garments are not shipped as fresh garments but treated as B or C grade garments causing huge financial losses to the company.

The A3 problem-solving technique was used to investigate, deep-dive, and control the alteration levels. Fig. 4.22 mentions a typical A3 report with a visual representation of the problem, root cause, countermeasures, and the effect confirmation. As illustrated in Fig. 4.22, the relevant data (monthly alteration% and defects%) of the last few months was collected from the shop floor to gage the actual condition and the real problem about the high alteration rate. Raw edge and Broken Stitch are the two most prominent defects (covering 46% of the total defects) that are observed after washing in the finishing & packing process. Raw edge (fraying),

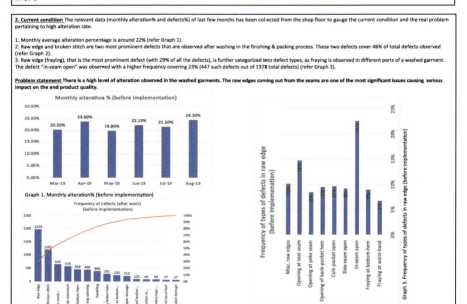

Figure 4.22 A3 problem-solving (a case of apparel manufacturing).

Lean problem-solving

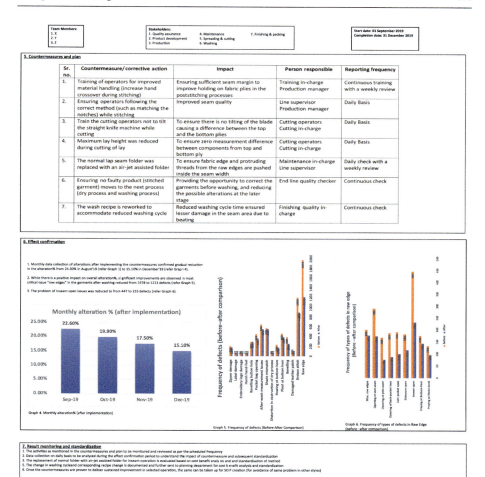

Figure 4.22 (Continued)

which was the most prominent defect (with 29% of all the defects) was further categorized into defect types, as fraying was observed in different parts of a washed garment. The defect "in-seam open" was observed with a higher frequency covering 23% (447 such defects out of 1978 total defects). The goals were set accordingly to keep the targets in measurable terms, so that progress can be assessed effectively in an objective manner. The RCA was undertaken to investigate the source of the problem, and accordingly, countermeasures (involving different areas including sewing section, quality, product development, cutting, washing, and maintenance) and plans were worked out. The fishbone diagram for only one of the defects "in-seam opening after washing" was shown in A3 sheet with problems not found to be present in this case was marked strikethrough. Postimplementation of the countermeasures as per the plan, the alteration and defect data were carefully collected, monitored, and analyzed. A critical review of the outcomes of each of the counter-measures was ensured, resulting in informed decisions for further fine-tuning. It can be observed from Fig. 4.22 that the countermeasures confirmed a gradual reduction in alteration% from 24.30% in August'19 to 15.10% in December'19. There was a positive impact on overall alteration%, significant improvements are observed in most critical issue "raw edges" in the garments after washing reduced from 1978 defects to 1213 defects. The problem of in-seam open issues was reduced to from 447 defects to 233 defects. The respective activities and processes were improvised with standardized parameters once the stability in improvements was achieved with the targeted outcome meeting the goals set. In the end, to make the improvements sustainable, adherence to the standardized processes/methods was ensured.

Here, it can be concluded that the A3 problem-solving involved logical, scientific, and data-driven solutions to address the issue and emerged as a key approach with the shared responsibility of all the stakeholders in a disciplined manner.

4.5 Standardized work

In a process of continuous improvement the processor work done by each individual must be standardized. Despite the best of industrial engineering (IE) practices, most of the apparel manufacturing organizations lack in documenting and standardizing the activities and processes. Industrial engineers put efforts for method improvement through work-study techniques, but still generally fail to standardize the methods. Even if the processes or work methods are standardized up to some extent, the scope is limited only to the stitching operations, and other processes are not given due attention. This results in high process and product variability, frequent creation of bottle-necks, issues of sudden breakdowns, the longer learning curve when the operator is replaced with other, increased training cost, higher setup times, and constant interrup-tions in the manufacturing process. The lean approach recommends a pull-based sys-tem where the production activities are initiated only after receiving a signal of demand from the next process. This ensures a lesser level of work-in-progress aiming to minimize different forms of wastages (Muda) in the process. To achieve an unin-terrupted pull-based flow, each of the work must be done in a standardized manner, at the right place, and should be done at the right time. Failing to any of such prereq-uisite, the pull-based flow system may be disturbed and may lead to process imbal-ances and unpredictability. Hence, for an accurate prediction of process cycles, the work must be standardized for each process (Choudari, 2002). Standardized work is also referred to as standard work, aims to establish precise work procedures to an operator. This is a lean tool used to define and document the work between people and their work environment and an important element of continuous improvement of a lean transformation. With an established practice of standardized work, the same processes can be performed every time in a repeated and identical manner. Such prac-tice can establish the "current best way" of doing the task (Charron, Harrington, Voehl, & Wiggin, 2015). Takt time, precise work sequence, and standard inventory are the three elements of standardized work (Lean Enterprise Institute, 2008). Process capacity sheet, standardized work combination table, and standardized work chart are three of the key documents used in the standardized work. Please refer to Figs. 4.23—4.27 for the application of standardized work for quality assurance, cutting process, label procedure, in-line quality, and for broken needle procedure.

Takt time: Takt is a German term (originated from the beats of a drum) that means rhythm or a signal at a particular interval. In lean manufacturing, takt is referred to the rate of producing a product to meet the customer demand on-time. In manufacturing setups, this is generally taken as the output rate of the finished product from the last operation or process. Takt time is determined using customer demand and the available time to produce the goods (McLean, 2015). The operations and processes should be balanced accordingly to meet the takt time, and the load leveling should be done to keep up an uninterrupted flow or the rhythm for standard work.

Let's assume, an apparel manufacturing setup operates in one shift of 8 hours per day with 40 operators, and the customer demand (production target) is 4800 garment per day to meet the delivery deadline. Then, the takt time will be as:

Total time available $= 1 \times 480 \times 40 = 19,200$ minutes per day
Demand $= 4800$ garment
Takt time $= 19,200/4800 = 4.0$ minutes/garment

Figure 4.23 SOP board: quality assurance (Standardized work). *SOP*, Standard operating procedure.
Source: Courtesy Corrigo Consultancy Pvt. Ltd., India.

114 Lean Tools in Apparel Manufacturing

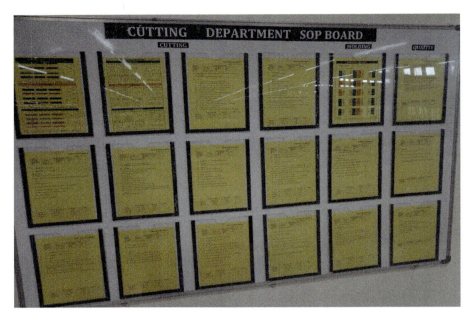

Figure 4.24 SOP board: cutting process (Standardized Work). *SOP*, Standard operating procedure. *Source*: Courtesy Corrigo Consultancy Pvt. Ltd., India.

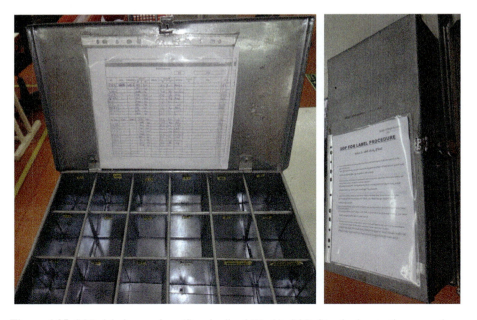

Figure 4.25 SOP: label procedure (Standardized Work). *SOP*, Standard operating procedure. *Source*: Courtesy Corrigo Consultancy Pvt. Ltd., India.

Lean problem-solving 115

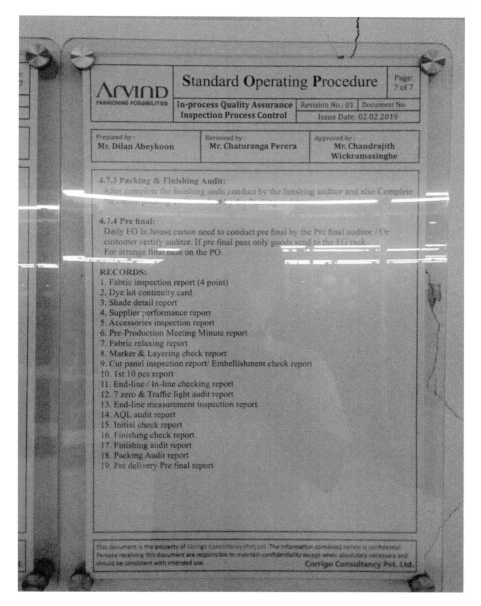

Figure 4.26 SOP: in-line inspection (Standardized Work). *SOP*, Standard operating procedure. *Source*: Courtesy Arvind Ltd. & Corrigo Consultancy Pvt. Ltd., India.

This means that to meet the demand, a garment needs to be completed every 4 minutes. Therefore every step or operation in assembly needs to be done/delivered every 4 minutes (or multiples of it).

116 Lean Tools in Apparel Manufacturing

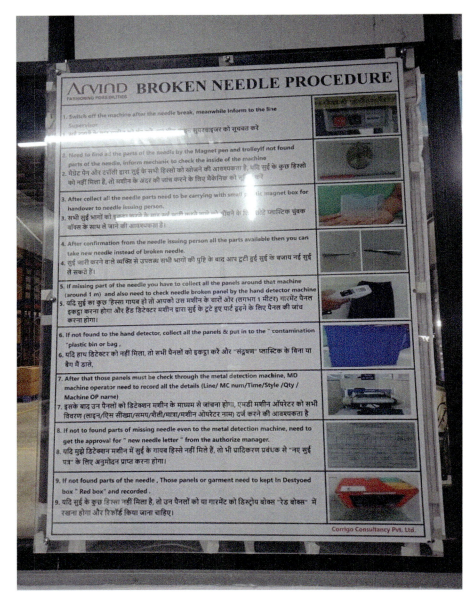

Figure 4.27 SOP: broken needle procedure (Standardized work). *SOP*, Standard operating procedure.
Source: Courtesy Arvind Ltd. & Corrigo Consultancy Pvt. Ltd., India.

Work sequence: The work sequence is the predetermined step of activities or operations that are performed in a specific order to complete the task. Work standardization is an important aspect of IE where with the help of method study and method improvement, the standard way of performing a task is established.

Standard work-in-process (standard WIP): In a standardized work setup where the process balancing is done as per takt time, and the work sequence of each of the tasks is predetermined, it becomes very critical to maintaining a minimum level of work-in-process (WIP). This is essential to maintain the uninterrupted flow of output by keeping a minimum amount of work between two upcoming processes or operations. Here, it is important to note that keeping an excess WIP is a waste as per lean philosophy while a WIP level (lesser than the minimum WIP level) may result in process or operation idleness due to martial run-out.

4.6 Plan−do−check−act cycle

PDCA, also referred to as PDSA (plan−do−study−act), is an important four-step process (refer Fig. 4.28) improvement approach of lean philosophy developed by Shewhart. Later in the 1950s practiced and promoted by Deming, hence this is also called the Deming cycle or Shewhart cycle (Hill, 2012).

PDCA is considered as an essential requirement of any lean transformation, as it provides an important opportunity to assess the improvements and necessary fine-tuning and act in the right direction to meet the objectives. In PDCA, there is always an opportunity to revisit the changes and set newer targets as a part of continual improvements (Gitlow & Gitlow, 1987).

As indicated in Fig. 4.28, there is no start or end of a PDCA cycle; hence, practically, an improvement can be initiated from any point. However, most of the time, any improvement effort starts from the planning of what we want to achieve or to carry out in the future. This can be referred to as setting up the benchmark targets or any small improvement. This also involves the planning of methodology for achieving the set target. The next step is the practical implementation of the things (Do) as per the plan to attain or meet desired results. The third step is checking and assessing the progress made so far, see if everything is on track as planned, and observe the challenges and obstacles. This step also involves collecting relevant information or data which helps in taking corrective informed decisions. The last step of the PDCA cycle is acting accordingly (based on the inferences of the third step—Check) to further fine-tune or resetting of the objectives. The process should be standardized and stabilized if the set targets or results

Figure 4.28 PDCA cycle. *PDCA*, Plan−do−check−act.

are achieved, else planning to achieve the set objectives should be done. As lean is a continuous journey, there is always a scope of improvement. Hence, the PDCA cycle should be repeated continuously seeking better always in pursuit of achieving perfection.

4.7   Look−ask−model−discuss−act cycle

LAMDA (look−ask−model−discuss−act) cycle is a problem-solving approach, used as a basic learning cycle of lean product and process development (Marchwinski, 2014). LAMDA cycle is a relatively new technique proposed by Allen Ward in 2002 at the University of Michigan's Japan Technology Education Program (Domb & Radeka, 2009). Before LAMDA, PDCA was used as a knowledge-driven decision-making tool. The PDCA was successfully applied, by established auto leaders such as Toyota and GM in their lean transformations. As observed by Ward (2002), PDCA was not sufficient enough for revealing the complete details of the activities undertaken by Toyota and other Japanese companies who practiced lean, as the same results were not achieved by American companies while following the PDCA approach. Eventually, LAMDA was proposed as an extension to PDCA with an effort to overcome the limitations of PDCA (Ward, 2002).

As a key feature of LAMDA, as well as a continuous improvement methodology, it is applied to develop the metaskill (which is the skill of skills development) for workforce skill development as a learning process to create knowledge. In this way, LAMDA is capable of revealing more insights into a problem (Stenholm et al., 2015). LAMDA focuses on the creation of knowledge and a true in-depth understanding of the cause of a problem before an action is taken based on the very first thought (Michael, Harmon, & Minnock, 2008). As there is an in-depth investigation, LAMDA also establishes the achieved results which are supported by the observations.

The basic LAMDA cycle includes five aspects, which can also be treated as levels or classes of learning and knowledge creation. These levels are:

Look: Focuses on first-hand primary observation-based approach where one needs to go to the floor or site and observe the things as it happens.

Ask: Focuses on the investigation approach which is based on critical questioning and probing. This aims to get deeper insights into the issue and the complete know-how. This approach of critical questioning also helps in identifying the potential root causes of the problem or issue.

Model: Focuses on the development of a hypothetical model, based on the learning and knowledge gained in the first two steps. The approach of the development of the model is based on scientific analysis, statistical modeling, and simulations to predict the expected performance of the proposed solution to the issue or problem being addressed.

Discuss: Focuses on receiving peer feedback on the model (solution) proposed. The solution may be presented and discussed with other teams, persons directly related to the issue (who understands the process and the situation thoroughly), and other supporting teams. It is advised to incorporate the feedback if it adds value to the model or solution proposed.

Act: Focuses on testing and deployment of the solution as an outcome of the first four levels. This also aims toward establishing (reconfirming or validate) the learning of the entire process.

LAMDA is also recognized as not just a tool but as a culture (LAMDA culture) where the key focus is on the importance of learning and sharing the knowledge which lays the ground for future improvement. LAMDA culture enables employees to naturally share their knowledge (Michael et al., 2008).

4.8 Define–measure–analyze–improve–control

The DMAIC (define–measure–analyze–improve–control) is a data-driven business improvement tools. It is a systematic and methodological approach. DMAIC is at the heart of Six Sigma (Samanta, 2019). The DMAIC methodology works on the problem which has been identified by the organization and uses a set of tools and techniques to arrive at a sustainable solution. The resultant solution(s) will minimize or eliminate the problem, placing the organization in a competitive position (Shankar, 2009). The detailed steps are explained in the following subsections.

4.8.1 Define

The "Define" phase ensures that the process/project, which will go through the DMAIC process, is linked to the organizational goals and has necessary management support to take it forward. Define phase starts with problem identification. It specifies the objectives, scope, and boundaries of the project and ends with the evidence of management support for the project (Samanta, 2019; Shankar, 2009).

Also, an approved project plan that includes detailed tasks to be accomplished for all the stages, namely, define, measure, analyze, improve, and control, is prepared while specifying the time schedules and resources required.

Typically, the design of the new process on a shop floor may be required to meet the following objectives:

- improve quality
- reduction in rejection
- reduce lead time
- implement a pull system
- increase production and productivity
- decrease inventories
- production line flexibility

The define phase may specify as the following steps:

- To define a project charter that outlines: the name of the project along with improvement goal, the problem statement or need of project, scope and significance of project, and process to be improved.

- To identify customers and stakeholders, both internal and external.
- To define the initial VOC and critical to satisfaction criteria from the customer perspective.
- To form a project team.
- To lunch the project.

4.8.2 Measure

Initially, baseline data is collected to understand and define a starting point. The baseline data are from the current process. The baseline data is also used to better understand what is happening in the process, customer expectations, and where the problem lies. The baseline data helps in identifying the areas or sections where improvements are to be made. The initial baseline data may be collected from quality reports, rejections reports, maintenance schedules, operation breakdowns, production reports, WIP report, etc. (Cudney et al., 2014; Shankar, 2009).

Further, the detailed customer data of both internal and external customers is collected. Usually, suitable tools, such as focus group discussion, brainstorming, benchmarking, and interviews with customers and stakeholders, are being used.

The data may be organized or synthesized in various forms. The raw data can be summarized using the various explorative and descriptive tools, such as mean, median, mode; charts; and graphs (Pyzdek, 2003; Shankar, 2009). For an instance, the baseline data obtained from time and motion study may be compiled in the form of a current state map chart and can be used to differentiate the VA activities or time from the NVA activities or time.

4.8.3 Analyze

Analyze phase helps to understand the cause-and-effect relationship in the process. It helps in filtering and identifying essential input factors from the list of factors identified in the measure phase (Shankar, 2009). The tools commonly used to identify the root cause are mentioned as follows:

- Cause-and-effect diagram
- Why-Why analysis for 5 Whys
- Waste analysis
- 5S analysis
- Kaizen plan using DMAIC
- Failure mode and effect analysis
- Customer survey analysis and hypothesis testing
- Defects per million opportunities (Cudney et al., 2014)

4.8.4 Improve

Generally, to a certain extent, the analyze and improve phases are conducted at the same time. Moreover, there is an improvement in every previous phase of the project. For instance, the activities conducted in define, measure, and analyze phases that help in identifying customer needs, how to measure these needs, and how the existing process may fulfill (Pyzdek, 2003).

In case if the performance falls short of the project goal the improvement recommendations to eliminate the root cause may be identified. For instance, the improvement recommendations could be various lean tools, such as 5S, poka yoke, Kanban, and Total Productive Maintenance (TPM) or Single-Minute Exchange of Die (SMED). Further, an action plan may also be suggested. Action plans may be of both short term and long term. It should specify project plans, resources, timeline, and risk analysis. Also, the team should develop and roll out training for all the stakeholders including the concerned project owner who will implement the change. The project team should identify the performance improvement target, and design scorecards to access the improvements achieved. The process owner should implement the changes and communicate to all stakeholders. The project team may use PDCA to implement recommendations (Cudney et al., 2014).

4.8.5 Control

In the previous phase of improvement the project goals may have been achieving. At last, the objective of the control phase is to ensure the gain achieved in the previous phase is permanent and to prevent problems with the new improved process (Pyzdek, 2003).

For instance, the evaluation of improvements may be determined by quantifying the actual gain achieved, monitoring the improvement continuously, and ensuing the gains are permanent. The benefits which are to be monitored may be as follows:

- quality level increased
- decrease in defects
- better feedback
- increase production
- increase flexibility
- decreased work-in-process (WIP)

4.9 Root-cause analysis

The complex work environment such as apparel manufacturing faces frequent challenges in its day-to-day operations. Many times, several issues are occurring at the same time and affect the efficiencies severely. The most common approach in such circumstances in apparel manufacturing is seeking immediate solutions to get quick relief. This is a kind of ad-hoc approach that can be considered as quick fix or bandage to the problems. Such practices may look good at the moment but not sustainable, and the same issues and problems occur again and again. This consumes a lot of time, effort, energy, and money in fire-fighting.

The core reason for this behavior of opting for covering up the issues using short-term solutions is the inability to see the bigger picture. People often give more value to immediate relief rather than a long-term permanent elimination of the problem. Slowly, this becomes the culture of the organization, and problems are treated casually. Most of the time in practical work environments, the scientific method with thorough observations and experiment are ignored over random adjustments to the system.

The RCA is a proven methodology for a permanent solution of a problem, through the identification of the real reason or the root cause (Ammerman, 1998; Andersen, Fagerhaug, & Beltz, 2009). The PDCA cycle may be combined with RCA, as it can help in the identification of the true state of nature or the root cause. Further, once the root cause is identified, the corrective actions or initiatives for further improvement can be taken as part of PDCA (Barsalou, 2015).

There are several tools and techniques used in RCA. There are seven classic quality tools for RCA as defined by the American Society of Quality in the 1960s: flowcharts, Pareto charts, Ishikawa diagrams, run charts, check sheets, scatter diagrams, and histograms (Geoffrey, 2006, pp. 59–61). Depending on the situation (according to the problem faced), appropriate tools can be applied in RCA by the practitioners. Apart from these seven classing tools, there are some additional tools used in RCA such as 5-Why analysis (also referred to as Why-Why analysis), cross assembling, is-is not analysis, parameter diagram, boundary diagram, and matrix diagrams (Andersen et al., 2009). In the context of apparel manufacturing, Ishikawa diagram (also known as cause-and-effect analysis and fishbone diagram) and 5-Why analysis are some of the most popular tools used in RCA. Given the same, only these two tools are discussed here in this chapter; for additional tools the readers may refer to other available literature in this area.

4.9.1 Ishikawa diagram

The Ishikawa diagram (or cause-and-effect diagram or fishbone diagram) was first created by Kaoru Ishikawa. An Ishikawa diagram (refer Fig. 4.29) is a visual presentation of the possible reasons and causes of a problem (observed as the effect of a cause) (Andersen et al., 2009; Okes, 2019, pp. 59–62). The problem or the effect is mentioned at the right-hand side with a horizontal line going toward the left side,

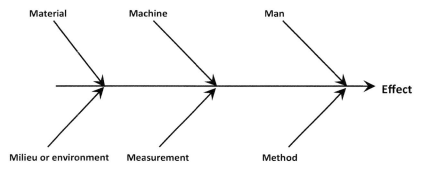

Figure 4.29 Ishikawa diagram.

while the causes (possible reasons) are mentioned in the form of lines at approximately a 45-degree falling on the horizontal line of effect. These lines represent possible reasons or the factors causing the problem. These factors are generally represented by six Ms: Man (people), Machine, Material, Method, Measurement, and Milieu (or Environment or nature of the working conditions) (Barsalou, 2015, p. 18; Charron et al., 2015, p. 418).

Let's try to understand the identification of a root cause in apparel manufacturing. Seam opening at the inside leg area is one common issue in jeans manufacturing. Most of the time, the jeans pants involve heavy washing with stones to get the desired washing effect. The cycle time varies from 2 to 4.5 hours depending on the washing recipe. In the case of heavy stone wash treatments the garments have to go through a rigorous washing. Many times, in the finishing sections, it is observed that the seams are found opened-up after washing. It is very difficult to repair such garments, as any correction or repair leaves marks of the previous seam, and also the effect of the new seam is not the same as the earlier seam, which was done in the stitching (before washing). Here, it is important to investigate the probable reasons for this problem of the seam open, and each of the factors (of 6Ms) needs to be examined critically. Each of these factors and associated probable reasons for the problem (refer Fig. 4.30) have been discussed below through the Ishikawa diagram.

Man—It should be checked and confirmed that the operators who are involved in the in-seam of jeans are following standard work instructions and they are possessing the necessary skills to undertake the tasks. The absence of proper training and lack of skills, and nonadherence to the instructions may result in the wrong panel attach where proper seam margins are not kept by the operators. Also, there are chances that the operator may not be following the notch marks of the leg panels.

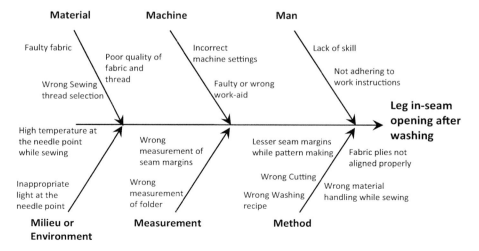

Figure 4.30 Ishikawa diagram for leg in-seam opening of jeans pants after washing.

Machine—The in-side leg seam of jeans pant is generally done using a feed-of-arm (FOA) machine, which is a double-needle chain stitch machine. During RCA, it should be ensured that all the machine settings and work-aids are done correctly. There can be problems due to faulty settings (such as stitches-per-inch settings of the machine) or not properly fitted seam folder (work-aid), and due to which the machine is not able to take the sufficient seam allowance (fabric margin) while stitching. If the fabric margin is lesser in the seam, it may come out during the washing process.

Method—The method of in-seam stitching should be checked properly. The manner fabric plies (leg panels) are aligned precisely, it is inserted in the seam folder properly, and the handling of the panels while stitching using the FOA machine is some key points to be checked while investigating the method. Also, it is important to check and confirm that all the operators are following the same method during stitching.

Apart from the stitching method, the method of pattern making, cutting, and washing methods can also be the source of problems, and each of these should be investigated thoroughly. For example, if the seam margins are not properly followed while cutting or the notches are not marked properly, the alignment of the panels during sewing may get disturbed and may lead to the opening of in-seam.

Material—The reason of defect may be a faulty material as well. For example, if the fabric is having some defects like skewness where the fabric is not balanced lengthwise and getting distorted. The other issues may be related to the quality of the sewing thread used, there can be chances where the sewing thread is not strong enough to sustain the rigorous washing treatment. Sometimes, if the fabric is stretchy or elastic it may also distort the seam responsible for seam open.

Measurement—This is related to the measurement- or dimension-related issues causing errors. There are chances that the seam margin is wrongly measured and mentioned, it may be due to a wrong scale used. Also, due to wrong measurements, the settings of the seam folder may be done incorrectly causing insufficient fabric margins in-side the seam.

Milieu (environment)—The environment may also be a factor that is responsible for the generation of the defect. The opening of the seam may be due to the high temperature (at the needle point) during stitching, where the sewing thread is losing its strength. Also, this problem of opening the seam, may occur due to the rise in the temperature (inside the washing machine) during washing, that may cause weakening of the seam. Further, inadequate light while sewing may act as the source of the problem, as the operator may not be able to see the fabric panels, and the seam line properly due to lower light intensity at the needlepoint.

4.9.2 5-Why analysis

The 5-Why analysis also popularly known as Why-Why analysis is one of the most used tools of RCA. The 5-Why technique can be applied in any environment, and many of the apparel manufacturers use this technique as a tool to identify the root cause of the problem, waste identified, or an issue that occurred. This approach suggests a continuous questioning or reasoning in a series to find out the reason, and eventually, it leads to the identification of the root cause of the problem (Andersen et al., 2009). It is believed that

Lean problem-solving

repeatedly asking (why) the questions for five times guide us to determine the root cause (Bhasin, 2015; Mussman, 2001). Let's try to understand 5-Why analysis with a real-life example of delayed shipment in an apparel manufacturing plant.

The initiation of the 5-Why reasoning may start from asking the first question:

1. *Why a particular shipment got delayed?*

    Because, the shipment got rejected on the day of shipment while final quality inspection by the buyer QA; however, all the garments were produced and packed according to the plan.
2. *Why the final quality inspection got rejected?*

    Because, the size labels in some of the garments were attached wrongly.
3. *Why the size labels in some of the garments were attached wrongly?*

    Because, the size labels got mixed in the sewing line.
4. *Why the size labels got mixed in the sewing line?*

    Because, all the size labels for the entire work-order quantity were issued at the same time.
5. *Why the size labels for the entire work-order quantity were issued at the same time?*

    Because, the work instructions regarding the size-wise issue of size labels were not given to the person responsible to issue the size labels.

Here it can be observed that the issue instructions (as indicated in the last question) are the real reason for the entire issue of shipment delayed. A very small issue at an earlier stage of manufacturing has led to a bigger loss at the later stage and causing significant losses (recall the discussion of "Rule of Tens" at the beginning of this chapter). To avoid such instances in the future, it should be ensured that the necessary instructions are given well in time, and a standard operating procedure may be prepared in this regard. Such an approach may also be applied to other issues and problems faced by the apparel manufacturers on a day-to-day basis, and accordingly, appropriate steps may be taken.

It is worth mentioning that some industries apply the Ishikawa diagram (fishbone diagram) and 5-Why (Why-Why) analysis collaboratively to identify the root cause of the problem. A car seat cover may have defects that include fabric defects and sewing defects. The fabric defects are generally from the fabric manufacturer's end. Loose stitch, skip a stitch, uneven seam margin, uneven notch, uneven stitching, faulty wire attach (or wire missed out), and notch mismatch are some of the major defects in a car seat cover. As a car seat is a major functional component directly linked with comfort and safety, quality control becomes vital. Lifter hole mismatch is one issue observed while sewing the car seat cover. The hole on the car seat panel should be aligned to the hole on the mirror panel (refer Fig. 4.31).

The cause-and-effect approach was applied to understand the possible reason for such a mismatch of the lifter hole (refer Fig. 4.32).

Margin out while sewing, wrong cutting, and incorrect punching of the hole due to the wrong template were three potential causes identified. On further investigation, seam margin and cutting accuracy were observed well within the acceptable limits. This left with one possible cause of using an incorrect template for punching the hole. In the next stage the Why-Why analysis was applied to identify the root cause.

1. *Why a lifter hole mismatch?*

    Because, the hole shifted vertically and horizontally while fitment.

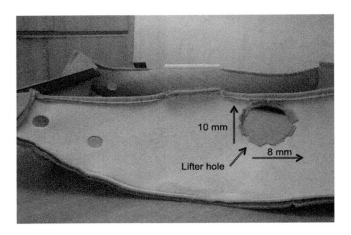

Figure 4.31 Lifter hole mismatch in a car seat cover.

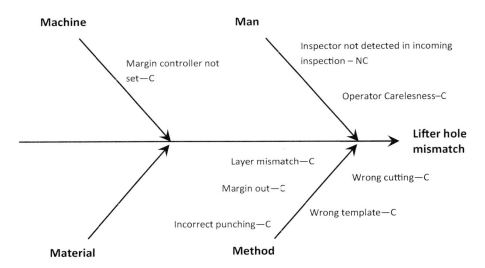

Figure 4.32 Cause-and-effect diagram (lifter hole mismatch) (Stitch World, 2013).

2. *Why the hole shifted vertically and horizontally while fitment?*
   Because, the hole punching was wrong.
3. *Why the hole punching was wrong?*
   Because, hole location in the template while punching was not correct.
4. *Why the hole location was incorrect?*
   Because, the punching template was incorrect.

As the complaint was received from the customer end, a separate Why-Why analysis (in a similar fashion) was conducted on why the lifter hole mismatch defect reached to the customer (Stitch World, 2013).

4.10 Conclusion

The path of sustainable growth and achieving excellence is challenging, but it results in sweet fruits of great success. The lean approach provides this opportunity to grow and excel in practical manufacturing environments, but the implementation of changes for improvement requires constant determination, dedication, discipline, and efforts. All these four qualities should be trickled down from the top management to every level of management to create a workforce of positive change. Change is always painful, and people always resist change. The top management should be ready for such situations and must put efforts in addressing and convincing people on why change is required, and how it is going to impact the organization as well as an individual. Further, the training & skill development (on lean initiatives, and applying lean tools effectively and efficiently) of the personnel should be given priority. The cost of training the employees properly is an investment, not a cost.

Any kind of improvement takes its own time, and one (especially the top management) must be patient in getting the results done. Also, one should not go too fast with the changes, as it may damage the existing systems. The transition should be smooth and seamless, and employees as well as systems should be given sufficient time to settle down and adjust with the new environment. All the lean tools are proven practices that have shown its worth again and again, provided these are implemented religiously with dedication and discipline.

References

Abernathy, F. H., Dunlop, J. T., Hammond, J. H., & Weil, D. (1999). *A stitch in time: Lean retailing and the transformation of manufacturing—Lessons from the apparel and textile industries*. Oxford University Press.

Aken, E. M., Farris, J. A., Glover, W. J., & Letens, G. (2010). A framework for designing, managing, and improving Kaizen event programs. *International Journal of Productivity and Performance Management, 59*, 641−667.

Ammerman, M. (1998). *The root cause analysis handbook a simplified approach to identifying, correcting, and reporting workplace errors*. Productivity Press.

Andersen, B., Fagerhaug., & Beltz, M. (2009). *Root cause analysis and improvement in the healthcare sector − A step-by-step guide*. ASQ Quality Press.

Barsalou, M. A. (2015). *Root cause analysis a step-by-step guide to using the right tool at the right time*. Boca Raton, FL: CRC Press.

Bassuk, J. A., & Washington, I. M. (2013). The A3 problem solving report: A 10-step scientific method to execute performance improvements in an academic research vivarium. *PLoS One, 8*(10).

Bhasin, S. (2015). *Lean management beyond manufacturing a holistic approach*. Coventry: Springer International Publishing.

Charron, R., Harrington, H. J., Voehl, F., & Wiggin, H. (2015). *The lean management systems handbook*. Boca Raton, FL: CRC Press.

Choudari, A. (2002). Lean manufacturing. In J. B. ReVelle (Ed.), *Manufacturing handbook of best practices: An innovation, productivity, and quality focus* (p. 189). Boca Raton, FL: CRC Press.

Cudney, E. A., Furterer, S. L., & Dietrich, D. M. (2014). *Lean systems: Applications and case studies in manufacturing, service, and healthcare.* Boca Raton, FL: Taylor & Francis Group.

Damelio, R. (2012). *The basics of process mapping book.* Productivity Press.

Development Commissioner Micro, Small & Medium Enterprises. (2010). *Guidelines for implementation of lean manufacturing competitiveness scheme.* New Delhi: Development Commissioner Micro, Small & Medium Enterprises, Government of India.

Domb, E., & Radeka, K. (2009). *LAMDA and TRIZ: Knowledge sharing.* Retrieved from www.triz-journal.com:https://triz-journal.com/lamda-and-triz-knowledge-sharing-across-the-enterprise/#Ref2. (Accessed January 2020).

Das, S., & Patnaik, A. (2015). Production planning in the apparel industry. In R. Nayak, & R. Padhye (Eds.), *Garment manufacturing technology* (p. 98). Sawston, Cambridge: Woodhead Publishing.

Farris, J. A., Aken, E. M., Doolen, T. L., & Worley, J. (2008). Learning from less successful kaizen events: A case study. *Engineering Management Journal, 20.*

Feld, W. M. (2000). *Lean manufacturing—Tools, techniques, and how to use them.* Boca Raton, FL: CRC Press.

Geoffrey, M. (2006). *Kaizen event implementation manual.* Society of Manufacturing Engineers.

Gitlow, H. S., & Gitlow, S. J. (1987). *The deming guide to quality and competitive position.* Prentice-Hall, Inc.

Hasan, K. M. (2016). Causes and remedies of batch to batch shade variation in dyeing textile floor. *Journal of Textile Science & Engineering, 6.*

Hill, A. V. (2012). *The encyclopedia of operations management.* New Jersey, NJ: Pearson Education, Inc.

Hirano, H. (2009). *JIT implementation manual—The complete guide to just-in-time manufacturing: Flow manufacturing—Multi-process operations and kanban* (Vol. 3). Boca Raton, FL: CRC Press.

ILO. (2017). *Lean manufacturing techniques for ready made garments industry.* Cairo: International Labour Organization.

Jana, P., & Tiwari, M. (2018). *Industrial engineering in apparel manufacturing.* New Delhi: Apparel Resources Pvt. Ltd.

Jones, D., & Womack, J. (2002). *Seeing the whole: Mapping the extended value stream.* Brookline, MA: The Lean Entreprise Institute.

Karrer, C. (2012). *Engineering production control strategies: A guide to tailor strategies that unite the merits of push and pull.* Berlin, Heidelberg: Springer-Verlag. (Management for Professionals, Ed.).

King, J. S., & King, P. L. (2015). *Value stream mapping for the process industries: Creating a roadmap for lean transformation.* CRC Press.

Lean Enterprise Institute. (2008). In C. Marchwinski, J. Shook, & A. Schroeder (Eds.), *Lean Lexicon a graphical glossary for lean thinkers* (4th ed.). Cambridge, MA: Lean Enterprise Institute (LEI).

Letens, G., Cross, J. A., & Aken, E. M. (2006). Development and application of a framework for the design and assessment of a Kaizen event program. In *International annual conference of the American Society for Engineering Management.* Huntsville, AL.

Locher, D. A., & Keyte, B. (2016). *The complete lean enterprise: Value stream mapping for office and services* (2nd ed.). Boca Raton, FL: CRC Press.

Locher, D. (2008). *Value stream mapping for lean development: A how-to guide for streamlining time to market.* New York: Productivity Press.

Marchwinski, C. (2014). *Lean Lexicon: A graphical glossary for lean thinkers.* Brookline, MA: Lean Enterprise Institute.

McLean, T. (2015). *Grow your factory, grow your profits.* Boca Raton, FL: CRC Press Taylor & Francis Group.

Michael, K., Harmon, K., & Minnock, E. (2008). *Ready, set, dominate: Implement Toyota's set-based learning for developing products and nobody can catch you.* CreateSpace Independent Publishing Platform.

Mika, G. (2006). *Kaizen event implementation manual.* Society of Manufacturing Engineers (SME).

Mussman, R. (2001). Lean manufacturing: A plant floor guide. In J. Allen, C. Robinson, & D. Stewart (Eds.), *Lean manufacturing—A plant floor guide* (pp. 357–370). Dearborn, Michigan: Society of Manufacturing Engineers (SME).

Okes, D. (2019). *Root cause analysis – The core of problem solving and corrective action.* ASQ Quality Press.

Ortiz, C. A. (2006). *Kaizen assembly designing, constructing, and managing a lean assembly line.* New York: CRC Press.

Pyzdek, T. (2003). *The Six Sigma project planner – A step-by-step guide to leading a Six Sigma project through DMAIC.* McGraw-Hill.

Roser, C. (2020). *Collected blog posts of AllAboutLean.com 2016.* Offenbach: AllAboutLean.com Publishing.

Rother, M., & Shook, J. (1999). *Learning to see: Value-stream mapping to create value and eliminate muda.* Cambridge: Lean Enterprise Institute (LEI).

Saad, N. M., Ashaab, A., Maksimovic, M., Zhu, L., Shehab, E., Ewers, P., & Kassam, A. (2013). A3 thinking approach to support knowledge-driven design. *International Journal of Advanced Manufacturing Technology, 68,* 5–6.

Samanta, M. (2019). *Lean problem solving and QC tools for industrial engineers.* CRC Press.

Saran, A. (2010). Matrix Clothing: Leading the change with collaborative Kaizen. *Stitch World, 8*(1), 24–27.

Shankar, R. (2009). *Process improvement using Six Sigma—A DMAIC guide.* American Society for Quality (ASQ).

Shook, J. (2008). *Managing to learn: Using the A3 management process to solve problems, gain agreement, mentor, and lead.* Cambridge, MA: The Lean Enterprise Institute.

Shopfloor Series. (2002). *Pull production for the shopfloor.* Productivity Press Development Team.

Sobek, D., & Smalley, A. (2008). *Understanding A3 thinking—A critical component of Toyota's PDCA management system.* CRC Press.

Stenholm, D., Mathiesen, H., & Bergsjo, D. (2015). Knowledge based development in automotive industry guided by lean enablers for system engineering. *Procedia Computer Science, 22,* 244–253.

Stewart, J., Charles, A., & David, R. (2001). *Lean manufacturing: A plant floor guide.* Dearborn, MI: Society of Manufacturing Engineers.

Stitch World. (2010). Madura Garments, Aditya Birla Group holds Kaizen Mela: Empowering people. *Stitch World* (5), 24–29.

Stitch World. (2013). *Manufacturing car seat covers quality control practices at Krishna Maruti Ltd. (A. R. Ltd, Producer).* Retrieved from www.apparelresources.com:https://in. apparelresources.com/business-news/sourcing/manufacturing-car-seat-covers-quality-control-practices-krishna-maruti-ltd/. (Accessed November 2019).

Taylor, R. E., & Vatalaro, J. (2005). *Implementing a mixed model kanban system the lean replenishment technique for pull production*. Productivity Press.

Tyler, D. (2008). *Carr and Latham's technology of clothing manufacture* (4th ed.). Oxford: Blackwell Publication.

Wang, J. (2011). *Lean manufacturing business bottom-line based*. Boca Raton, FL: Taylor and Francis Group, LLC.

Ward, A. C. (2002). *The lean development skills book*. Ward Synthesis, Inc.

Womack, J. P., Jones, D. T., & Roos, D. (2007). *The machine that changed the world — The story of lean production—Toyota's secret weapon in the global car wars that is now revolutionizing world industry*. Free Press.

Visual management

Ankur Makhija[1], Chandrajith Wickramasinghe[2] and Manoj Tiwari[3]
[1]Department of Fashion Technology, National Institute of Fashion Technology, Gandhinagar, India, [2]Corrigo Consultancy Pvt. Ltd., Noida, India, [3]Department of Fashion Technology, National Institute of Fashion Technology, Jodhpur, India

> *The worst kind of waste is the waste you don't even see.*
> — Shigeo Shingo

5.1 Introduction

Since the beginning of human civilization, we have been using visual symbols (in the form of drawings, colors, flags, fire, fumes, etc.) for fast and effective communication of significant information to others (Ortiz & Park, 2011). In present times also, every day we do interact through several visual tools (through signs, images, symbols, colors, etc.), but generally, we do not pay attention to it. The usage of visual tools has become a part of our regular life. (Some simple examples are communication using emojis in an online discussion, or mobile phone showing signal of lower battery or a lesser space available in the laptop memory.) After all, these tools make anything easy to tell just by looking. Researches have confirmed that 50% of our brain function is dedicated to find and interpret visual information. This ability is in-built in the human brain, and our brain does not need any instructions to seek out visual information. It happens naturally and automatically without any decision making by an individual. This is the reason that our sense of sight dominates over other senses, and what we see and observe visually becomes impactful to us, it can make us happy, fires our thoughts or imagination, or may lead to deep thoughts as well (Galsworth, 2017).

Visual management (VM) (or visual management system known is VMS) is a popular lean pillar while lean deployment. VM has been an integral part of Toyota Manufacturing System and lean production since the 1940s. Kanban and Andon were one of the initial and the most popular VM tools that have been widely used worldwide in lean transformations. The workplace management tools like 5S (now treated as 6S as well with addition to the sixth S as Safety), and poka-yoke (mistake proofing) were also gained popularity. Since the inception of the Toyota Production Sys, the VM tools have been used primarily for production control; hence, the VM tools are also known as visual control (Monden, 1998; Ohno, 1988; Schonberger, 1986;

Shingo, 1989). According to Niederstadt (2014), VM may be referred to as visual methods or tools applied to enhance productivity, efficiency, quality, and safety of a process or processes without much verbal interaction. Keeping these visual tools simple, clear, and easy to understand is the prime objective to ensure the best results enabling the same understanding of the message, matter, or issue to all concerned. As these tools are simple, clear, and easy, it does not lead to any confusion, problem, abnormalities, and deviations in understanding. This helps in taking accurate and quick corrective actions (Niederstadt, 2014).

The VMSs have gained popularity with time and treated as an essential requirement in lean transformations for continuous improvement (Kurpjuweit, Reinerth, Schmidt, & Wagner, 2018). The VM practices have been developed primarily through the practitioner(s) efforts. It is different from the other tools and techniques of lean philosophy, which are mostly driven by theoretical concepts. This makes VM more dependent on Gibson's concept of affordance (1977) as proposed for VM by Beynon-Davies and Lederman (2017). The theory of affordance in VM focuses attention on human cognitive abilities and corresponding features of visual devices. This has resulted in a lack of theoretical guidelines for visual design, and in the scholarly field, VM is defined too narrowly (Koskela, Tezel, & Tzortzopoulos, 2018).

According to Beynon-Davies and Lederman (2017), VM refers to "a way of making work actions visible to improve the flow of work." There are several terms used interchangeably such as the visual workplace, visual control, visual factory, shop-floor management, or visual tools (Kurpjuweit et al., 2018). Galsworth (2017) defines a visual workplace as "a work environment that is self-ordering, self-explaining, and self-regulating, where what is supposed to happen does happen, on time, every time, day or night, because of visual devices." VM is referred to as a measurement approach that monitors organizational results and then displays those results in flat formats (Galsworth, 2017).

Beynon-Davies and Lederman (2017) propose a five-point prescription for developing VM solution, as:

- Visual devices should be aimed to utilize all senses or multiple senses.
- Visual devices should facilitate action-taking.
- Physical structures (e.g., whiteboards) should be utilized as performative structures (how to communicate and the outcome of such communication).
- While designing the VM, a holistic approach should be applied, means one should not just focus on an individual device, but focus on the entire physical environment.
- The patterns of action (from the VM devices) should aim to embrace current status (as-is), targeted status (as-if), and change (to-be) (Beynon-Davies & Lederman, 2017).

VM has evolved through disconnected and unintegrated efforts, mainly by the practitioners. It mainly focuses on developing some visual aids or visual tools to address an issue or a specific problem. Most of such solutions are simple (in terms of functioning) and include intuitive design features. The superficial understanding of lean as common sense also shifted the focus of such VM solutions from "what" (conceptual understanding) to "how" (practical understanding) (Farris, Van Aken, Doolen, & Worley, 2009; Langstrand & Drotz, 2015; Saurin, Ribeiro, & Vidor, 2012; Sorge & van Witteloostuijn, 2004).

In many of the cases, VMS can be deployed or implemented independently from other lean solutions. The VMS brings much-needed transparency of process performance. It helps in choosing the right metrics and targets, and eventually, it creates the initial level of empowerment where people where can influence the process performance (Thys, 2011). In a manufacturing environment, an interaction about the process and its performance is vital, here "what" is happening and "why" it is happening become important. For an effective interaction, it needs to be designed in a structured and constructive manner with ease and clarity. It helps in facilitating people with quick and relevant information, based on that further understanding may be developed. According to Tezel, Koskela, and Tzortzopoulos (2016), the application of VM tools enables production processes comprehensible for shop-floor personnel with clear visualization of the work activity responsively and consistently (Tezel et al., 2016).

In the context of operation management, VM tools play a vital role in the manufacturing environment. Traditionally, VM was limited to visualize the information on the shop floor. Applications of VM in this context were limited to signs, markers, labels, or color codes (Jaca, Viles, Jurburg, & Tanco, 2014). In the last few years the VM has been used with a wider scope and a holistic approach to continuous improvement, and VM is equipped with more strategic tools with additional quality features. For example, these days the VM is also being used in real-time data monitoring of production data (Lee, 2018), quality (defect level) data even at an individual workstation level. It is also used in electronic display boards that alert for the required production rate to achieve the target on a real-time basis. Several studies suggest that VM should not be treated merely as a supporting feature, but a lean management practice can be implemented undependably from other lean practices (Bateman, Philp, & Warrender, 2016; Eaidgah, Maki, Kurczewski, & Abdekhodaee, 2016).

According to Tezel et al. (2016) treat VM as a close-range communication strategy based on cognitively effective information conveyance. It is also observed that the available literature on VM is fragmented, and it lacks and integrated focus and cohesion between theory and practice. Also, there is a lack of clarity on the potential benefits of VM to an organization (Tezel et al., 2016).

5.2 Gemba

In Japanese, Gemba (also spelled "genba" with an n) means actual place. Gemba is where the action is and where the facts may be found. News reporters can refer to the location of the story as Gemba. Doctors can refer to the place where they are interacting with patients as Gemba. Police call the crime scene as Gemba. In business, Gemba is the place where value is created. In manufacturing, Gemba is the shop floor. In the service sector, *Gemba* is where the customers come into contact with the services offered. In the hotel business, for instance, *Gemba* is everywhere: in the lobby, the dining room, guest rooms, the reception desk, the check-in counters, etc. At banks, the tellers are working in *Gemba*, as are the loan officers receiving applicants. Thus *Gemba* spans a multitude of office and administrative functions (Imai, 2012).

Gemba as referred by Japanese Companies such as Toyota is "Genchi Genbutsu." "Genchi" means the real place and "Genbutsu" means the real thing. These terms focus on reality in the workplace (Womack, 2011).

We should see Gemba from the customer's point of view. It is a tool to experience the process where it happens. Instead of making the decisions in the board room meetings, you make decisions on the shop floor. Gemba is not simply the area of value-added (VA) work; it is also the originator of critical data that can be used to analyze and determine problems. Once upon a time, good managers used to avoid Gemba. They treat it as a dirty workplace where all sin happens. According to Imai (2012), "In Japan, production-related work is sometimes referred to as 3K, signifying the Japanese words for "dangerous" (kiken), "dirty" (kitanai), and "difficult" (kitsui)." Today, in contrast, many companies organized Gemba trips to learn the kaizen process with a benchmarking.

5.2.1   Gemba walk

Gemba walk is a powerful improvement tool. The objective is to eliminate Mudas and observe Genchi Genbutsu. With the help of Gemba walk, you can have a deep understanding of the process of what is happening in the organization. It provides a structured approach for accessing process performance and brings key issues to the surface.

Gemba walk is typically defined as going to where the action is. It is a key element of the Toyota Production System. Every Gemba walk is also a teaching exercise. As you go through Gemba, you have the opportunity to look under the surface with your own eyes and understand more deeply what is happening inside your organization. A Gemba walk can save your job and your business. You need to go at least 2 hours a day to observe Gemba. The time spent in Gemba makes you closer to the customer, and it makes your matrix comes alive (Kirchner, 2013). Gemba walk can be used to encourage people to create an environment of the ideas, futuristic approach, and experimentation to reach a level of accomplishment (Bremer, 2015).

Gemba walk is generally performed by senior managers, outside people, or consultants, whereas support departments such as accounting, human resource, and purchase do not show their keen interest in Gemba or they are not encouraged to do so. Sadly, this is the root cause of the weak association between support functions and the VA work of the manufacturing plant to understand the customer requirements.

To manage an adhesive system the work done in these support roles should be well tied with what occurs at the Gemba, so that they adapt themselves with how the Gemba operates and also instruct Gemba about customer requirements. The vision obtained by visiting the shop floor helped the support departments to give better support to the production team, and thus this depicts how the support roles can make a positive change for the Gemba (McQuade, 2017).

Gemba walk can be organized at three stages:

Stage 1: Gemba walk is conducted to understand the process, how it operates, identifying the problems, root-cause analysis (RCA), making an action plan for improvements.
Stage 2: Gemba walk is conducted for routine checking, to check the deviation between the standards and the process metrics.
Stage 3: It is conducted to incorporate a radical change in the process when there is business process reengineering or technological advancement.

In Gemba, a more intense communication between management and operators results in a much more effective two-way information flow between them. Workers had a much clearer understanding of management expectations and their responsibilities in the whole kaizen process. The resulting constructive tension on the work floor made the work much more challenging in terms of meeting management expectations and giving workers a higher sense of pride in their work (Imai, 2012). Best practices for successful Gemba walks are:

- Focus on the process, *not* the people.
- Be a student of the process.
- Do *not* correct the interviewees during the walk.
- Do *not* bring the written procedure on the walk.

5.2.2 Gemba centric management

Gemba centric management is a concept where management should focus on Gemba while taking decisions. With this, the management shares the vision with the employees who work together to accomplish performance improvement and make the organization more successful. Following are the features and benefits of this approach (Imai, 2012):

- Gemba management must accept accountability for achieving quality, cost, and delivery (QCD).
- In Gemba, there should be enough room for Kaizen.
- Management should be accountable for every outcome in Gemba.
- Process owners, supervisors, and workers at Gemba can easily identify the needs of the Gemba and thus encourage solutions.
- Management should be ready for change and they should inspire people for solutions that are grounded to reality and low-cost.
- Kaizen awareness and work efficiency are enhanced simultaneously.
- It is not always necessary to gain upper management's approval to make changes.

While going to Gemba is essential, it is equally important to remove the people immersed in the day-to-day activities and away from the constant frenzy so they can better see their situation from a new perspective. After the work team gains a new perspective of how its work is a process and how the process flows and affects others, it's much easier for them to come up with improvement ideas (Nestle, 2013).

5.2.3 Rules of Gemba management

Most managers prefer their desk as their workplace and wish to distance themselves from the events taking place in the Gemba. They take reporting from their subordinates, and in this way, most information originating from the Gemba becomes increasingly abstract and remote from reality as it goes through many layers before reaching them. It is therefore important for the top management to stay in close contact and understanding the Gemba effectively. Hence, the five golden rules of Gemba management are:

- When a problem arises, go to the *Gemba* first.
- Check the genbutsu (relevant objects).
- Take temporary countermeasures on the spot.
- Find the root cause.
- Standardize to prevent a recurrence.

Managers who do not take initiative to standardize the work procedure, and instead let supervisors run the show, are not doing their job of managing Gemba (Imai, 2012). Gemba is not a hall pass to roam freely in the work area and ask awkward questions. Gemba needs to be followed by Kaizen (small changes) or Kaikaku (radical changes). Gemba-Kaizen constitutes three principles: reducing Mudas, good housekeeping, and standardization. Every Gemba should have a Kaizen action list so that the specific problem is resolved with subsequent Gembas. Your Gemba should be always Customer Ready (Jansson, 2017).

5.2.4 Best practices for successful Gemba walks

To conduct successful Gemba walks, the set of rules defined in the following subsections should be kept in mind.

5.2.4.1 Focus on the process, not the people

The Gemba walk focuses on the process flow, not the people who work in the process. The management needs to create a blame-free environment during the walk which should make assure that Gemba walk is not a search for guilty. The Process Walkers should *not* be pointing fingers during the walk. If the focus is to correct the process, then process owners will also come freely sharing their experience.

5.2.4.2 Be a student of the process

The facilitator should ensure that Process Walkers should maintain a student perspective. With this perspective, they would have an open approach. Walkers are trying to understand and build profound knowledge of the current state of the process. Walkers should stay curious about the process design and seek to learn. The purpose is not to "fix" the process during the walk or brainstorm solutions. The purpose is to *see* and understand the process first.

5.2.4.3 Do not *correct the interviewees during the walk*

The purpose of the walk is not to see if people are following the procedure. The walk is not about "correcting" people. Sometimes, we find that a step is done incorrectly, and this should be handled later. A Process Walk is *not* the forum to correct the workers. The point of the walk is to understand the current state. If anyone is publicly shamed, they will be less likely to share their process knowledge.

5.2.4.4 Do not *bring the written procedure on the walk*

Again, the goal is to find out what is happening, not what is supposed to happen. Workers may not be following the written procedure and there could be many legitimate reasons for this. The procedure might be outdated, inaccurate, or poorly written. Often procedures are rewritten after a Process Walk because the team gains valuable insight on how to change the process.

Gemba is not a hall pass to roam freely in the work area and ask awkward questions. Gemba needs to be followed by Kaizen (small changes) or Kaikaku (radical changes). Gemba-Kaizen constitutes three principles: reducing Mudas, good housekeeping, and standardization. Every Gemba should have a Kaizen action list so that the specific problem is resolved with subsequent Gembas. Your Gemba should be always Customer Ready (Jansson, 2017).

5.3 Visual factory

A visual factory embeds the details of operations into a living landscape of work through visual devices and visual systems. They help to meet daily performance goals, impeccable safety, vastly reduced lead times, dramatically improved quality, and an accelerated flow that you control at will.

According to Galsworth (2007), "A visual workplace is a self-ordering, self-explaining, self-regulating, and self-improving work environment—where what is supposed to happen does happen, on time, every time, day or night, because of visual solutions."

In a visual factory the necessary information is converted into simple, universally understood signs and language and installed at the Gemba near to the point of use. This information is quickly comprehensible and easily accessible to those who need to understand the status of a process. The Gemba itself speaks fluently and precisely, about how to use it effectively and efficiently. The floors, boards, signage, and metrics help to check the abnormalities, do the work with precision. They become performance partners.

The visual factory is an integral part of lean manufacturing which intends to develop quality culture and effective communication. Visual language enables the organizations associated to quickly distinguish between the desired (standard) situation and abnormalities in the manufacturing process. In a visual factory workplace, every item has a defined location and remains there except when in use. Each activity in an organization has a defined standard process which is well communicated and clear through visualization. The underlying concept behind the visual factory is tools aimed to provide shop-floor employees with the unambiguous information needed to best perform their jobs (Borkar & Rai, 2019).

The concept of the visual factory which is also known as a visual workplace or VM aims to put critical information in physical workspaces by using signs, labels, posters, display, and other mediums. These visuals help to create a safer work environment and sufficient to eliminate the need for repetitive training and ongoing supervision. Information deficits can have a major impact on all performance indicators such as defects, reworks, changeover time, performance, efficiency, safety-related issues, cycle time, and overall manufacturing lead time (Borja, 2016).

The visual system has a significant impact in reducing the learning curve time of new and experienced employees. The visual factory becomes an instructive work area where information is strategically located in the environment to help employees to learn and work effectively. Visual factory plays an important role in implementing lean manufacturing principles such as 5S, TPM, Standard work, Kanban, Andon directly, and others indirectly by displaying their principles at prominent locations in Gemba.

A visual factory should also be incorporated with a smart layout with all the departments to be linked with the value flow. For example, the highest priority needs to be given to the production process to finish the product in the given takt time with less Mudas such as waiting, transportation, and motions. Then direct support departments such as maintenance, quality assurance, planning, and supervisory personnel are placed. Finally, indirect support systems such as sales, plant management, engineering, safety, and environment which are considered as overhead functions are placed (Ortiz & Park, 2011).

5.3.1 Characteristics of visual tools

Visual tools are integrated and generally kept as openly exposed in the work environment. According to Kattman, Corbin, Moore, and Walsh (2012), developing a visual tool for day-to-day requirements by an employee is not a difficult task (Kattman et al., 2012). Being simple and easy to see is no doubt the key quality of any visual tool; however, there are four common characteristics of VM tools have been recommended by the experts (Berkley, 1992; Galsworth, 2005; Greif, 1991; Harris & Harris, 2008; Suzaki, 1993):

1. A VM tool works as an aid to provide information to create information fields, from where concerned people can easily extract information.
2. A VM tool presents the information in advance (ahead of time) to avoid deficiencies.
3. The information display is integrated into process elements such as man, machine, material, method, and space as a direct interface between people (e.g., operators) and the process elements in a work environment.
4. A VM tool works as an aid of simple communication, with little or no dependency on verbal and textual information.

Tezel et al. (2016) suggest nine conceptual VM functions: transparency, discipline, continuous improvement, job facilitation, on-the-job training, creating shared ownership and the desired image, management-by-facts, simplification, and unification. The researchers also identify the mainstream practices that can be improved using VM tools. An adapted version of the same has been mentioned in Table 5.1.

5.3.2 Classification of visual tools

There are several visual tools available that are applied in lean transformation at workplaces; visual tools can be classified into four broad categories: information giving tools; signaling tools; controlling tools, and response guaranteeing tools (Galsworth, 1997). As there are several tools available, it may create confusion on which tool does what. The most common VM formats include such as KPI (key performance indicator) dashboards: LCD/LED screens for data monitoring, glass walls, vision and mission statement boards, hourly/weekly tracking charts. The flat figures of VM may include papers, charts, boards, posters, or screens. All these tools have the same common goal that is making it easy to see and understand what is going on, at a more abstract and summative level (Galsworth, 2017).

Tezel et al. (2016) discuss visual tools taxonomy with classification, definition, and practical implications of different visual tools (Tezel et al., 2016). Please refer to Table 5.2 for the adapted form of the same.

5.3.3 Basics of visual factory system

There are three main elements in a visual factory system:

Element 1: Workplace organization and standardization.
Element 2: Visual displays for sharing information.
Element 3: Prevention of abnormalities through error proofing.

There are six levels of visual displays and control (refer to Table 5.3), Levels 1 and 2 pertain to visuals displays, and Levels 3−6 cover visual control (Borkar & Rai, 2019; Greif, 1991).

5.3.4 Components of visual management systems

With the literature discussed, it can be summed up that a visual workplace should be self-explaining, self-ordering, and self-improving. The application of VM should be that the response of concerned personnel is immediate and obvious, and employees should be empowered to correct them easily. Any visual workplace should result in a clear understanding of 5W and 1H (5W—what, why, when, where, and who and 1H—how) about the particular issue or area. As an extension of the same, the visual workplace should aim to the "eight zero" in different aspects including zero waste, zero customer complaints, zero downtime, zero defects, zero injuries, zero delays, zero loss, and zero changeovers. In a VMS, comprising 6S (5S + sixth

Table 5.1 Visual management (VM) functions.

VM function	Key characteristics	Status before VM	Possible VM tool (indicative only)
Transparency	Increased information flow and its easy and open access to all concerned Communication capacity of a production process with concerned people	Information or knowledge in people's minds or files	Dashboard
Discipline	Habitually maintaining correct procedures Process standardization through by visualizing process requirements, work instructions, work specifications, and process flows	Lack of SOPs People taking actions as per their understanding Warning, scolding, inflicting punishments, dismissing, etc.	Kanban 5S Standard Operating Procedure (SOP) boards Andon
Continuous improvement	Process of sustained incremental improvement	Static organizations or big improvement leaps through considerable investment	Idea board A3 Sheet Plan-Do-Check-Act (PDCA) VSM Superstar boards (line of the month, an employee of the week, etc.)
Job facilitation	Making people's routine tasks easy aided by relevant visual aids Easing cognitive perception and physical execution	Expecting people to perform their tasks perfectly without providing them any aids	Coding in terms of color, shape, texture, size, location, and label Shadow boards for hand tools at workshops
On-the-job training	Systematical dissemination of information and acquiring tacit knowledge through the visual display on the shop floor	Conventional training practices or offering no training	One Point Lessons (OPL) Ishikawa diagrams Define, Measure, Analyze, Improve and Control (DMAIC)

(Continued)

Table 5.1 (Continued)

VM function	Key characteristics	Status before VM	Possible VM tool (indicative only)
Creating shared ownership and the desired image	Conveying the message of a caring and supportive workplace culture Encourage a sense of belongingness and ownership	Management dictation for change efforts, vision, and culture creation	Look-Ask-Model-Discuss-Act (LAMDA) cycle 5-Why analysis Yamazumi charts Boards with vision and mission statements Display boards with internal and external customers Display boards with various certifications, global standards achieved, and best practices followed Team building displays Posters related to social compliances Display of Corporate Social Responsibility (CSR) activities
Management-by-facts	Encouraging objective fact data-driven decision making Developing information-based monitoring and control mechanisms	Lack of informed decisions and subjective actions	Display of various kind of graphs and chart Real-time data monitoring board Dashboards
Simplification	Bringing simplicity and clarity in displaying information enabling ease of grasping	Expecting people to acquire, analyze and monitor data on their own Mismanaged information with information overloads	Visual display of Key Performance Indicator (KPI) Display of quality levels (from example sewing line-wise or style-wise Defects per Hundred Units (DHU) levels)

(Continued)

Table 5.1 (Continued)

VM function	Key characteristics	Status before VM	Possible VM tool (indicative only)
Unification	Enhancing awareness about working conditions of different departments and the organizational environment Encouraging systematic information sharing within and outside the organization	Working and acting in silos	Andon Kanban Visual indicators in machines Process flow diagrams Display of SOP at a common area Knowledge management display boards

KPI, Key performance indicator; *OPL*, One Point Lessons *VSMs*, Value Stream Mapping.
Source: Adapted from Tezel, A., Koskela, L., & Tzortzopoulos, P. (2016). Visual management in production management: A literature synthesis. Journal of Manufacturing Technology Management, 27(6), 766–799.

S as safety), visual displays, and visual controls (refer to Fig. 5.1) lead to the systematic achievement of these eight zeros.

Visual systems can be referred to as a group of visual devices designed to share information at a glance with an aim that the visual systems should communicate to everyone.

5.3.4.1 Visual display

A visual display may be referred to as a method to visually communicate a statement of essential information such as SOP (Standard Operating Procedure), SOE (standard of engagement), storage locations, quality points, equipment needs, and priority. Several types of visual display are possible to communicate the information, for example, location indicators and labels, checklists, worksheets, flow diagrams, signboards, status boards, product displays, area maps, process maps, and product calendars. The recommended steps to implement a visual display are:

- Identify the categories to focus on based on factory or office objectives—What?
- Identify the problems associated with those categories—Why?
- Identify who needs to see the display—Who?
- Determine the best location—Where?
- Develop and test the display—When and how?
- Finalize and implement the display.

Visual management

Table 5.2 Taxonomy of visual tools.

Visual tools	How it is used	Application areas (related to apparel manufacturing)	Impact (practical implications)
Signals, labels, name tags, and direction lines Border shadows and coding	Systematic workplace structuring and housekeeping efforts (i.e., the 5S)	Workplace management, inventory management, safety management, maintenance management, process management, production management	Improved workplace orientation for employees Reduction in learning curve/ training time for new employees Reduction in process wastes Improved workplace safety Reduction in process set-up times Higher equipment availability Easier identification of problems and deviations
Graphs, photos, posters, drawings, models	Communicating performance, lessons learned, mission statement, goals, change programs best practices, and internal/external marketing efforts	Change management, performance management, quality management, knowledge management, human resources management	Improved employee behavior for better and raise commitment among employees Building a positive image of the or organization to its stakeholders
Pareto charts, sticky boards, decision trees, A3 sheets	Visual tools and systems supporting continuous improvement	Process management, change management	Provide training for employees on critical issues Facilitate problem-solving Summarize and communicate a process
Performance cantered obeya rooms	Visual performance figures, process information, and KPIs grouped in designated locations in the workplace	Performance management, process management, change management	Increased focus and efficiency Encourage employee participation through group discussions, coordination, and problem-solving

(Continued)

Table 5.2 (Continued)

Visual tools	How it is used	Application areas (related to apparel manufacturing)	Impact (practical implications)
Control tables	Visual tracking boards	Production management (production control), inventory management, human resources management	Facilitate identifying improvement opportunities Facilitates visual production control through increased transparency Visual communication of production plans for increased awareness in employees Improved material flow Encouraging employee participation
Samples and prototypes	Demonstrating a real sample or a prototype of the end product	Quality management, knowledge management	Improved understanding of quality aspects Effective training
SOSs	Visual instructions of operational steps, approximate durations, critical points, Work in Process (WIP) amounts, etc.	Process management, quality management, maintenance management, safety management	Process and procedure standardization Waste reduction
One Point Lessons (OPLs)	Visual one-page-sheets (short) to disseminate new ideas, new knowledge, and critical points on a specific topic OPLs can be basic information sheets, problem case study sheets, and continuous improvement sheets	Knowledge management, safety management, maintenance management, quality management, workplace management, process management	A glance at new or better knowledge improved understanding of process functions Effective on-the-job training

(Continued)

Visual management 145

Table 5.2 (Continued)

Visual tools	How it is used	Application areas (related to apparel manufacturing)	Impact (practical implications)
Value Stream Mappings (VSMs)	Visual documentation of the flow of information and materials required to produce a product or service	Process management, change management (communicating improvements)	Visually summarizes the entire process chain for a given product or process family Facilitates the identification of bottlenecks and wastes in the Current state map, and improved process through a Future state map
Andon—electronic display	Audio—visual signaling boards to communicate the status of a process	Quality management, change management, production management	Effective display of production status and enhances transparency Effective shop-floor supervision and control Enhances employee participation and accountability Enhanced quality and safety It helps in communicating the deviations from the target
Heijunka boards	Visual leveling boards often linked with kanban	Production management (production planning and leveling), maintenance management	Process balancing through load leveling Improves the flow of material
Kanban systems (kanban card, kanban light, etc.)	Visual signals used to "pull" a product or service from preceding work units or other functional departments	Production management, maintenance management, safety management	Controls different types of Muda due to the demand-driven approach Improved flow with effective WIP control

(*Continued*)

Table 5.2 (Continued)

Visual tools	How it is used	Application areas (related to apparel manufacturing)	Impact (practical implications)
Mistake proofing (poka-yoke)	Electromechanical systems used to warn operators in case there is a chance of any mistake/error	Safety management, quality management, and process management	Effective supervision and control Reduction in the need for quality control (waste) Quality enhancement Improved safety of personnel and equipment Minimized production set-up and change-overtime

KPI, Key performance indicator; *OPLs*, one point lessons; *SOSs*, standard operating sheets; *VSMs*, value stream maps.
Source: Adapted from Tezel, A., Koskela, L., & Tzortzopoulos, P. (2016). Visual management in production management: A literature synthesis. *Journal of Manufacturing Technology Management, 27*(6), 766–799.

5.3.4.2 Visual metrics

Visual metrics may be referred to visually communicate a key measurement (KPI) indicating the changing status of any process, for example, visual display related to supply chain, support functions, manufacturing, product development, and quality. Roles of visual metrics include the usage of data appropriately, keys to effective reporting and analysis, facilitating people in decision making, and communicate about the characteristics of "new" measures. Please refer to Table 5.4 for key features of visual metrics roles.

5.3.4.3 Visual controls

Visual controls are referred to as the methods and devices that alert employees about the specific actions that need to be taken. Visual controls may be applied in several of situations such as communicating about an undesirable condition in the process, schedule production activities, safety issues (SOE), inventory status, routine maintenance interval, refill office supplies, toner needs to be replaced, and message light is on (phone, email). Please refer to Table 5.5 for the recommended seven-step methodology for visual control in the work environment.

Table 5.3 Levels of visual display and control.

Level 1
Visual displays involve results of daily performance control activities to assess its confirmation to expectations. For example, graphs indicating performance, efficiency, etc.
Level 2
Visual displays involve the sharing of standards/Standard Operating Procedure (SOP) at the workplace which everyone needs to follow. For example, SOP, work instruction sheets, manuals, etc. near the workplace.
Level 3
Visual control involves building primary controls in the workplace. For example, status board with lights.
Level 4
Visual control involves stage two controls which warn of abnormalities with an alarming. For example, machines with sensors to detect abnormalities.
Level 5
Visual control involves stage three controls that help to stop the line in case of abnormalities with a device. This is like a machine shutdown. For example, Machines with Jidoka.
Level 6
Visual control involves stage four control which prevents abnormalities occurrence through automatic detection For example, A smart system with control of light, power consumption, excess heat.

5.3.5   Steps in creating a visual factory

A visual factory works on the principle of self-service communication which makes it ready for the employees and visitors across hours. The messages in a visual factory are systematically placed with a larger and more tangible context. Any communication from the management is easily conveyed across all the employees without any adulteration. There are six steps (refer to Fig. 5.2) involved in creating a visual factory (Galsworth, 2007).

5.3.5.1   Visual order

The first step is visual order where we need to prepare the physical environment to hold visual information. It includes sorting out the junk, housekeeping, repainting, ensuring safety, set locations for work areas and tools, and finally following customer-driven Gembas.

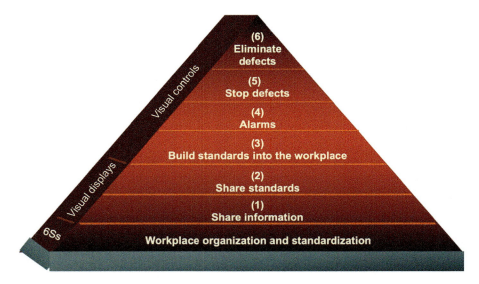

Figure 5.1 Visual management system.

Table 5.4 Roles of visual metrics.

Visual metrics role	Objectives
Use data appropriately	To keep management and operators informed of progress
	To drive and fine-tune improvement activities
	To educate the factory or office staff
	To display at the point of use
	To communicate across functions
Effective reporting and analysis	Involve operators in measurement
	Provide results to people who can take action
	Present results in the appropriate form for your audience
	Post results in a place all the operators can see them
	Visually link results with specific improvements
	Report data in at the frequency required for timely decision making such as hourly, daily or weekly
Visual metrics in measurement	Makes data easy to understand and react
	Exposes and explains the waste
	Shows improvements made and those left to implement
	Does not rely on written language to communicate information
	Everyone should understand how they have done and what is expected of them

Visual management 149

Table 5.5 Methodology for a visual control.

Sr. no.	Step	Key points
1.	Define the problem	Workplace environment Safety Storage Location Production/process Equipment Quality Priority First In First Out (FIFO)
2.	What needs to be controlled?	For the category selected, decide what needs to be controlled Write the items on a card Post them on the development chart
3.	What is the desired result or target?	Set a target result for what needs to be controlled Be specific; quantify if possible Ask: When we achieve the desired control level, what will be the tangible benefit? Write the target on a card Post it on the development chart
4.	Identify facts/information—gather data	Information needed problem-solving by the team Collect information from the problem area employees Identify current standards or lack of standards Evaluate adherence to standards Gather feedback on operations and activities
5.	Evaluate improvement ideas	Brainstorm ideas about how to control each activity and post ideas on a chart Evaluate the control level and potential impact of each idea
6.	Test ideas for implementation	Decide which ideas to use Determine time-line Assign responsibilities Get authorization Check guidelines for effective implementation Test Ideas
7.	Implement and measure results	Compare actual results to expected outcomes (visual metrics) Implement improvement as part of daily work and include in Standard Operating Procedures (SOPs)

5.3.5.2 Visual standard

The second step is to create visual standards which could be technical and procedural. Standards define what is supposed to happen at the workplace includes the

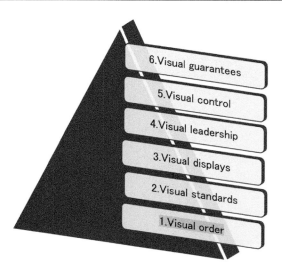

Figure 5.2 Visual workplace pyramid.

data of planned, expected, and normal. If standards are weak, there are chances of the occurrence of abnormalities.

5.3.5.3 Visual display

The third step is visual displays wherein a concise format, supervisors, and managers should have a range of visual answers such as where, what, when, who, how many, and how. It should be placed at a strategic location where it can influence the most.

5.3.5.4 Visual leadership

The fourth step is visual leadership in the form of visual metrics, visual problem-solving (DAMIC, A-3, 7 quality tools), and policy deployment.

5.3.5.5 Visual control

The fifth step is visual control where information is deeply installed into the landscape so that people should do the right things. It includes concepts such as Kanban and Heinjunka.

5.3.5.6 Visual guarantee

The sixth and last step is the visual guarantee which accomplished the goal of all workplace visuality and avoids doing the wrong things. The Gemba at this stage is close to perfect performance with concepts such as poka-yoke, and fail-safe devices. It involves the latest technology, visual machines with sensors, and limit switches.

5.3.6 Visual factory: zoning approach for plant layout

A plant floor worker normally works on a horizontal floor that specifies a particular field of work that can be conveniently represented in a design view, the primary sense of contact, vision, and the physical activities of the worker are expanded and performed vertically on vertical surfaces. The information must, therefore, be spread according to the regions, not arbitrarily but in a coordinated way, using the prescribed method. According to the visual factory principle, the production area is separated into three task zones "A," "B," and "C" (Bilalis, Scrobelos, Antoniadis, & Koilouriotis, 2002).

5.3.6.1 Zone A—activity zone

This is the zone where the major activities take place; this zone typically occupies the operational facility's central portion.

Visual displays in this zone show the result of the control activities and help to examine how closely the performance matches expectations, for example, graphs indicating performance and efficiency. Visual display with appropriate infographics would allow everyone to follow the standard specification and method, such as SOP, work instruction sheets, and manuals near the workplace. It also helps in building standards into the workplace through Andon Board.

This includes activities such as cutting, sewing, and finishing.

5.3.6.2 Zone B—boundary zone

This is where materials of auxiliary services are spread out to feed zone "A" and it incorporates:

- staging areas for raw and packaging materials
- safety equipment
- auxiliary/support workstations (production/warehouses)
- supplementary workstations (laboratories)
- support equipment parking areas (warehouses/production)

Visual controls in this zone can warn about the abnormalities with something alarming like a bell, for example, machines with sensors to detect abnormalities. Visual control such as machines with Jidoka can stop abnormalities. This is like a machine shutdown. Visual control can further also prevent abnormalities, such as with a smart system with control of light, power consumption, and excess heat.

5.3.6.3 Zone C—circulation zone

This is the zone that is utilized for administrator or vehicle flow; this zone isolates either zones "A" and "B" or partitions huge territory zones "A" or "B" into subzones or both; this zone should as a rule, by standard structure, stay free from any, even transitory, stockpiling of fixed material or equipment. Generally, this zone must stay free of all, however temporary, storage of permanent content or machinery by normal arrangement. The zoning allows us to determine the visual contact and regulation that will be shown at each location.

5.3.7 Visual garment factory

Apparel manufacturing is a dynamic mechanism involving supply management, quality assurance, and the health and compliance of employees (refer to Fig. 5.3). Visual administration aims to explain what to do, what is the normal approach, how to do with any single task. Working rule visuals in front of the worker decrease the risk to make an error. Having a goal and putting outcomes on the board will encourage the employee to maximize productivity, and competitive rivalry for better success starts across all divisions. Typically, three types of data can be used in a visual garment factory.

5.3.7.1 Process metrics

Metrics are installed at the machine or operating unit for real-time delivery of information. When information is quick, adjustments can be made to a process

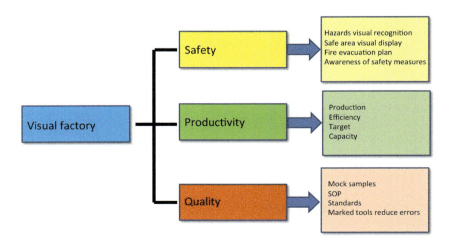

Figure 5.3 Visual garment factory parameters.

Visual management

immediately. Refer to Figs. 5.4—5.7 for different examples of process metrics in apparel manufacturing.

5.3.7.2 Work instructions

These provide individuals with information on what to do and how to do it. Visuals and depictions are recommended (refer to Figs. 5.8—5.12 for different examples of work instructions in apparel manufacturing) instead of verbal instructions since they provide precise instructions and prevent output errors.

5.3.7.3 General plant information

This is normally posted in a focal area and animates two-way data exchanges. A footprint is a framework of the things required at work territories demonstrating where the things ought to be placed (refer to Figs. 5.13—5.17). Machines have regions that need to be accessed for maintenance, arrangement, or activity, and these zones of access must be kept clear. Lines on the floor can outwardly impart the need to keep regions clear. The lines can be made utilizing tape (e.g., the standard yellow-dark striped "Alert" tape) or paint if the earth is particularly harsh. The standard color coding should be followed as indicated in Fig. 5.18. This information raises awareness, alerts about changes, and posts warnings about how to handle potentially dangerous manufacturing

Figure 5.4 Process metrics in a garment factory.
Source: Courtesy Method Apparel Consultancy, Gurugram.

154 Lean Tools in Apparel Manufacturing

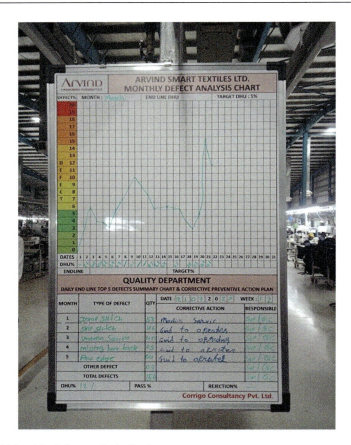

Figure 5.5 Monthly defect analysis display.
Source: Courtesy Arvind Ltd. - Corrigo Consultancy Pvt. Ltd, India.

5.3.8 Benefits of visual factory

This leads to a concept of poka-yoke: Visual factory accredits early warning and faster recovery of abnormalities. When standard specifications and failures are shown together, it is easy to comprehend the mistakes. Further, the concept of RCA and Jidoka can lead to a mistake-proof factory; poka-yoke.

Ease of exchange of information: Visual factory makes the information in the public domain so that every employee can participate. Every department comes out with the idea of Kaizen with the shared information to remove abnormalities. It also enables employees to participate in creating standards. Please refer to Figs. 5.19 and 5.20 for practical examples of the exchange of information in apparel manufacturing.

Promote employee independence: Visual factory operations help to build informal contacts outside the hierarchy. Process owners of different departments are given roles and responsibilities to operate visual controls. It reorients inspection functions toward observation of facts and problem-solving, instead of monitoring individuals and seeking to blame.

Visual management 155

Figure 5.6 Operators allocation analysis display.
Source: Courtesy Arvind Ltd. - Corrigo Consultancy Pvt. Ltd, India.

Thus it determines the need for cultural change toward employee independence. Please refer to Fig. 5.21, where a product help desk has been installed on an apparel manufacturing floor which can be easily operated by any person keen to get relevant information or status.

Improve performance metrics: It helps to improve performance metrics such as efficiency, reworks percentage by visible standards and targets, and motivating them to reach the goal. The visual factory makes abnormalities in the 4Ms (man, machine, material, and method) process inputs visible to everyone for prompt corrective action. Please refer to Fig. 5.22 for skill matrix display board installed on an apparel manufacturing floor.

Self-communicating Gemba: It makes the whole Gemba self-communicating across the hours without any assistance. Reduce the learning curve time by making operation standards quicker and easier to understand. It also helps to reduce Mudas and align the value stream.

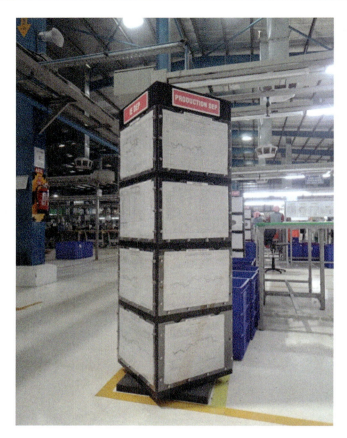

Figure 5.7 Production analysis display.
Source: Courtesy Arvind Ltd. - Corrigo Consultancy Pvt. Ltd, India.

5.4 5S

5S is a Japanese Management Philosophy that was developed by Hiroyuki Hirano for a Safe Working Environment. It is a technique for organizing a workplace, especially a collective workplace (like a shop floor, an office space, or a factory production floor). The methodology originates from a Japanese housekeeping idea named because of the five Japanese words each designed with a letter Se or Shi. They are seiri, seiton, seiso, seiketsu, and shitsuke. If these Japanese words are translated to English give the best explanation in itself, that is, sort, set in order, shine, standardize, and sustain (Khan & Islam, 2013). The 5S approach is geared toward eliminating waste, providing visual control, and preparing the workplace to enable improvements to be effective and implemented efficiently (Delisle & Freiber, 2014).

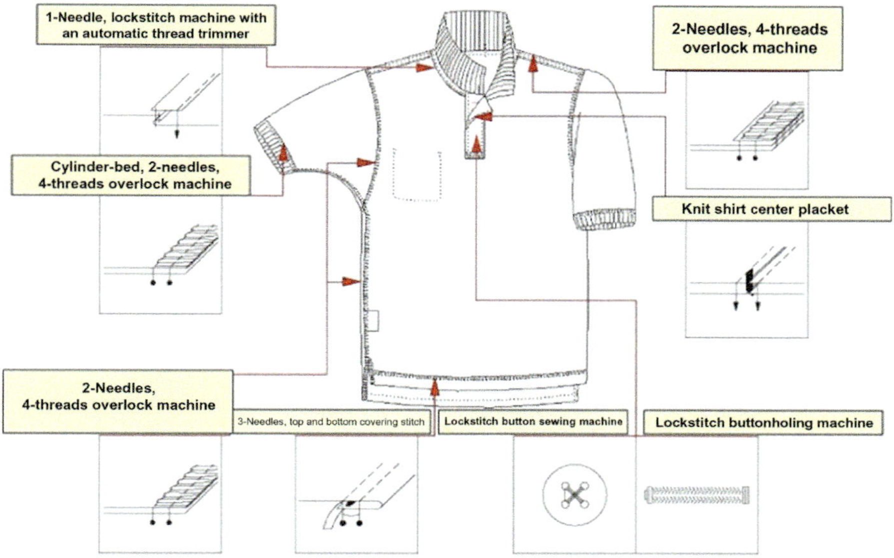

Visual management 157

Figure 5.8 Work instruction sheet.

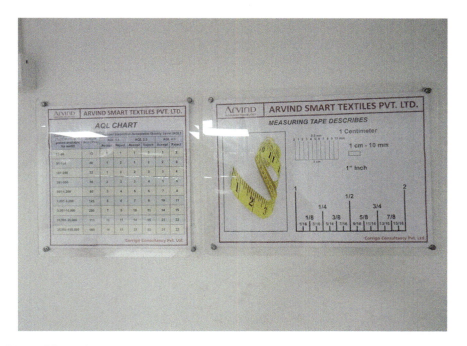

Figure 5.9 Quality instructions display.
Source: Courtesy Arvind Ltd. - Corrigo Consultancy Pvt. Ltd, India.

Figure 5.10 Garment measurements instruction display.
Source: Courtesy Arvind Ltd. - Corrigo Consultancy Pvt. Ltd, India.

Figure 5.11 Fabric and sewing defects display.
Source: Courtesy Arvind Ltd. - Corrigo Consultancy Pvt. Ltd, India.

Visual management

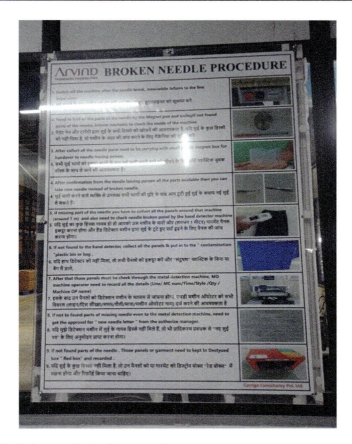

Figure 5.12 Broken needle handling procedure.
Source: Courtesy Arvind Ltd. - Corrigo Consultancy Pvt. Ltd, India.

According to Hirano, "Behind all workplace successes and failures are the 5S." The 5S method is a fundamental lean manufacturing tool, the aim of which is to support the management process at workplaces, and more specifically to create a well-organized, effective, orderly, and safe workplace. According to Hirano, lean culture requires a change in people's mentality as well as it is also the basis and the first step to implementing other Lean tools such as just-in-time (JIT), Kanban. The implementation of 5S involves all areas of the company from top management (which should become the driving force of the entire implementation process) to employees at the lowest levels, building a culture of continuous improvement of the organization. The 5S method helps to facilitate in-house communication, development of communicativeness, and efficiency of information flow. 5S is not only a method of rigid organization of the company, but it is also primarily building employee awareness and fighting with habits,

Figure 5.13 Input area layout.
Source: Courtesy Corrigo Consultancy Pvt. Ltd, India.

Figure 5.14 Signages for process and module.
Source: Courtesy Corrigo Consultancy Pvt. Ltd, India.

Visual management 161

Figure 5.15 Fabric storage.
Source: Courtesy Corrigo Consultancy Pvt. Ltd, India.

Figure 5.16 Trims storage.
Source: Courtesy Corrigo Consultancy Pvt. Ltd, India.

Process-wise audit layout   Process-wise carton storage

Figure 5.17 Process-wise audit layout.
Source: Courtesy Corrigo Consultancy Pvt. Ltd, India.

Color	Description
YELLOW	Aisle ways and traffic lanes: paths of progress work cells
WHITE	Production or racks, machines, carts, benches, and all other equipments that does not fall under any other colour guidelines
RED	Defects/scrap area ; red tag area
ORANGE	Materials or production inspection or energized equipment
GREEN	Raw materials or first aid related locations
BLUE	Work in progress
BLACK	Finished goods
BLACK-YELLOW	Area which presents physical or health risks to employees. Indicates that extra caution is to be exercised
BLACK-WHITE	Areas to be kept clear for operational purpose (nonsafety related)
RED-WHITE	Areas to be kept clear for safety reasons (around emergency access points)

Figure 5.18 Color coding standard for footprint.

which are often the biggest problem in building a well-organized work environment (Cichocka, 2018).

One of the main purposes of the 5S is to prepare the work environment to hold visual information. From that perspective, 5S is a method, while creating a visual workplace is the goal. Therefore 5S and VM (Visual Management) go hand in hand. Each S has its impact and needs to be implemented stage-wise through the

Figure 5.19 Visual board (daily plan vs achieved).
Source: Courtesy Corrigo Consultancy Pvt. Ltd, India.

effective PDCA (Plan, Do, Check & Act) cycle. It is a continual process and success depends on daily workplace discipline with everyone following it in their work areas (Marascu-Klein, 2015). Proper implementation of 5S guarantees the real benefits that increase the productivity and efficiency of the work as well as improve the quality and safety requirements.

The Japanese 5S system is very essential for ensuring systematic discipline. Moreover, this is a worldwide-acceptable formula that helps in a great deal in solving the managerial level problems. 5S can be considered a philosophy, a way of life, which can raise morale and create a good impression to customers and enhance efficiency (Mridha, 2020). The 5S methodology is a very suitable way to start the process of continuous improvement. It is one of the best tools to generate a change in attitude among workers and serves as a way to engage in improvement activities in the workplace (Bartnicka, 2018). 5S is a strategy that delivers results by a systematic approach to planning and organizing the activities, thus ensuring the compliance to standards.

The 5S system shapes the organizational culture of the company. However, some companies contemplate themselves as too busy to rearrange the workplace believing that it would take ample time cleaning and rearranging the workplace. But if these companies follow the 5S procedure properly, it will not only help the company to have a smooth operation as the developed standards through 5S would allow for easy identification of problems, making it possible to start the analysis of processes currently carried out, and on the other hand, it would also keep the

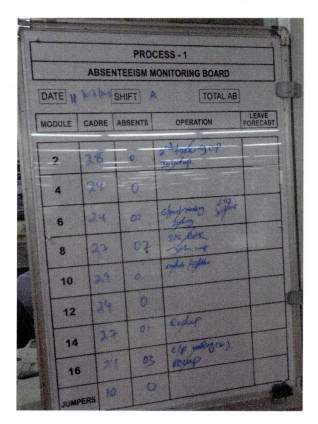

Figure 5.20 Absenteeism monitoring board.
Source: Courtesy Corrigo Consultancy Pvt. Ltd, India.

employees happy with the new process of working in an improvised environment (Bartnicka, 2018).

5.4.1 Benefits of 5S

The perceived benefits arising from successful 5S implementation like less searching, improved cleanliness, easier recognition of defects, decreased walking and motion add up to overall improvements in productivity, quality, cost, delivery, safety, and morale (Hirano, 1995).

5S operation model brings a lot of benefits. Some of those are economical and some improve human capital. The results of strategic 5S implementation can be seen in staff, environment, quality, production, and offices. The significant measurable benefits realized through the 5S program are depicted in Table 5.6 (Hirano, 1995; Visco, 2015).

Visual management 165

Figure 5.21 Production help desk.
Source: Courtesy Corrigo Consultancy Pvt. Ltd, India.

The following can be summarized as some important benefits from implementing a 5S process:

- Orderliness (seiri and seiton)—defect rate will reduce and efficiency will increase.
- Cleanliness (seiso and seiketsu)—a company can ensure a safe and healthy environment if its workplace is clean and equipment is regularly maintained.
- Discipline (shitsuke)—by training and educating employees their morale will improve, quality of work will enhance (Gapp, Fisher, & Kobayashi, 2008).

166   Lean Tools in Apparel Manufacturing

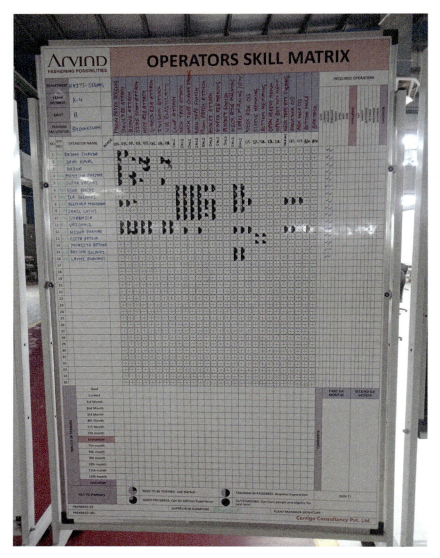

Figure 5.22 Skill matrix display board.
Source: Courtesy Arvind Ltd. - Corrigo Consultancy Pvt. Ltd, India.

5.4.2 Implementing 5S

Before implementing 5S, any organization needs to be ready with the following roadmap. It includes four phases under PDCA as discussed in the following subsections (refer to Fig. 5.23).

Visual management

Table 5.6 Tangible and intangible benefits of 5S.

Tangible benefits	Intangible benefits
Productivity and quality improvement, low	Improved company image
Reduction in Work in Process (WIP) and shorten lead time	To create an enthusiastic/motivating work environment
Safe shop floor and offices due to reduced accidents and unsafe situations	Reduced non−value adding activity
Reduction in rework and rejects which brings higher quality	Better working circumstances and rising comfort
The decrease in search time, improve cleanliness	Improved safety at workplace
Improvement in equipment reliability and maintainability	Improved coordination and teamwork among employees
Easier recognition of defects fewer mistakes, decreased walking and motion, improved flow	Helps lay the foundation for kaizen, continuous improvement culture in the organization
Improved workplace visual management and better utilization of space.	Greater employee participation
Reduced downtime, fewer safety hazards, and accidents	A way to avoid blaming people for defects
Reduced unnecessary human motion and transportation of goods	More focused employees in an organized workplace

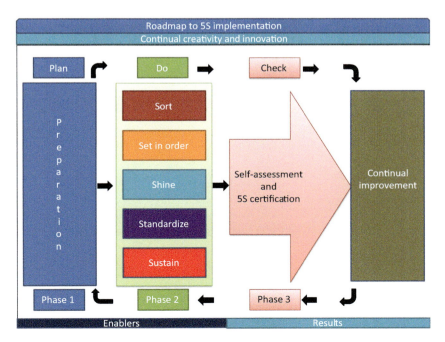

Figure 5.23 Phases in implementing 5S.

5.4.2.1 Phase 1: 5S preparation

Formation of the 5S council: To enhance total participation at all levels of employees and develop a continuous improvement culture and best performance spirit in the teams, a 5S council is formed which comprises:

- 5S Steering Committee:
  - Managing Director as Advisor
  - General Manager as Chairman
  - Head of Departments as Facilitators
- 5S Training Committee
- 5S Promotion Committee
- 5S Audit Committee

The 5S Steering Committee is responsible for developing the implementation plan, and the selection of team members based on ability, organizational representation, and expertise. 5S implementation responsibilities are to be distributed throughout the organization. Every member must know their 5S responsibilities and perform accordingly. The Chairman has to administer accountability for each of the responsibilities. The steering committee is also required to appoint 5S coordinator, 5S facilitators, and 5S leader from each department.

5S coordinator is required to communicate with everyone involved. He is responsible to facilitate workgroup implementation activities. He is required to motivate and monitor implementation activities ensuring total organization participation. On behalf of the organization, the coordinator acts as a resource for information.

5S facilitators support 5S implementation by communicating with everyone involved, motivating workgroups, ensuring employee implementation plan, and finally monitoring measurement systems.

5S leaders participate in the workgroup implementation process by communicating with everyone involved and monitoring the progress of group activities (Kumar, 2012).

Set up 5S zones: 5S coordinator demarcates the zones. 5S facilitators assign responsibilities and divide activities into manageable tasks. This involves obtaining the layout of the entire work area and dividing each section into small zones, assigning one team to each section, determining the number of people per team, and displaying the names of team members and their areas.

5S preliminary training: Training is done to disseminate 5S methodology and prepare the workforce for meaningful participation in 5S activities (Kumar, 2012).

5.4.2.2 Phase 2: 5S implementation

5S implementation is an important activity when things happen and we see a physical significant change on the floor. As the name indicates 5S involves five S (refer to Fig. 5.24): seiri/sort, seiton/set, seiso/shine, seiketsu/standardize, and shitsuke/sustain.

Seiri—sort

Segregating the wanted and the unwanted materials, removing the unwanted materials systematically through a red tag area

The sort is the first pillar of 5S, it corresponds to the JIT principle of "only what is needed, only in the amounts needed, and only when is needed." During this step,

Visual management

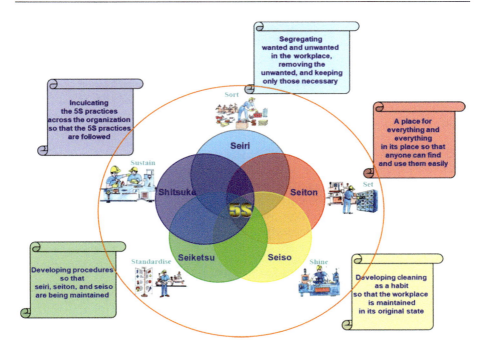

Figure 5.24 5S and its principles.

one will go through the chosen area and remove everything not needed for the current production. This leads to fewer hazards and less clutter. The goal is to eliminate nonessential items from the workplace. Items are "red-tagged" and stored in the red tag area for a specific period before disposing of them (Hirano, 1995). Refer to Fig. 5.25 for sorting methodology.

Action steps for sorting
1. The 5S project area is identified.
2. A local red tag area is created.
3. Separate needed and unneeded items.
4. Red tag the unneeded items and transfer them to the red tag area.
5. Discard all truly unneeded items.
6. As per the disposal policy, dispose of unwanted items from the red tag area (Hirano, 1995; Visco, 2015).

Target outcome
- A neat workspace
- Improves workflow
- Reduced WIP inventory
- Improvement in floor space utilization
- Converted unwanted things into cash (Visco, 2015)

Figure 5.25 Sorting methodology.

Red tagging A red tag area is a brief holding period of items that are no longer needed before their disposal. The items accumulating in the red tag area for more than 7 days have to be moved to a common red tag area along with the red tag card (refer Fig. 5.26) by the person/section who kept it. Items that are not taken would be reviewed for disposal by the team consisting of a member from the concerned department, 5S Champion, and the Head of the Purchase Department, quarterly (Hirano, 1995; Visco, 2015).

Seiton—set in order
A place for everything, and everything in its place

Set in order is the second pillar of 5S. According to Hiroyuki Hirano, it can be defined as, "arranging needed items so that they are easy to use and labeling them to make their storage sites easily understood by anyone." Each item should be arranged properly so that anyone can see where it is kept, can easily pick it up, use it, and return to its proper place. It minimizes searching waste, waste due to difficulty in using items and waste due to difficulty in returning the item (Hirano, 1995). Please refer to Figs. 5.27—5.30 for seiton (set in order) examples in shop floor in apparel manufacturing.

Action steps for set in order
1. The current and future state workplace diagram is created.
2. Determine how and where to lay out your tools and equipment and get this step done.
3. Label the locations of equipment and materials (Visco, 2015).

Target outcome
1. Minimal waste and human errors.
2. The workspace is properly organized.
3. Items are easily located, stored, and retrieved (Visco, 2015).

Examples of set in order (Figs. 5.27—5.30)

Seiso—shine
Cleaning with meaning

It is the third pillar of 5S. It emphasizes keeping the workplace clean. According to Hiroyuki Hirano, shine can be defined as, "keeping everything swept and clean." Develop cleaning as a habit so that the workplace is maintained in its original state. Treat cleaning as an inspection process inspects equipment and facilities to identify abnormalities and prefailure conditions (Hirano, 1995). Refer to Fig. 5.31 for an example of a shine calendar in apparel manufacturing.

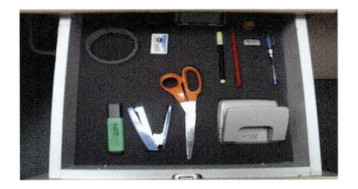

Figure 5.26 Red tag.

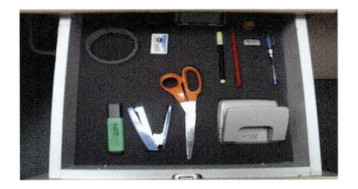

Figure 5.27 Workplace drawer.

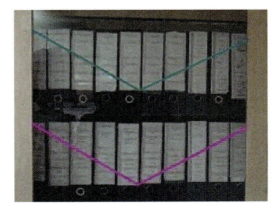

Figure 5.28 Color lines on files in a pattern.

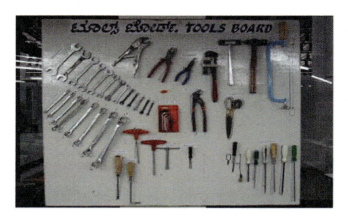

Figure 5.29 Maintenance tool board.

Figure 5.30 Workplace zoning in a factory.

Action steps for shine
1. The total area is divided into zones.
2. Responsibility for cleaning of each zone is allocated.
3. Cleaning points, cleaning aids, etc. are decided.
4. The cleaning schedule is displayed.
5. The workspace is cleaned along with inspecting and maintaining tools and equipment (Visco, 2015).

Target outcome
1. Improved equipment lifespan.
2. Clean and safer work environment.
3. Lower maintenance cost.
4. Fewer product defects (Hirano, 1995; Visco, 2015).

Seiketsu—standardize
Developing procedures so that first 3S are maintained

Shine calendar			
Zone: 2			Section: Sewing floor
Work area	**Responsibility**	**Apparatus**	**Timing**
Walls, floors, ceilings, notice boards	Housekeeping helper	Housekeeping tools	5:30–5:40 pm
Sewing and other machines	Operators	Autonomous maintenance	5:30–5:40 pm
Chairs	Operators	Cleaning cloths	5:30–5:40 pm
Dust bin	Housekeeping helper	Cleaning cloths	Hourly
Side storage shelves	QC	Cleaning cloths	5:30–5:40 pm
Checking tables	Checker	Cleaning cloths	5:30–5:40 pm
End of line checking table	End of line checker	Cleaning cloths	5:30–5:40 pm
Feeding rack	Feeding helper	Cleaning cloths	5:30–5:40 pm
Lighting fitting busbar	Electrician and HK helper	Cleaning cloths	Daily before shift beginning

Figure 5.31 Shine calendar (in a sewing floor).

According to Hiroyuki Hirano, it is defined as "the state that exists when the first three pillars are properly maintained." The purpose of the fourth S is to develop standardized procedures to make sure that everyone in the company follows the same method, the same names of items, etc. It helps to build a rapid learning curve for new employees (Hirano, 1995; Visco, 2015). Please refer to Fig. 5.32 for shop-floor examples of work instructions in apparel manufacturing.

Action steps for standardize
1. Brainstorm ideas for making the 5S changes standard operating procedure.
2. Update documentation to reflect changes.
3. Aware all stakeholders of the new standards—inform and educate (Visco, 2015).

Target outcome
1. Consistency in the work practices.
2. Clear work instructions.
3. Fewer mistakes.
4. Allows the workers to know what is expected.
5. With better visual and transparency management, work efficiency will improve (Visco, 2015).

Shitsuke—sustain
Inculcating the 5-S practices across the organization so that the 5-S practices are followed.

Figure 5.32 Standardize examples in the form of signage, work instructions; safety levels.

It is the fifth pillar of 5S. It incorporates disciplines needed to maintain the improvements. According to Hiroyuki Hirano, discipline is defined as "making a habit of properly maintaining correct procedures." To maintain discipline standard audits are developed and implemented. Employees are trained to practice 5S system so that it becomes ingrained in the company's culture. Many companies give 5S awards to motivate employees (Hirano, 1995; Visco, 2015). Please refer to Figs. 5.33 and 5.34 for shitsuke/sustain examples in apparel manufacturing.

Action steps to sustain
1. Create awareness and publicize the system by developing 5S news, 5S posters, 5S slogans, 5S day, etc.
2. Create a structured way to implement 5S activities.
3. Formulate guidelines for audit/evaluation of 5S implementation.
4. Develop and maintain 5S status boards showing percentage improvement and team members along with "before and after" pictures and activities.
5. Reward the best performers.
6. Follow the PDCA cycle to plan a systematic approach (Visco, 2015).

Target outcome
1. Improves organizational efficiency and profits.
2. Improves health and safety.

Visual management

Figure 5.33 Sustain examples in the form of 5S training, 5S success stories.

Figure 5.34 Lean room.
Source: Courtesy Corrigo Consultancy Pvt. Ltd, India.

3. Reduces Muda (waste).
4. Improves employee morale.
5. Higher quality with greater consistency (Visco, 2015).

Further 5S audits can be conducted with scores under each S at respective zones. The scores can be assessed for further improvements and winners for each zone can be rewarded with 5S rolling trophies.

Understanding the sixth S—safety

Some companies have added S which stands for safety; hence, the approach is also called 6S. The goal of this step is to ensure safety by removing workspace hazards and including personal protective gear in all the departments. However, many organizations believe that safety is an integral part of every step of the 5S system. By marking the dangerous areas and equipment, consistently maintaining equipment, and developing and implementing safety standards, one can ensure safety in the workspace. It is an ongoing practice that should be incorporated in both workspace's culture and people's mindset (Osakue & Smith, 2014; Visco, 2015).

5.4.2.3 6S audit

6S (5S + sixth S as safety) in an important lean tool that is widely used in apparel manufacturing. Let us discuss the 6S implementation at XYZ Clothing Ltd. To gage the potential of workplace improvement and to identify the need to lean interventions, a 6S audit was conducted as a part of the diagnostic study. 6S audit sheet indicates a checklist (refer Table 5.7) with several checking parameters for each of the S (seiri/sort, seiton/set, seiso/shine, seiketsu/standardize, seitsuke or shitsuke/ sustain, and safety) is mentioned. The current state of the floor under each parameter was assessed on a four-point scale, namely, "Not satisfactory," "Average," "Good," and "Very good." For each parameter, -1, $+1$, $+3$, and $+4$ points were awarded to "Not satisfactory," "Average," "Good," and "Very good," respectively. Subsequently, a score for each of the head (the 6S) was calculated, and a final audit score (which was sum up of individual scores of the 6S heads) was calculated. From Table 5.7, it may be noted that there are a total of 32 audit points in the checklist. A maximum score of 128 may be awarded if each of the audit points is observed as very good (means awarding $+4$ points to each of the audit points). In this case the final audit score observed is $+9$, which makes the current state of shop floor at a 7% level.

The 6S audit score for different S has been mentioned in the radar chart (refer to Fig. 5.35). It provides a comparative status of the current scenario before 6S implementation. It can be observed from Fig. 5.35 that sort, set, shine, and standardize were the areas of concern as these had relatively low scores, while sustain and safety looked good. However, there is a possibility that after the improvement (due to lean interventions) when a score of sort, set, shine, and standardize improves, it may have an impact on improvement potential of sustain and safety due to addition of some newer parameters which are not part of the audit before 6S implementation.

A necessary plan may be worked out based on the outcome of the 6S audit. The areas of improvement and required corrective actions may be decided under each S after a critical analysis of the results. Please refer to Figs. 5.36–5.38 for 5S shop-floor implementations in apparel manufacturing.

5.4.2.4 Phase 3: 5S self-assessment and 5S certification

The objective is to ensure that the organization can assess its strength as well as the areas for improvement and where the organization stands in the

Table 5.7 6S audit sheet.

6S Audit Checklist

Department / Location:

S.NO	AUDIT PARAMETERS	NS	Avg	G	VG
	Auditors:	Date:			
I	**SEIRI**				
1	Is the area free of clutter?	✓			
2	Is there a system of detecting non moving items, unused files, unused machines, etc. ? (Red tag area)	✓			
3	Is there a regular clearing of storage areas/ files etc.as per the Disposal Policy	✓			
4	Are notice Boards / White Boards cleared of outdated information?			✓	
5	Are the Drawers, Cupboards, Shelves, Racks & Storage Areas arranged neatly?		✓		
		-3	1	3	
II	**SEITON**				
6	Is there an updated Master File Index?				
7	Is there an updated Master Key Index?				
8	Is there a specified place clearly demarcated for trolleys, tools, stationery, spares, consumables in the office/ store / work area?	✓			
9	Are all Cupboards / Shelves / Racks / Drawers / Keys / Equipments numbered and labelled? (Equipments List to be maintained in the Master List of the Department)	✓			
10	Are all files coded with proper identity?		✓		
11	Do all cupboards have a list of contents neatly pasted?	✓			
12	Are X-Axis, Y-Axis alignments evident in the placement of all documents, equipments, notices etc.?		✓		
13	Is color coding used for effective identification? (files / keys etc)?	✓			
14	Is there an ability to find a document / item within 2 minutes?	✓			
		-5	2		
III	**SEISO**				
15	Are Updated cleaning responsibility maps, Schedules & Cleaning Checklists displayed?	✓			
16	Are dustbins present at regular intervals and not overflowing?	✓			
17	Is use of and appropriate cleaning tools evident and are they stored well and in accessible places? (Cleaning cloth, etc.)	✓			
18	Is "cleaning with meaning", daily self cleaning (inspect and clean) and "Original Condition Policy" (as good as new, careful usage) followed?	✓			
19	Are the following areas dust free ?(Floor, Tables/Chairs /Conveyors /Drawers /Shelves /Cupboards /Equipments/Walls / Windows / Notice Boards/ Ceilings)	✓			
20	Are all cables / Wires etc. dressed?			✓	
		-5		3	
IV	**SEIKETSU**				
21	Are all standardized 6S procedures being practiced	✓			
22	Is there a 6S Manual?				
23	Are SOP's displayed in relevant operations and are they followed	✓			
		-2			
V	**SHITSUKE**				
24	Evidence of Training?		✓		
25	Evidence of follow up?			✓	
26	Point of Work Reminders concepts are in practice? (Check of work reminders in forms viz. Work schedule/Things to do/Control sheets to avoid deviation in work)		✓		
27	Are there any evidence of steps taken on the previous audit findings?		✓		
			3	3	
VI	**SAFETY**				
28	Are Aisles, Fire exits, Fire Extinguishers blocked?		✓		
29	Are all safety equipments in good condition and in proper use?			✓	
30	Are fire exits, escape route, first aid location marked?		✓		
31	Employees awareness of the emergency preparedness plan			✓	
32	Evidence of improvement on the previous audit findings		✓		
			3	6	
	Total	-15	9	15	

NS -	Not Satisfactory	-1	
Avg. -	Average	1	9
G -	Good	3	
VG -	Very Good	4	

Source: Courtesy Corrigo Consultancy Pvt. Ltd, India.

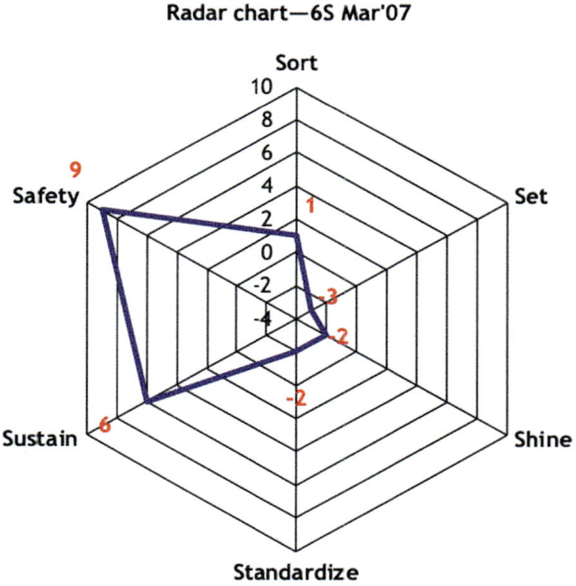

Radar chart—6S Mar'07

Figure 5.35 Radar chart (6S audit).
Source: Courtesy Corrigo Consultancy Pvt. Ltd, India.

5S movement. After activities from Phase 2, 5S self-assessment is done, monitored, and documented through 5S audit methodology. Organizations that have successful 5S activities measure their performance through weekly or monthly audits using 5S checklists, audit summary sheet, and 3S improvement stickers. The 5S certification is a public declaration of commitment to higher quality culture to meet changing customer needs. Please refer to Fig. 5.39 for shop-floor example of self-assessment based on KPI score in apparel manufacturing.

5.4.2.5 Phase 4: key performance indicators for continual improvement

In employing the 5S management techniques, every organization needs its own set of KPIs to enable it to monitor its progress. The organizations must examine their work process to develop the KPIs which describe how they influence productivity across the key areas. The KPIs most commonly used are productivity, inventory levels, inventory cost, lead time, reworks, machine breakdown, reject rate, etc. The organization discusses and prioritizes the usage of KPIs to foster a productive work culture through benchmarking. Please refer to Fig. 5.40 for shop-floor example of management KPI board in apparel manufacturing.

Visual management 179

Figure 5.36 Before 6S implementation.
Source: Courtesy Corrigo Consultancy Pvt. Ltd, India.

Figure 5.37 After 6S implementation.
Source: Courtesy Corrigo Consultancy Pvt. Ltd, India.

180　　Lean Tools in Apparel Manufacturing

Figure 5.38 5S in office.
Source: Courtesy Corrigo Consultancy Pvt. Ltd, India.

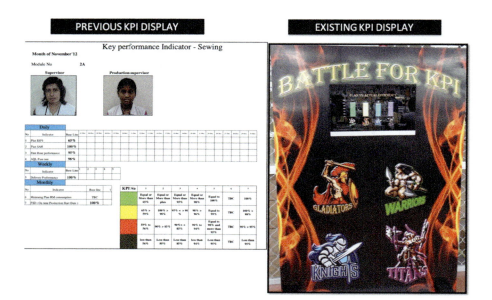

Figure 5.39 Self-assessment based on KPI score. *KPI*, Key performance indicator.
Source: Courtesy Corrigo Consultancy Pvt. Ltd, India.

Visual management

Figure 5.40 Management KPI board. *KPI*, Key performance indicator.
Source: Courtesy Corrigo Consultancy Pvt. Ltd, India.

5.4.2.6 5S implementation in the garment industry

5S implementation is considered the first step toward lean implementation, which helps to organize workstations, leading to the formation of a clean, safe, and quality workplace. It is also one of the commonly used tools in lean manufacturing. Seven manufacturing units from Okhla Garment & Textile Cluster (OGTC), in Northern Capital Region (NCR) in India reported up to 20% improvement in on-time delivery and saving of 30 lakhs to 1.5 crores annually through 5S and Kaizen implementation (OGTC, 2016). 5S is being implemented in raw material stores, cutting, sewing, finishing, and finished goods store by numerous organizations. 5S implementation in the cutting department of a garment manufacturing company has resulted in an organized and clean workplace with a reduction in transportation distance by 19.97% and an improvement in quality by 4.5% (Sharma & Jha, 2014). In another case, organization was facing major problems regarding handling stains on products and unorganized workplaces. 5S solution with a designated place for everything and usage labeled racks, trollies, and trash bins to have a smooth workflow and an organized workplace resulted in a 4.5% decrease of monthly DHU (Defects per Hundred Units) related to stains (Sharma & Jha, 2014). Implementation of 5S in Jacket Trim store reported a reduction of searching time up to 17 minutes per order and the process improved from 40% to 84% (Patil & Tewari, 2014).

5.5 Andon

Andon is a Japanese word which means "fixed paper-enclosed lantern, paper-covered wooden stand housing an (oil) lamp." It emerges in the Edo Era between the 17th and 19th centuries.

The concept of Andon in the industrial application was first started in the book Ford Men and Methods (Norwood, 1931) as

Along the back wall of this room stands an instrument board, studded with signal lights. Aside from their visitor, two men are present. One, seated on a stool, is drawn close to a shallow desk that extends from the board. The other, a trouble mechanic, is intent upon that constant motion to be seen through the windows. As you watch there comes the whir of a bell fixed to the top of the panel. The man at the desk moves a switch. The bell is silenced but in the same instant, a green light glows in the face of the board.

It is also said to be an extension of Deming's speech on Statistical Product Quality Administration, in 1950, where he quoted that for quality "Appliances and techniques can be used to a higher degree" (Deming, 2012).

It was later briefly described in the book Toyota Way (Liker, 2004)

In the case of machines, we build devices into them, which detect abnormalities and automatically stop the machine upon such an occurrence. In the case of humans, we give them the power to push buttons or pull cords called andon cords which can bring our entire assembly line to a halt. Every team member has the responsibility to stop the line every time they see something that is out of standard. That's how we put the responsibility for quality in the hands of our team members. They feel the responsibility they feel the power. They know they count.

The Andon initially surfaced within the Toyota Production System as a VMS (Visual Management System) with the principle of Jidoka. It allow the supervisors or even operators to stop production due to part shortages, defects, machine problems, and more simply by alerting mechanics and process owners through three-colored lights colleagues to problems through the three-colored lights (green for normal operations, yellow to indicate that changes are needed, and red to indicate production must stop) (Liker, 2004).

"Andon" has become a gesture to uncover a disorder. This signal is used to detect potential defects in quality. When a defect is suspected, a signal board may indicate a problem for a particular workstation. The signal event indicates that the system has shut down due to an error and is waiting for the problem to be resolved. This process is called Jidoka. The idea behind Jidoka is that shutting down the system will bring an immediate opportunity for the source of improvement by RCA instead of blaming it further down the line (Willis, 2015).

The Andon response system depends on two factors WIP and cycle time. When a mechanic takes an Andon call, he has that time to fix the problem until the line should not disturb. This should be equal to the cycle time of the previous operation or depends on WIP or flexibility of the system. This is done under yellow light. If the problem is not fixed in the stipulated time, the whole line needs to be stopped with a red light following the principle of Jidoka.

Organizations need to build a culture of stopping to fix problems, to get the right quality for the first time. Andon lights are used for short repetitive cycles, repetitive jobs where immediate help is needed and even seconds count. In case of any irregularity, the Andon light would come on, and an alarm would sound. Andon system would be only effective if operators followed a standardized work. In a way, it is a *poka-yoke* device (Liker, 2004).

5.5.1 Types of Andon systems

The Andon system is usually a two-way system consisting of inputs and outputs which can either be done manually or automatically (Roser, 2015). When talking about manual Andon input systems they can of various types.

5.5.1.1 Andon cords

Taichi Ohno introduced the system of *Andon cords* as a part of Jidoka. Toyota implemented the Andon cord as a physical rope in the assembly line and could be pulled to stop the manufacturing line at any time. Any worker on the assembly line can, and should, pull the Andon cord. This signal to the team leader that there is a problem, if the worker and the team leader cannot find a resolution in a predefined amount of time, then the assembly line is stopped.

5.5.1.2 Andon switch

Andon switch or an *Andon button* is used for smaller workspaces, a button or switch on the reach, and workers need not stand or walk away to pull the cord.

Andon buttons can be classified into different colors (refer Fig. 5.41) based on the severity of the problems and the required action to be taken.

5.5.1.3 Automatic Andon input system

This is an intelligent system that can detect if the operations are completed in time or if the workers are falling behind, a signal is sent automatically. Similarly, for an automatic or semiautomatic machine, sensors and programmed logic can detect a potential upcoming slowdown or stop. This information is then also forwarded to the Andon system. Many Andon systems can also automatically measure things such as production speed, actual and target quantities, and other parameters relevant to the performance of the line. Please refer to Fig. 5.42 for the shop-floor example of the Andon System in apparel manufacturing.

In terms of output, various visual tools are used as indicators to alert the team leader about the problems in the line. The Andon system obtains productivity and quality-related data across the line. This data can be visually illuminated or displayed in numbers or through audio signals and mobile phone networks. That is why Andon is ultimately a part of VM.

Color cord	Condition	Action
🟢 Green	Production is normal or smooth	Proceeds to net level
🟡 Yellow	Problem appeared	Operator takes help of concerned authority to fix the problem
🔴 Red	Production stopped	When problem is not identified and need further investigation

Figure 5.41 Andon switch.

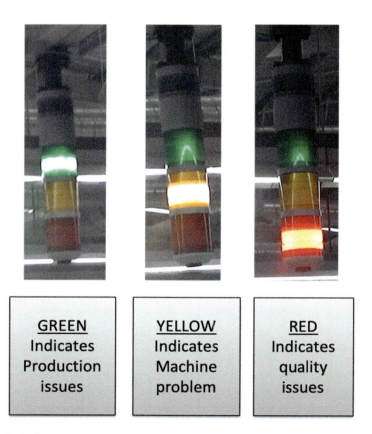

Figure 5.42 Andon system.
Source: Courtesy Corrigo Consultancy Pvt. Ltd, India.

5.5.1.4 Stack light

The simplest Andon output is a stack light or industrial light tower on top of a machine. Stack lights (refer Fig. 5.43) are used in a variety of machines and process environments with a specific color coding which indicates the status of the machine.

Visual management

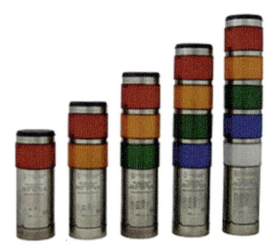

Figure 5.43 Stack lights.

Red: Failure conditions such as an emergency stop or machine fault
Yellow: Warnings such as overtemperature or overpressure conditions
Green: Normal machine or process operation
Blue: External help request, where an operator might be requesting raw materials, scheduling or maintenance personnel assistance
White: User-defined conditions to a specific machine, often related to productivity monitoring

Optionally an audible alarm buzzer, typically in the range of 70–105dB, may be added to alert machine operators to high priority conditions.

5.5.1.5 Andon boards

Andon boards summarize the information about the status of the system on one board using different lights and LED number displays. This board displays the relevant information from the Andon system such as Takt time, Overall Equipment Effectiveness (OEE), an actual and target production value, defect quantities, efficiency, rejects, and downtime (refer Fig. 5.44).

5.5.2 Implementing the light and sound in visual factory
5.5.2.1 Color scheme

Plant workers utilize the colors of the Andon lights to convey, without vocal prompts, the condition of the industry line. To make a successful visual framework the colors of the Andon lights should be effectively interpreted by line operators (Bansal, Jankins, & Duran, 2017).

Figure 5.44 Andon boards.

The first model of the Andon light system for the intercept line was comprised of four stacked colors as mentioned in Table 5.8. Refer to Fig. 5.45 for a practical example of Andon lights and switches installed in an apparel manufacturing floor.

5.5.2.2 Adding sound

When there is an issue on a sequential assembly line, a sound framework could head out to alert close by mechanics or managers about the issue. The quick reaction would help spare assets and time, two of the losses of lean assembling.

With the execution of the sound framework, every assembly line would need to receive its distinguished sound example so the alerts do not get mistaken for issues on different lines or the signal for move changes. Optionally an audible alarm buzzer, typically in the range of 70–105 dB, may be added to alert machine operators to high priority conditions (Bansal et al., 2017).

5.5.3 Working of Andon system

Whenever there is a problem in the machine or manually by the operator, a light or alarm is triggered to call for help and immediately a planned team reaches the location of the problem.

The time required to solve the problem depends on the complexity of the same:

- The problem is solvable at the moment, then the dedicated team solves the problem and the production continues.
- Or the problem is not-solvable at the very moment if any part of the machine is broken, then he particular task is set aside and the production continues or the line is stopped and the other workers also come in support (e.g., for quick transport of material from the warehouse to get things going again).

Andon working flowchart is indicated in Fig. 5.46; whenever the problem is resolved, it is updated in the Andon. A backup is done on the activity taking into account the time of intervention and the type of problem. In case the same problem

Visual management 187

Table 5.8 Andon light system stacked colors.

Red light	Stop. Warn the workers that there is a machine-related problem and that their tasks had to be halted. A sign of Jidoka.
Yellow light	Caution. Would warn the workers that there might be machine-related problems.
	An example would be if one item of clothing did not pass quality checks but the ones who followed and ended did.
Green light	Go. The all-inclusive sign for go would guarantee the operators that everything is running easily on the line and that they can keep on working.
Blue light	Signs that the machine isn't yet prepared to work, however, is starting up.

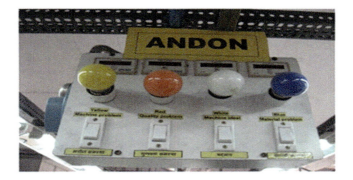

Figure 5.45 Andon lights and switches.

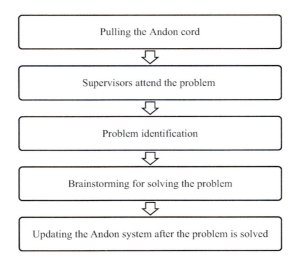

Figure 5.46 Andon working flowchart.

occurs in the future, backup data can reduce the time to solve the problem thus increasing the efficiency of the process.

5.5.4 Benefits of Andon system

Expedite information flow: Visual factory with Andon system improves the information flow in case of abnormalities. It has fewer problems, can fix them faster, and tries to eliminate the root cause of the problem. Andon can help in process standardization and accelerating the information flow.

Empowering the employees: Usage of this system empowers the employees to identify the status and takes action quickly. Operator/supervisor is empowered and informed to pull an Andon cord or press the button in the manufacturing line when abnormalities occur.

Mistake proofing: Andon is associated with the concept of Jidoka; stopping the line in case of abnormalities and further find the root cause. It creates an environment to produce right at the first time with no defects. It blocks defects from reaching customers and creates an opportunity to improve the system to prevent future defects.

Improve culture: In Andon culture the team leader thanks to the team member for pulling the Andon cord as it saved a customer from receiving the defect. This improves the cooperate culture of participation and removes the fear of stopping the line. Further, they can mutually participate in finding the root cause to prevent future occurrences.

Quick response time: It also reduces downtime by improving the ability of supervisors/ operators to quickly identify and resolve manufacturing issues. Andon also saves costs and time as it provides a simple and consistent mechanism for communicating information on the plant floor.

Improve transparency: Andon also promotes transparency and encourages immediate reaction to various problems in an assembly line such as downtime, quality, and safety problems. This also results in increased productivity and flexibility.

Promote standardization: To avoid defects occurrence, every process moves toward standardization. It involves creating SOP, understanding it, and pulling Andon in the case of abnormalities.

5.6 Yamazumi charts

Manufacturers must respond correctly and quickly to the volatility in this competitive market. One way factories can deal with evolving customer demands is by using Yamazumi charts to build flexibility in their production processes (Harris & Harris, 2009).

Yamazumi is a Japanese word that means stacking up. "Yamazumi Chart (or Yamazumi Board) is used to represent processes for optimization purposes in a graphical manner" (Kelly, 2019). It is also called operator balance sheet and operator loading diagram.

Yamazumi charts use vertical bars to represent the total amount of work that each operator is expected to do in contrast to the takt time. For each operator the vertical bar is built by stacking small bars representing individual work elements, with the height of each element proportional to the amount of time required. These charts help with the crucial task of redistributing work elements among operators.

Visual management

This is essential for minimizing the number of operators needed and is done by making the amount of work for each operator nearly equal to the takt time (Lean Enterprise Institute, 2008).

5.6.1 How to make a Yamazumi chart?

A Yamazumi chart can be made in three steps, as follows:

Step 1: Work elements
The first step is to gather all the elements of the work needed to complete the product. A work element can be defined as the smallest work increment that can be transferred from one individual to another. For example, a work element would be the opening of bundles and arranging the components.

Step 2: Classifying the work elements
The production process comprises three types of work. The first type, and the lowest percentage of the work, is typically VA work. More than 90% of the work elements are non-VA (NVA). The definition of a VA work element is any work element that directly adds value to the product. Attaching a button in a shirt, for example, is VA work.

The two remaining work elements are both NVA. The first is what we call necessary NVA (NNVA) work. NNVA is the work that has to be done but does not directly add value to the product—transporting the material (such as cut panels) from one work station to another, for example.

The third work element is waste (W). Waste can be defined as a work element that does not add value to the product being produced and isn't necessary to produce the product like waiting for the bundles.

For instance, the following chart (refer to Fig. 5.47) is a Yamazumi chart of a casual shirt, VA is represented by green, NNVA by orange, and waste is represented by red.

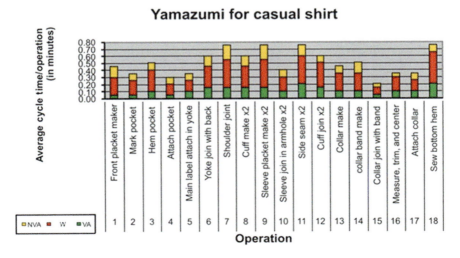

Figure 5.47 Yamazumi charts (Jana, 2010).

The goal is to eliminate all the waste work elements and reduce the NNVA work elements as much as possible. This to VA work being a higher percentage of the work-content in the process, ensuring that manufacturing associates use a greater portion of their time and resources to add value to the company.

Step 3: The paper kaizen

The paper kaizen method is relatively basic, and it is a significant part of the improvement effort. It is a process where the information gathered is discussed for process improvement. The process examines the classification of each element of work, removes the elements of work classified as waste, and determines which NNVA work elements can be reduced. Please refer to Fig. 5.48 for kaizen display board on an apparel manufacturing floor.

Once the paper kaizen is complete, a new list of work elements is available that is needed to produce the product. This list often gives the opportunity and need to rearrange the production area, so that the process can be completed without the waste work elements of walking and waiting (Harris & Harris, 2009).

5.6.2 Benefits of Yamazumi chart

- It's easy and visual. As the Yamazumi Board tells you the story at a glance, the staff can see where the delays come from instantly and intuitively.
- It is inescapable. Since the board hangs above the production line, it is a constant, perpetual inspiration for continuous improvement or Kaizen.

Figure 5.48 Kaizen display board.
Source: Courtesy Corrigo Consultancy Pvt. Ltd, India.

Visual management 191

- It is public. The Yamazumi Board is, glaringly so, in the open. This is a great motivator for constructive performance improvement of competing work teams.
- It visually underlines key constraints. It reveals the "vital few" opportunities that can change the game, much like the Pareto diagram (Jana, 2010).
- A Yamazumi chart is an image of the distribution of work among the operators with takt time. This reflects the rate at which the product should be manufactured to prevent the significant waste of overproduction.
- It helps define and prioritize opportunities for change (i.e., reduce the NVA elements).
- Yamazumi Charts help eliminate bottleneck operations in an assembly line by identifying the most time-consuming element and equally redistributing it among the operators.
- It increases the operator's work efficiency by eliminating the waste elements from the process.
- The time of the assembly station is considerably reduced and it also helps in aligning with the next assembly station (Sabadka, Molnar, Fedorko, & Jachowicz, 2017).

5.7 Turtle diagram

According to NQA Certification Body, "A Turtle Diagram is a visual tool that can be used to detail, in a very precise manner, all of the elements of any given process within an organization" (NQA, 2017). Another definition is given by Thompson as "It is a schematic visual representation of the key elements that make up a single process. It is used to illustrate the 1-page plan for a single process including the resources needed to achieve the process's purpose" (Thompson, 2013).

Turtle diagrams are helpful as they can replace various lengthy multipage, complex documented procedures with a clear and easy to understand visual illustration of what is expected from a process. This tool can also work as a contract between cross-functions, illustrating where product, communication, and other hand-offs begin and end. The tool can be used to showcase the interactions between various processes and define the inputs and outputs involved clearly.

The Turtle diagram was first introduced by Philip Crosby. Crosby's Process Worksheet builds on the traditional SIPOC diagram (supplier, input, process, output, customer) by adding performance requirements for input and output as well as training, equipment, and procedures (documentation). Sometime around the year 2000 as the automotive QS 9000 standard was being replaced by TS 16949, certified registration auditors were taught the use of the Turtle Diagram as an auditing tool. The purpose was to provide a tool for process auditing as compared to the audit of "shalls" or of "clauses" (Ambrose, 2016).

5.7.1 Components of a Turtle diagram

The components of a Turtle diagram comprises of inputs, outputs, what, who, how, and the measures respective to the process or activity. The turtle chart can fill in as a "contract" between the process owner and administration of the association. On the off chance that desires (outputs) change, at that point there ought to be thought given to particular resources expected to meet these new expectations. Please refer to Fig. 5.49 for an illustrative example of the fusing process in apparel manufacturing through the turtle diagram.

Input

- shell fabric cut panels (for fusing)
- fusible interlinings cut panels

Output

- fused cut panels of consistent quality

Who?

- fusing machine operator
- maintenance engineer
- helper (for material handling)

Criteria/measures

- machine speed (meters/minute)
- time (seconds)
- temperature (°C)
- pressure (bar)

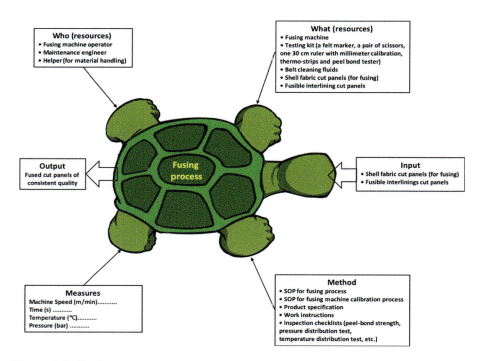

Figure 5.49 Turtle diagram.

Visual management 193

With what?

- fusing machine
- testing kit (a felt marker, a pair of scissors, one 30cm ruler with millimeter calibration, thermo-strips, and peel-bond tester)
- belt cleaning fluids
- shell fabric cut panels (for fusing)
- fusible interlining cut panels

How/method?

- SOP for the fusing process
- SOP for fusing machine calibration process
- product specification
- work instructions
- inspection checklists (peel-bond strength, pressure distribution test, temperature distribution test, etc.)

Inputs: Inputs to the process should be described in this category which may include internal and external customer specifications, part numbers, and other relevant information.

Outputs: The output side of the turtle diagram represents the process's results. That includes information, reports, or data for a system process. The outputs for a manufacturing process include the finished product, statistical data, and other records related to the production. The actual output should be as per customer requirements.

What: These are nonhuman resources needed to perform the process, that is, material, infrastructure, equipment.

Who: These are all the persons who add value to the process in an organization directly or indirectly.

How: This section is dedicated to finding any relevant documents within the management framework that tell the people in charge of implementing the VA steps on how to complete them effectively within the best practice of the organization. This includes processes, guidelines for the job, action plans, failure mode, and effects analysis, standards, environmental specifications, etc.

Measuring: This is a list of the performance indicators which will show the process's success or failure. These metrics or KPIs are also the driving force behind organizational progress and preparation for companies (Granowicz, 2011).

In the center, one can detail all the value-adding steps that fall with the scope of the process, plus their sequence. Please note that any given process may involve multiple employees and departments within the organization (NQA, 2017).

5.7.2 Benefits of Turtle diagram

Turtle diagram correlates the information in one simple to read diagram. There are many benefits of using a Turtle diagram, here are just a few of those:

- Turtle diagrams can be used in conjunction with process flowcharts, procedures, work orders, standard operating procedures (SOPs) to align and link all of them in one simple diagram (McDonalds Consulting Group, 2018).

- An added reward of making a turtle chart for each procedure can be found during the audits, regardless of whether it's an internal or third-party audit. During an internal audit, the auditor can utilize the turtle outline to comprehend the procedure better. This permits the inspector more opportunity to investigate the procedure instead of spending time figuring out what the procedure is and how it fits into the general plan (Granowicz, 2011).
- Understandable onboarding document, these are ideal for offering to new workers, so they comprehend their job role in the organization easily and also can understand the value-adding tasks they aid in completing.
- It is helpful to align the organizing processes to the organization goals. They serve as a benchmark for current best practices for each process and make processes more visible and easier to understand. The efficiency and efficiency of the turtle diagram can help visualize the quality of the finished product, turn it over time, and pave the way for any defects or risks associated with the process. It helps to prepare and implement changes that improve quality and can bend multiple times while reducing the number of defects and accidents.
- The separated "legs" of the turtle diagram make it easy to understand the path taken from first to last in a process. Identifying not only raw roles but also resources such as warehouses, machine software, and databases, and what resources are needed. It identifies the support process, from staff training to IT support and maintenance of used equipment.
- The focus of the turtle diagram is the VA steps in a process and the dimensions of the effectiveness and efficiency of those steps. Where waste occurs in the process, quickly, management can identify value-reducing steps in the process, or if the VA steps go wrong. These can be addressed and changes implemented to improve the overall process (Granowicz, 2011).

5.7.3 Applications of Turtle diagram

ISO 9001 and ISO/TS 169-49 expect associations to show how every one of their interior procedures connects with others. One well-known apparatus being utilized to portray process cooperation graphically is the turtle diagram. It is not required for compliance with clause 4.1 of ISO 9001 (Granowicz, 2011).

Note: The turtle diagram is not the only way to show objective proof to a third-party auditor that you are using the process approach, and neither ISO/TS 16949 nor ISO 9001 requires its use. But the diagram is one method being employed regularly by organizations to demonstrate having used the process approach.

The combined turtle flow diagram can be utilized by the operations and quality teams in the everyday working of the procedure, just as by internal and third-party auditors as a beginning stage to assemble data and follow the procedure.

To address the quality control issues of a product-service system, including job shop manufacturing and customization of orders, incapacity, and default of process quality control, one can fabricate a model of procedure quality control framework, utilizing improved turtle outline and assessment technique dependent on VDA (VDA is an acronym in German, Verband der Automobilindustrie, in English: Association of the Automotive Industry), for the greater accomplishment of sustainability.

5.8 Affinity diagram

KJ method also known as affinity diagram was developed by Jiro Kawakita in the 1950s. Kawakita was a professor at the Tokyo Institute of Technology. He directed several different scientific expeditions to Nepal and India to collect Himalayan Ethnographic data. He developed analytical techniques to comprehend qualitative data of Himalayan Ethnographic in the 1950s. It took another 15 years to systematize the method and finish the guidelines for the development of the program to teach this method which was known as the KJ method in Japanese business. It is a useful creative and brainstorming technique to organize complex, immeasurable, idiosyncratic, nonrepetitive, behavioral, qualitative, and heterogeneous data collected in the field.

Kawakita facilitates creative bridging synthesis between ethnographic fieldwork and more structured hypothesis testing. Initially, he developed the W-shaped problem-solving model, upon which the contemporary KJ method is based upon (refer to Fig. 5.50). In Kawakita's formulation A−D−E−H represents the "armchair sciences," E−F−G−H represents the "laboratory sciences," while A−B−C−D represents the "field sciences."

> A−B−C−D "*field sciences*" *route*: If a person encounters a problem at point A on the thought level, he explores the conditions surrounding the problem between points A and B, and next collect all relevant and accurate data through observations between points B and C. By this data, he next formulates several hypotheses between points C and D.
>
> E−F−G−H "*laboratory sciences*" *Route*: In the next level, at point D, he evaluates hypotheses and decided which one hypothesis to adopt. Between D and E, he infers and revises the adopted hypothesis through deductive reasoning. Next, he plans an experiment

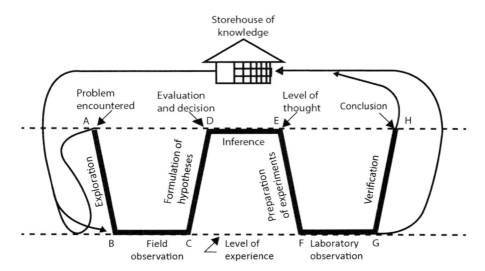

Figure 5.50 Outline of basic steps in W problem-solving method.

for testing the adopted hypothesis between E and F and observes the experiment between F and G. After viewing the result of the experiments, he can verify the hypothesis between points G and H and can finally acquire a correct conclusion at point H. If they falsify the hypothesis, they go back to point A and start the problem-solving again.

There are many methods of inference, experimental design, and hypothesis testing for performing the basic steps from points D to H, but there is a little systematic methodology for illuminating the steps from A to D, in particular the step from points C to D. The KJ method was developed to bridge the gap between unstructured field observations, induction, hypothesis adoption, and experimental laboratory methods. The KJ method serves to unite two general approaches to problem-solving: the laboratory method that starts from definitions of hypotheses and tests them under controlled laboratory conditions, and field method of making observations in an unstructured environment and generating hypotheses based on them (Kunifuji, 2016; Scupin, 1997).

The affinity diagram is used to organize a large number of ideas into their natural relationships. This tool can be used to produce, segregate, and integrate information related to a product, process, complex issue, or problems that have a lot of data. After generation, the ideas are grouped according to their affinity, or similarity. This tool is a quick way to get a group consensus after a brainstorming session. It is an easier way to get honest feedback limiting the impact of peer pressure and group politics. It is generally used at a definition or ideation stage of a project.

The affinity diagram process is a good way to get people to work on a creative level to address difficult issues. It may be used in situations that are unknown or unexplored by a team, or when people in a team are with diverse experiences with incomplete knowledge of the area of analysis. It encourages a group of people to come out with ideas that were mulling at a gut level.

The team should consider all ideas from all members without criticism. This stimulus enables the team to enhance their creativity and intuition (Oakland, 2014). This tool helps us make decisions about the future. It is often used to plan events that have not occurred yet. They help people analyze relationships between ideas and attributes-tangibles that are difficult or impossible to quantify. They help in structuring ideas and map their interrelationships (Cohen, 1988). It is used in situations like when a group of people is confronted with many facts or ideas in apparent chaos when issues seem too large and complex to grasp when group consensus is necessary. It can be used after a brainstorming exercise, to analyze verbal data such as survey results, to collect and organize large data sets, to develop relationships or themes among ideas

5.8.1 Creating an affinity diagram

This process lets a group move beyond its habitual thinking and preconceived categories. This technique accesses the great knowledge and understanding residing untapped in our intuition. Affinity diagrams tend to have 40−60 items; however, it is not unusual to see 100−200 items.

Visual management

Materials needed: Sticky notes or cards, a group of people (your team), marking pens, and a large work surface (wall, table, or floor) (Cleary, 2018; Kunifuji, 2016). Refer to Fig. 5.51 for the basic steps of affinity diagram.

Step 1: Label each idea with a sticky note or card

In the first step, label making step, record each idea with a marking pen on a separate sticky note or card. A group of people familiar with the issue phrase the issue. Each member is given a sticky note or label where they should write one thought or one concept related to the problem on each label. This is used to generate ideas through brainstorming. In the end, they need to display it on a common board.

Step 2: Label grouping

After writing all the labels the team needs to stick all the labels on a common board. The process owner must carefully consider what the labels are saying. Labels that appear to belong together should be arranged close to each other and kept at a distance from other labels to form a group. Kawakita emphasizes a nonlinear−nonlogical method where it should not be grouped simply based on similarity (similar words being used) but rather mental association. Labels that do not seem to be related to any other labels (called "lone wolves") might become key concepts, or be merged into another group at a higher level of label grouping. In the end, you should be ready with 5−10 related groups.

Step 3: Naming group label

After about two-thirds of the labels have been grouped, you can start making one-line headers as titles for each group. Identifying titles or more generalized concepts are subsequently used to order the data into larger groups. Arranging the teams into larger groups and placing titles on the new groups and posting them together helps classify the data. This intuitive process may be repeated as many times as necessary to reduce the labels to less than 10 families.

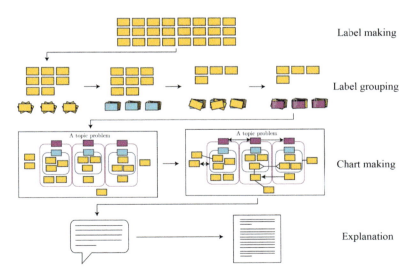

Figure 5.51 Basic steps of affinity diagram.

Step 4: Spatial arrangement

In this step, place all the groups and labels in a spatial arrangement. Discover a relationship between two or more groups and combine them under supergroups. The output of this exercise is a compilation of a maximum number of ideas under a limited number of major headings. This data can then be used with other tools like an interrelationship diagram.

Relationship symbols

The following relationship symbols are used in KJ Ho chart making between objects:

1. *Cause and effect*: One is a predecessor or a cause of another.

2. *Contradiction*: Objects are conflicting with each other.

3. *Interdependence*: Objects are dependent on each other.

4. *Correlation*: Both objects relate to one another in some way.

Step 5: Verbal or written explanation

The last step is to explain the affinity diagram in written or verbal form making it smooth clear and concise. It can start with any part of the chart and then proceed adjacent until the whole chart is covered. This step helps the people to understand the interrelationships among components of the problem thoroughly.

5.8.2 Advantages of the affinity diagram

Speed—Imagine a group discussing how to compile hundreds of bits of data. The affinity exercise forces the team members to use instinctive reactions rather than lengthy discussions.

Acceptance—Reading, thinking, and moving the data build buy-in from the team. They become more familiar with issues and relationships.

New thinking—Because of the spatial movement, this exercise forces people to use the creative right side of the brain, in addition to the analytical left side of the brain. This combination can open people's eyes to new ways of seeing issues (Plain, 2007).

5.9 Dashboard

Many organizations are not able to take time-bound decisions because of a lack of information or they need to blog down themselves into endless spreadsheets or outdated reports. They need to implement a performance management system that translates the organization's strategy and metrics visually to each individual in the organization in time.

According to Eckerson (2005), "Dashboards are the most visual elements of a performance management system that merges the functionality of Business Intelligence and performance management. Besides displaying metrics visually, these full-fledged information systems let users drill into detailed data to identify the root causes of problems and intervene while there's still time." It is a tool that helps organizations to collect, display, understand, deploy, and use information.

Visual management

The purpose of a dashboard is to help operators, supervisors, and managers to assess their performance and take quick decisions based on visual analysis. It enables the organization to communicate strategic objectives and enables employees to measure, monitor, and manage the key activities and processes needed to achieve their goals. It is a visual display of the most important information needed to achieve one or more objectives; consolidated and arranged on a single screen so the information can be monitored at a glance. Thus it can be used to facilitate quick decision making an important part of a lean management system (Few, 2006).

A dashboard can be defined as an integrated display of data and infographics in such a way that provides at-a-glance views of relevant and much-needed information to a particular process and its KPIs and performance metrics (Biggs, 2013). Refer to Figs. 5.52–5.54 for different shop-floor examples of dashboards installed in apparel manufacturing set-ups.

A dashboard is a type of graphical user interface that often provides at-a-glance views of KPIs relevant to a particular objective or business process. In another usage, "dashboard" is another name for "progress report" or "report."

A dashboard is a powerful agent of organizational change that enables organizations to measure, monitor, and manage business performance more effectively. It is linked to a database that allows the report to be constantly updated. A performance dashboard lets business people to:

- monitor critical business processes and activities using metrics of business performance that trigger alerts when potential problems arise;

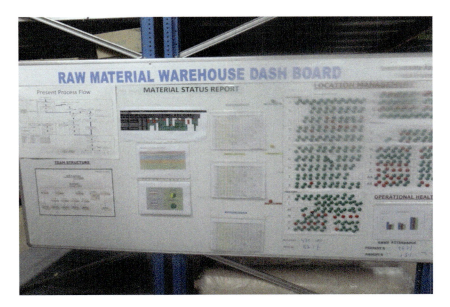

Figure 5.52 Raw material warehouse dashboard.
Source: Courtesy Corrigo Consultancy Pvt. Ltd, India.

Figure 5.53 Process-wise trims allocation display board.
Source: Courtesy Corrigo Consultancy Pvt. Ltd, India.

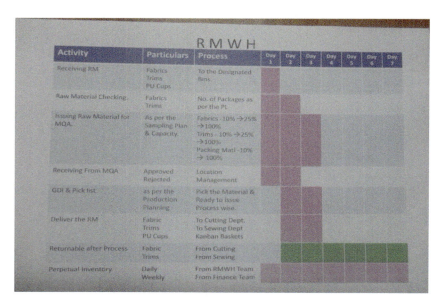

Figure 5.54 Raw material warehouse process flow.
Source: Courtesy Corrigo Consultancy Pvt. Ltd, India.

Visual management

- analyze the root cause of problems by exploring relevant and timely information from multiple perspectives and at various levels of detail; and
- manage people and processes to improve decisions, optimize performance, and steer the organization in the right direction (Eckerson, 2005).

5.9.1 Types of dashboards

Dashboards are often linked to a database that allows the report to be constantly updated. There are different types of dashboards (refer to Table 5.9) based on the usage and information sought (Koch, 2018).

5.9.2 How to make a dashboard?

Monitoring your operations with real-time dashboards can help ensure that you're seeing data clearly, correcting inefficiencies, identifying areas for optimization, and communicating with team members effectively (Wolny, 2014). Refer to Figs. 5.55 and 5.56 for examples of digital dashboards installed in apparel manufacturing.

Step 1: Ask the basic questions: To start designing a dashboard, basic questions to be answered like the look of the dashboard, type and number of metrics, how data can be tailored at different levels, and who will be using the dashboard.

Step 2: Tool selection and dashboard design: Before investing heavily in a fancy dashboard, it is important to draw your metrics on a whiteboard. The dashboard design should be a single one-page display screen with three to six metrics with minimal texts that can be affected by a larger audience.

Table 5.9 Type of dashboards.

Dashboard	Objective
Strategic	Strategic dashboards provide a quick overview of high-level measures of performance, and forecasts to take key decisions of the business. It focuses on static snapshots of data (daily, weekly, monthly, and quarterly) that are not constantly changing from one moment to the next.
Analytical	Analytical dashboards include more context, comparisons, and history, along with subtler performance evaluators. It involves drilling down data to find underlying details.
Operational	Operational dashboards are used for monitoring operations and activities that include real-time performance display and require attention on the spot.
Informational	Informational dashboards combine different elements such as metrics and KPIs on one single graph to give an overall picture. The data contained in this dashboard is archived, derived, summarized, and historical.

KPI, Key performance indicator.

Figure 5.55 Dashboard.
Source: Courtesy Corrigo Consultancy Pvt. Ltd, India.

Figure 5.56 Dashboard in apparel industry.
Source: Courtesy Methods Apparel Consultancy, Gurgaon.

Step 3: Choose your audience: This is an important step that the metrics are visible to which audience and who should be taking an action. Dashboard metrics on the production floor should be of maximum benefit for real-time action. Thus it should be designed based on the audience on the floor.

Step 4: Choose your metrics: With the target audience in mind, the next step is to define the metrics that would be best suited for driving improvements on the floor. Various brainstorming sessions, kaizen, and other tools can be used in this regard. The metrics can

Visual management 203

include line efficiency, cycle time, rework percentage, target, OEE, WIP, maintenance and repair costs, etc.

Step 5: Choose the type of dashboard and level of data: The next step is to choose the type of dashboard from strategic, analytical, operational, or informational based on audience, level of data drill, and the time of action.

Step 6: Designing the display VM: The dashboards are generally placed overhead and should be visible from a distance. It is necessary to design the display and graphics based on the principle of VM, metrics importance, and alert data. The design should be simple and made so that it can be acted upon quickly by the intended audience. Further, the size of the board depends on the location, several metrics, and location.

Step 7: Training: The next is to provide training regarding the features, metrics, levels of data, and how to use it effectively by a different set of audiences. The adoption of a dashboard is quite complex as it is difficult to overcome the habit of using the old method of data. It should be kept in mind that the dashboard should make the data handling easy and comprehensive avoiding its redundancy.

5.9.3 Benefits of dashboard

The purpose of a dashboard is to help workers, stakeholders, and management makes data-driven decisions without getting bogged down in endless spreadsheets and outdated reports.

One of the most important aspects of a dashboard is "drill-downs", simply put in words, a data drill-down allows you to dig deeper into top-level data, gaining an even more detailed view of core information and related metrics. In complex operations, such as lean production, every component of the process will involve different metrics, and every worker involved in those stages may need slightly different sets of data. With drill-downs, dashboards use high-level metrics to guide workers (Eckerson, 2005).

Communicate strategy: Dashboards translate corporate strategy into measures, targets, and initiatives that are customized to each individual in an organization. Each morning when managers log in to the dashboard, they get a clear picture of the organization's strategic objectives and what they need to do in their areas to achieve these goals. They can also refine their strategy based on issues observed.

Increase visibility: Dashboards give greater visibility into daily operations and future performance by collecting relevant data and forecasting trends based on past activity. This helps the organizations to assess their performance on a specific day.

Increase in coordination: Dashboards encourage each department to work closely and discuss performance results and shortcomings.

Increase motivation: It has been said, "You can't improve until you measure it." By the visual display of performance and abnormalities, it tells where you stand today and how far you have to go. It motivates employees to reach specific goals.

Give a consistent view of the business: Dashboards consolidate and integrate corporate information using common definitions, rules, and metrics. This creates a single version of business information that everyone in the organization uses, avoiding data hiding giving a consistent view of the business.

Reduce costs and redundancy: By consolidating information, dashboards eliminate the need for redundancy of the data. A single dashboard can help an organization to remove various independent reporting systems, spreadsheets, data marts, and data warehouses.

Empower users: Dashboards empower users by giving them self-service access to analyze and act on information.

Prompt action: Dashboards provide information in a concise manner that lets users take prompt action to fix a problem, help a customer, or capitalize on a new opportunity before it is too late. A dashboard prevents users from wasting time searching for the right information or report.

In short, dashboards deliver the right information to the right users at the right time to optimize decisions, enhance efficiency, and accelerate bottom-line results.

5.9.4 Application of dashboards in apparel industry

Apparel manufacturing process generates a lot of data at each level for its operation. It requires some level of analytics to run the business. Dashboards in the apparel industry should be designed keeping in mind the following key metrics.

- Productivity—Productivity and performance for each department.
- Quality control—Monitoring of inspected pieces, rejected pieces, reasons for rejection, rework percentage.
- Cost structure—Analyzing the fixed and variable costs per department.
- Safety—Keeping a track of the accident-prone areas, safety precaution information.
- Logistics service—Information about transportation responsibility, quality, and customer complaints and feedback.
- Human resource—Hiring new people, the record of employee turnover, training programs.
- Research and development—Tracking the area or processes which may need improvisation using the research of better alternatives, their results after implementing it.

5.10 Conclusion

With the detailed discussion of 5S and VM philosophies, it is confirmed that adopting these practices is not just a choice but a necessity for an organization. Application of such tools and techniques help immensely in developing efficient and effective shop-floor management by controlling or eliminating different types of wastes. More importantly, it helps in people's empowerment, where they are updated with the relevant information and can handle issues and situations on their own. This also boosts employee morale, teamwork, and a sense of belongingness to the organization. It is always emphasized in lean approach, that being lean or becoming lean is not just about following some tools and techniques, but it is all about how people are adapting to it as a habit or living the lean way. In this context, 5S and VM play a vital and decisive role. We see a growing number of organizations adopting VM practices in apparel manufacturing in recent years. These days, VM that is powered by technology (in the form of real-time data monitoring of production process, quality management systems, maintenance management systems using RFID, Bar-codes, QR codes, and artificial intelligence) is transforming VM

Visual management 205

into knowledge management. The application of VM is up to the extent that these tools and techniques are being practiced independently, irrespective of applying other lean tools. This is a good indication, as it helps immensely in informed decision making based on real-time accurate data.

References

Ambrose, P. (2016). *Turtle diagrams — myths and misconceptions.* SystemsThinking.Work. Retrieved from https://systemsthinking.works/post-turtle-diagrams. (Accessed 30 May 2020).

Bansal, M., Jankins, S., & Duran, W. (2017, July 21). Implementing A Visual Factory Check in Silver Line.

Bartnicka, C. (2018). The effects of implementing 5S as the foundation for work improvement on the workplace. *Multidisciplinary Aspects of Production Engineering, 1*(1), 451−455.

Bateman, N., Philp, L., & Warrender, H. (2016). Visual management and shop floor teams — Development, implementation and use. *International Journal of Production Research, 54*(24), 7345−7358.

Berkley, B. (1992). A review of the kanban production control research literature. *Production and Operations Management, 1*(4), 393−411.

Beynon-Davies, P., & Lederman, R. (2017). Making sense of visual management through affordance theory. *Production Planning & Control: The Management of Operations, 28* (2), 142−157.

Biggs, J. (2014). *Management reports & dashboard best practice.* Target Dashboard. Retrieved from https://www.targetdashboard.com/site/kpi-dashboard-best-practice/default.aspx#KPI-Dashboard-Design (Accessed 31 May 2020).

Bilalis, N., Scrobelos, G., Antoniadis, A., & Koilouriotis, D. (2002). Visual factory: Basic principles and the 'zoning' approach. *International Journal of Production Research, 40* (15), 3575−3588.

Borja. (2016). *What is visual factory?* Arrizabalagauriarte Consulting. Retrieved from https://arrizabalagauriarte.com/en/una-fabrica-visual/ (Accessed January 2020).

Borkar, S., & Rai, A. (2019). A review paper on visual factory. *International Research Journal of Engineering and Technology (IRJET), 06*(07).

Bremer, M. (2015). *Quality progress-ASQ.* ASQ. Retrieved from http://asq.org/quality-progress/2015/03/lean/walk-the-line.html. (Accessed 31 May 2020).

Cichocka, M. (2018). A practical appliance of the 5S method in the work organization of the manufacturing company. *Journal of Positive Management, 9,* 41−54.

Cleary, B. A. (2018). *ASQ.* (American Society for Quality). WHAT IS AN AFFINITY DIAGRAM? Retrieved from https://asq.org/quality-resources/affinity. (Accessed 11 December 2018).

Cohen, L. (1988). Quality function deployment—An application perspective from digital equipment corporation. *National Productivity Review, 7*(3), 197−208.

Delisle, D. R., & Freiberg, V. (2014). Everything is 5S: A simple yet powerful lean improvement approach applied in a preadmission testing center. *The Quality Management Journal, 21*(4), 10−22.

Deming, E.W. (2012). *Speech by Dr. Deming to Japanese business leaders in 1950.* The Deming Institute. Retrieved from https://blog.deming.org/2012/11/speech-by-dr-Deming-to-Japanese-business-leaders-in-1950/#: ∼ :text = You%20don't%20hear%20the,power%2C%20forests %2C%20and%20railroads (Accessed December 2019).

Eaidgah, Y., Maki, A. A., Kurczewski, K., & Abdekhodaee, A. (2016). Visual management, performance management, and continuous improvement: A lean manufacturing approach. *International Journal of Lean Six Sigma, 7*(2), 187–210.

Eckerson, W. W. (2005). *Performance dashboards: Measuring, monitoring, and managing your business*. John Wiley & Sons, Inc.

Farris, J., Van Aken, E., Doolen, T., & Worley, J. (2009). Critical success factors for human resource outcomes in Kaizen events: An empirical study. *International Journal of Production Economics, 117*(1), 42–65.

Few, S. (2006). *Information dashboard design*. O'Reilly.

Galsworth, G. D. (1997). *Visual systems: Harnessing the power of visual workplace*. New York: AMACOM.

Galsworth, G. D. (2005). *Visual workplace: Visual thinking*. Portland, OR.: Visual-Lean Enterprise Press.

Galsworth, G. D. (2007). *Visual workplace visual thinking*. Boca Raton, FL: Taylor & Francis Group.

Galsworth, G. D. (2017). *Visual workplace: Visual thinking*. Boca Raton, FL: CRC Press.

Gapp, R. P., Fisher, R. J., & Kobayashi, K. (2008). Implementing 5S within a Japanese context: An integrated management system. *Management Decision, 46*.

Granowicz, G. (2011). Best of both worlds. *Quality Progress* (4), 48.

Greif, M. (1991). *The visual factory: Building participation through shared information*. Portland, OR: Productivity Press.

Harris, C., & Harris, R. (2008). *Lean connections: Making information flow efficiently and effectively*. New York: CRC Press.

Harris, C., & Harris, R. (2009). Lean decision making. *Manufacturing Engineering; Dearborn, 143*(3), 83–89.

Hirano, H. (1995). *5 Pillars of the visual workplace*. Cambridge, MA: Productivity Press.

Imai, M. (2012). *Gemba kaizen: A commonsense approach to a continuous improvement strategy* (2nd ed.). McGraw-Hill.

Jaca, C., Viles, E., Jurburg, D., & Tanco, M. (2014). Do companies with greater deployment of participation systems use visual management more extensively? An exploratory study. *International Journal of Production Research, 52*(6), 1755–1770.

Jana, P. (2010). *Yamazumi charts red, yellow and go!*. Apparel Resources Pvt. Ltd. Retrieved from https://in.apparelresources.com/business-news/manufacturing/yamazumi-charts-red-yellow-go/. (Accessed 29 May 2020).

Jansson, K. (2017). *A guide to successful Gemba walk*. Retrieved from https://blog.kainexus.com/improvement-disciplines/lean/gemba-walks/11-steps-to-an-effective-gemba-walk. (Accessed 31 May 2020).

Kattman, B., Corbin, T., Moore, L., & Walsh, L. (2012). Visual workplace practices positively impact business processes. *Benchmarking: An International Journal, 19*(3), 412–430.

Kelly, R. (2019). *The myths and truths of lean transformations: How to successfully make the transition from theory to effective deployment*. New York: NY Routledge/Productivity Press.

Khan, A. M., & Islam, M. M. (2013). Application of 5S system in the sample section of an apparel industry for smooth sample dispatch. *Research Journal of Management Sciences, 2*(7), 28–32.

Kirchner, M. (2013). *Two hours a day in the "gemba"*. Products Finishing. Retrieved from https://www.pfonline.com/columns/two-hours-a-day-in-the-gemba (Accessed January 2020).

Koch, C. (2018). *Introduction to information technology*. Edtech Press.

Koskela, L., Tezel, A., & Tzortzopoulos, P. (2018). Why visual management? In V. González (Ed.), *26th Annual conference of the international* (pp. 250–260). Chennai: Group for Lean Construction (IGLC).

Kumar, K. (2012). Steps for implementation of 5S. *International Journal of Management, IT and Engineering, 2*, 402–415.

Kunifuji, S. (2016). A Japanese problem solving approach: the KJ Ho Method. In A.M.J. Skulimowski, & J. Kacprzyk, *Knowledge, information and creativity support systems: recent trends, advances and solutions* (Vol. 364, pp. 165–170). Switzerland: Springer International Publishing.

Kurpjuweit, S., Reinerth, D., Schmidt, C. G., & Wagner, S. M. (2018). Implementing visual management for continuous improvement: Barriers, success factors and best practices. *International Journal of Production Research, 57*(17), 5574–5588.

Langstrand, J., & Drotz, E. (2015). The rhetoric and reality of lean: A multiple case study. *Total Quality Management & Business Excellence, 27*(3–4), 398–412.

Lean Enterprise Institute. (2008). *Lean Lexicon: A graphical glossary for lean thinkers.* Cambridge: Lean Enterprise Institute, Cop.

Lee, H. L. (2018). Big data and the innovation cycle. *Production and Operations Management, 27*(9), 1642–1646.

Liker, J. (2004). *The Toyota way: Fourteen management principles from the world's greatest manufacturer.* McGraw-Hill Education.

Marascu-Klein, V. (2015). The 5S lean method as a tool of industrial management performances. In *IOP conf. series: Materials science and engineering.* Brasov: IOP Publishing.

McDonalds Consulting Group. (2018). *What are the benefits of a Turtle Diagram?* The McDonalds Consulting Group. Retrieved from https://mcdcg.com/blog/quality/what-are-the-benefits-of-a-turtle-diagram/. (Accessed 30 May 2020).

McQuade, L. (2017). Gemba in the workplace. *Cost Management, 31*, 22–27.

Monden, Y. (1998). *Toyota production system: An integrated approach to just-in-time.* Norcross, GA: Engineering and Management Press.

Mridha, J. H. (2020). Contrivance of 5S system to effectuate higher productivity in apparel industries. *Global Journal of Researches in Engineering: J General Engineering, 20*, 21–28.

Nestle, M. (2013). Gemba is gold. *ASQ Six Sigma Forum Magazine, 13*(1), 32–36.

Niederstadt, J. (2014). *Kamishibai boards: A lean visual management system that supports layered audits.* Boca Raton, FL: CRC Press.

Norwood, E. P. (1931). *Ford Men and Methods Doubleday.* Doran.

NQA. (2017). *How important are turtle diagrams for an organization?* NQA. Retrieved from https://www.nqa.com/en-US/resources/blog/june-2017/turtle-diagrams. (Accessed 30 May 2020).

Oakland, J. (2014). *Total quality management and operational excellence: Text with cases.* Abingdon: Routledge.

OGTC. (2016). *Lean manufacturing in OGTC cluster.* New Delhi: OGTC.

Ohno, T. (1988). *Toyota production system: Beyond large-scale production.* Portland, OR: Productivity Press.

Ortiz, C. A., & Park, M. R. (2011). *Visual controls: Applying visual management to the factory.* New York: Productivity Press.

Osakue, E. E., & Smith, D. (2014). A 6S experience in a manufacturing facility. In *121st ASEE annual conference & exposition.*

Patil, A., & Tewari, A. (2014). *Implementation of lean tools in Trims Store.* Gandhinagar: National Institute of Fashion Technology.

Plain, C. (2007). Built an affinity for K-J method. *Quality Progress, 40*(3), 88.

Roser, C. (2015). *All about Andon.* AllAboutLean. Retrieved from https://www.allaboutlean.com/andon/. (Accessed May 2020).

Sabadka, D., Molnar, V., Fedorko, G., & Jachowicz, T. (2017). Optimization of production processes using the Yamazumi method. *Advances in Science and Technology Research Journal*, *11*, 175–182.

Saurin, T., Ribeiro, J., & Vidor, G. (2012). A framework for assessing poka-yoke devices. *Journal of Manufacturing Systems*, *31*(3), 358–366.

Schonberger, R. (1986). *World class manufacturing*. New York: The Free Press.

Scupin, R. (1997). The KJ method: A technique for analyzing data derived from Japanese ethnology. *Human Organization by Society for Applied Anthropology*, *56*.

Sharma, B., & Jha, G. (2014). *Quality improvement and enhanced workplace using the 5S—Survey cum case study approach*. New Delhi: National Institute of Fashion Technology.

Shingo, S. (1989). *A study of the Toyota production system from an industrial engineering viewpoint*. Portland, OR: Productivity Press.

Sorge, A., & van Witteloostuijn, A. (2004). The (non)sense of organizational change: An essay about universal management hypes, sick consultancy metaphors, and healthy organization theories. *Organization Studies*, *25*(7), 1205–1231.

Suzaki, K. (1993). *The new shop floor management: Empowering people for continuous improvement*. New York: The Free Press.

Tezel, A., Koskela, L., & Tzortzopoulos, P. (2016). Visual management in production management: A literature synthesis. *Journal of Manufacturing Technology Management*, *27* (6), 766–799.

Thompson, J. (2013). *Turtle diagram*. Concentric Global. Retrieved from https://www.concentricglobal.co/blog/turtle-diagrams. (Accessed 30 May 2020).

Thys, M. (2011). *Business excellence partners blog*. Business Excellence Partners. Retrieved from https://businessexcellencepartners.wordpress.com/. (Accessed March 2020).

Visco, D. (2015). *5S made easy: A step-by-step guide to implementing and sustaining your 5S program*. Productivity Press.

Willis, J. (2015). *The Andon cord*. IT Revolution. Retrieved from https://itrevolution.com/kata/ (Accessed 10 February).

Wolny, T. (2014). *Build a visual dashboard in 10 steps*. iSixSigma. Retrieved from https://www.isixsigma.com/tools-templates/metrics/build-a-visual-dashboard-in-10-steps/. (Accessed 31 May 2020).

Womack, J. (2011). *Gemba walks* (1st ed.). Lean Enterprise Institute, Inc.

Rapid setup

Chandrark Karekatti[1] and Chandrajith Wickramasinghe[2]
[1]Ananta Garments Ltd., Dhaka, Bangladesh, [2]Corrigo Consultancy Pvt. Ltd., Noida, India

6.1 Introduction

Before we go into the deeper details of rapid setup, let us understand the analogy of order quantity and setup.

Company A is dealing with working with some global buyers with relatively stable order quantities and no or minimal diversification in the product. The sewing lines are set for longer runs with huge inventories [in form of work-in-progress (WIP)] in the lines ensuring no idleness due to shortage of work. Profit margins are relatively low in such kind of functioning. Company B is working with some high-end fashion brands and dealing with lower order quantities with diversified products. The changeovers are frequent, and there is always a process of setup going on in one or other lines almost every day. Profit margins are relatively higher in this way of operations. Here, when we say smaller order quantity, medium order quantity, and large order quantity, it is referred to as orders with less than 500 units, between 501 and 5000 units, and more than 5000 units, respectively.

Whenever an Industrial engineer, Production Manager, or the Factory manager is asked about the biggest challenge of his/her work environment, the responses primarily revolve around the challenges of handling smaller order quantities and tackling the frequent changeovers (due to diversification in the products). And for the very reason, people want to go work with the larger order quantities, as it involves lesser headache. People generally have an opinion that the setup time (ST) is more or less the same, irrespective of the size of the order. It means, ST (e.g., 4 hours) shall remain the same for 500 units order as well as for 5000 units order. If the order size is bigger, the ST per unit shall be very small, and it may have a negligible impact. Even if the STs are higher (in apparel manufacturing, there are instances where a typical setup takes even 2 working days or 16 hours), it does not impact much due to the large order quantity. On the other hand, the same (ST per unit) may be high (substantial) for a small order quantity. Several times, it is observed that the ST is more than that of the processing time (sewing time is referred here), or the style is so small that the peak of the learning curve in sewing line is not achieved and there is time for a fresh setup for the next style, thus the line was never stable. Most probably, this is the reason, where larger order quantities with fewer diversifications are preferred over diversified smaller orders. It is interesting to note that in the traditional approach of manufacturing, to avoid the production loss, efforts were put

Lean Tools in Apparel Manufacturing. DOI: https://doi.org/10.1016/B978-0-12-819426-3.00012-6
© 2021 Elsevier Ltd. All rights reserved.

toward minimizing the frequency of changeovers, but in modern times, smaller order quantities and product diversities demand frequent changeovers to be responsive.

Another myth is that because the order quantity is large (high volume), there have to be larger lots as well (in the context of keeping a higher level of inventories in form of WIP). According to Shingo (1985), inventory is treated as a necessary evil. This badly affects the flow and the throughput time and creates different types of Muda in the process. Ideally, in such cases, larger order quantities should be handled in smaller lots for effective control. The creation of internal factory PO (purchase orders) based on cut-wise (one cut as one lay) or internal PO based on size and color are some of the practices followed by the apparel manufacturers who understand the importance of handling smaller lots.

Here, it is assumed (by default) that ST is a constant...!!! If we see the other side of the coin, that an improvement in ST may drastically impact the line performance and may enhance our capabilities to handle smaller orders more efficiently and effectively. The ST and the lot size are two different and independent aspects. With a reduced ST (by making setup process efficient, effective, and standardized), style changeovers can be handled at ease resulting in the smooth run of the production order irrespective of order quantities, which may eventually lead to Just-in-Time. This can make life easy for several production personnel who struggle almost every day whenever there is a style change.

The other important aspect to think in this area is questioning if the setup (for a new style) is a value-added activity? The answer to this is NO, which means it is a nonvalue-added activity as the buyer has nothing to do with how we are executing the orders. But we can see that setup is a necessary activity while any style changeover, hence it is a necessary nonvalue-added activity (NNVA). There are many operations in the manufacturing environment, which do not add value to the product but are indispensable to the production. Adaptation to change becomes critical to meet the production requirements depending on the product order quantity and product diversification (Kiran, 2016). Frederick Taylor, who is also popularly known as the father of Industrial Engineering, referred changeover as "unproductive" in his famous book "Shop Management" in year 1911. Further, a need for precise and repetitive changeover or set work has been discussed. Similarly, Henry Ford in his autobiography found changeover as an unproductive work and advocated for never performing changeovers (Henry, 2013). As the setup is an NNVA activity, to become lean, we must put efforts to eliminate or reduce the ST. Ideally, there should not be any setup required whenever there is a style changeover. Single Minute Exchange of Dies (SMED) and One-Touch Exchange of Dies (OTED) are some means of achieving faster or rapid setup. There are success stories reported in the area of automobile manufacturing, where the STs were reduced drastically from hours (4.0 hours) to 1.5 hours to finally up to 3 minutes (Shingo, 1985; Womack, Jones, & Roos, 2007). SMED is a set of techniques in lean manufacturing aiming for rapid setup. SMED enhances line flexibility as it improves the machine changeover (De la Vega-Rodríguez, Baez-Lopez, Flores, Tlapa, & Alvarado-Iniesta, 2018).

According to Shingo (1985), SMED is not merely a technique but a way of thinking about the production itself. A focus on understanding theory and practice is important to get the best results. In SMED, understanding "why" is more important than the "know-how" (Shingo, 1985). The setup process may be simplified logically with a categorization of Common setup and Similar setup aspects. Irrespective of the diversification in the styles to be produced it shares some common features which result in the usage of some common machine settings, tools, and parts. The products which share common setup features should be identified. In the context for apparel manufacturing, for example in a jean manufacturing plant, the inseam to be done using a Feed-of-Arm machine irrespective of size, color, or the buyers. Similarly, there are good chances that the side seam is done using a five thread Overlock machine. Also, there may be some style groups that require the same type of waist-band attachment work-aid may be with the same Stitch per Inch (SPI). Another category is similar setup, where the styles share similar requirements while style changeover; however, some elements differ due to product variation. For example, a jean style may have different waist-band construction, a different back pocket profile, or the pockets having flaps, etc.; in such situations, we need to categorize such styles separately and prepare for setup accordingly.

6.2 Tool changeover

Godina, Pimentel, Silva, and Matias (2018) discuss the latest trends in SMED while reviewing the literature in this area. The researchers emphasized the need of adopting SMED practices to reduce the nonvalue-added activities (such as setup) and machine downtime for enhanced quality, cost, variety, and delivery deadlines in the competitive manufacturing environment (Godina et al., 2018). Rapid tooling is a critical economical factory for a company's profitability. Production time and product diversity are two factors responsible for making rapid tooling critical to the success of a company. As frequent interruptions (may be due to style changeover, which is a planned breakdown or may be due to unplanned breakdowns) cause Muda and are obstacles in achieving higher levels of efficiency (Godina et al., 2018).

6.2.1 What is tool changeover time

A tool changeover time (COT) is a time elapsed between the last conforming output of the previous style, and the first conforming output of the next style, at the maximum output rate (when the line is at its peak) (Shingo, 1985). It is a process of making a machine (which is producing a particular product type) ready to produce another type of product. According to Henry (2013), "*Changeover is the total process of converting a machine, line, or process from running one product to another.*" A rapid changeover is an objective of SMED. A changeover can be performed quickly by focussing on transporting, positioning, and fixing a tool. Achieving the tool changeover in record time is the ultimate goal, SMED targets for changeover within 10 minutes (that is a single-digit minute exchange). Completing the changeover between the manufacture of different products quickly and precisely is one of the components of modern manufacturing

(Tu, Vonderembse, & Ragu-Nathan, 2004), (Spencer & Guide, 1995). That is the reason, changeover is also called Quick changeover (QCO) as well. However, according to Henry (2013), the word "Quick" may result counterproductive, as people involved in the changeover may take in the context of doing or performing the same activities or process but in a faster or quicker mode. This may result in errors or inefficiencies in the changeover process. Eventually, the outcome may be not as expected, not sustainable, and annoying. According to Feld (2001), the primary objective of applying SMED is to enhance flexibility in the process. The focus is not on reducing the ST but to achieve the ability to perform more setups in the same amount of time (Feld, 2001). Given the same, it is suggested to use the word "lean" instead of on quick, and the QCO should be called Lean changeover (Henry, 2013). We can recall the changeover activities in the Formula 1 racing cars, the fastest tire change in the history of Formula 1 is just 1.82 seconds, with all four tires changed for RB15 at the Brazilian Grand Prix (Perkins & Silvestro, 2019). Such interventions are not only quick but also methodological and precise.

6.2.2 Design for changeover

The changeover improvement techniques involve equipment modification (hardware modification) through the application of knowledge and control of information, and the management, skill enhancement, and motivation of personnel. Design for the changeover is popularly known as DFC is the methodology to carry out the changeover efficiently and effectively. Reik, McIntosh, Owen, Mileham, and Culley (2006) advocate the need of rapid changeover in time-based manufacturing environment where companies have to respond swiftly (in an economical manner) to the dynamic demand patterns. DFC transforms a manufacturing process into a highly flexible and responsive system (Reik et al., 2006). According to McIntosh, Culley, Gest, Mileham, and Owen (1996), there are two primary improvement drives in a changeover (McIntosh et al., 1996), as:

> Organizational-led changeover: With the involvement of skilled personnel, where people perform changeover activities in a disciplined manner using the appropriate tools. It also involves alteration in the sequence of activities performed while changeover.
> Design-led changeover: With a focus on physically altering the design of the manufacturing equipment used in the changeover. In design-led changeover, sometimes the design of the product itself may be beneficially altered.

Reik, McIntosh, Owen, Mileham, and Culley (2005) propose changeover improvement under the "4P" model of People, Practice, Process, and Product (refer to Fig. 6.1). These four attributes influence changeover performance, and there is a potential for improvement in each of these. People involved in the changeover process may be provided with training to make them more skilled and competent. They may be given a better orientation and motivational training to carry out the work. The practice and process may be reengineered, revised, or re-sequenced for better results. Also, the tools and equipment used in changeover, work-aids, machines, as well as the product itself may be modified to make the changeover smooth and faster (Reik et al., 2005).

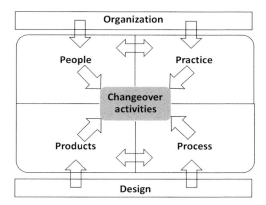

Figure 6.1 4P model for changeover improvement.

Henry (2013) classifies changeover tasks or activities into two domains, as Mechanical and Operational. The mechanical tasks are directly the part of a changeover process, whereas the operational tasks are in supporting roles (Henry, 2013). Such tasks include documentation work, material or spare parts reconciliation, after changeover activities, an inspection of the changeover, planning, and execution of the next changeover. An improvement in the operational tasks may result in a significant impact on reducing the COT. In the absence of efficient and effective execution of operational tasks, the mechanical tasks may not result in time-saving not matter how well and timely these are performed.

- Improve the design of manufacturing machine and equipment: Investigate the use of clamps instead of fasteners wherever feasible. For example, use magnetic edge guides in place of screw fastener type.
- Improve the design of the manufactured product: Exercise the scope of changes in product construction parameters in consultation with buyers to facilitate a reduction in setup time. For example, the margin for stripe matching at sleeve joint for yard dyed knits, increasing/decreasing SPI by a unit as the specified SPI cannot be produced on some machines, or to increase the width of J stitch so that it can be easily performed on J Stitch machine. However, since such changes require buyer's consent, which often is a time-consuming process, they have limited application in the apparel industry.

It is observed that in comparison to design improvement, organizational improvement is cheaper, easy to implement, and produces significant reductions in COT.

6.2.3 Changeover methodology

Changeover may be divided into three components, Cleanup, Setup, and Start-up. Cleanup is related to the removal (or clearing the space) of the material from the site of the previous changeover. Setup (most of the time, this word is used interchangeably with changeover) is referred to the physical change (modification or setting up) of the machinery or equipment and making it ready to produce the next product. Start-up is the last phase of the changeover, also referred to as ramp-up or run-up. This is the phase when the machine or line has finally produced the next product and the entire process is settled-down and the machine or line can work at a normal pace.

Shingo (1985) proposed a concept of SMED. The SMED concept got such high popularity across the industries all over the globe, that SMED is now known as "Changeover improvement." According to Herrmann et al. (2004), SMED is established as the most retrospective changeover improvement tool (Herrmann et al., 2004).

Shingo (1985) recommends a five-step methodology for QCO, as indicated in Fig. 6.2.

Close analysis of changeover process shows that changeover process comprises two sets of activities namely internal activities and external activities (refer Fig. 6.3 for a graphical depiction of the changeover activities).

Internal activities: The tasks or activities during changeover for which the machine needs to be stopped or kept idle.

These are activities which are performed when the sewing machines in the line are stopped, and no product is being manufactured in the line. These are activities performed for sewing machines that are already in the line, for example, gauge set, SPI change, and folder adjustment, as per the requirement of a new style. The time taken to complete all the internal activities constitutes the setup phase.

External activities: The tasks or activities for which the machine does not need to be stopped or kept idle. Such activities are generally performed before the changeover (Shingo, 1985).

These are the activities that can be performed outside the line without disturbing the production inline. For example, presetting of a sewing machine for a new style or early loading of cut parts without disturbing the production of running style.

Fig. 6.4 depicts the Shingo's SMED methodology for an improved changeover performance, in three stages as (1) Separate internal and external setup, (2) Convert internal into the external setup, and (3) Streamline all aspects of the setup operation (Shingo, 1985, McIntosh, Owen, Culley, & Mileham, 2007).

Henry (2013) proposed ESEE (pronounced as Easy) model for the changeover. Here, the alphabets "E, S, E, and E" stand for Eliminate, Simplify, Externalize, and Execute.

Eliminate (E): Elimination of unnecessary and nonvalue-added tasks or activities.
Simplify (S): Simplification of the process or activities performed while changeover.
Externalize (E): Taking out the tasks or activities which can be performed without stopping the machine or can be performed while production is ON.
Execute (E): Execution of performing changeover tasks repeatedly with minimum variations (Henry, 2013). This shall bring efficiency and precision in the changeover process through skill development.

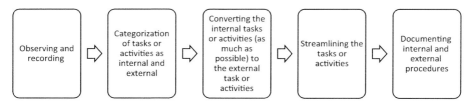

Figure 6.2 SMED steps.

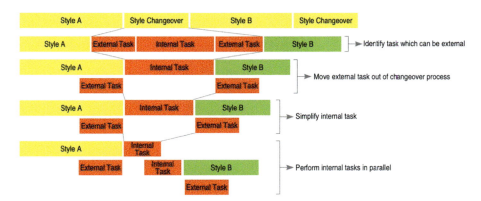

Figure 6.3 Rapid Setup changeover process, identifying internal and external activities (King, 2009).

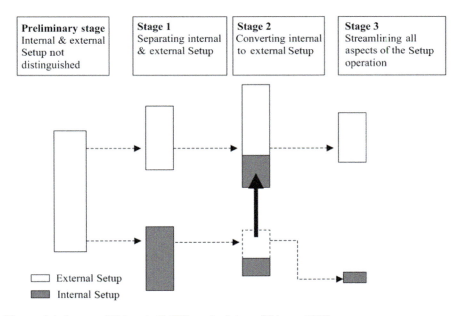

Figure 6.4 Stages of Shingo's SMED methodology (Shingo, 1985).

McIntosh et al. (2007) reinterpreted Shingo's SMED methodology in the context of two fundamental mechanisms for improvement of the changeover. According to McIntosh et al. (2007), the first fundamental is a better allocation of tasks to the resources necessary to conduct, and here, the tasks remain unchanged. The second mechanism is seeking structural change to existing tasks and enabling them to be completed quickly (McIntosh et al., 2007).

6.2.4 Prerequisites for SMED implementation

For a successful SMED implementation, it is vital to have a conducive manufacturing environment. The prerequisite requirement for a successful SMED implementation has been discussed and proposed as SMED-ZERO by Moxham and Greatbanks (2001). The researchers recommend SMED-ZERO as an additional stage to Shingo's three-stage SMED methodology and place it before Stage 1 (of Shingo's methodology) as an improved version of the same. The research was conducted at a small to medium-sized textile enterprise based in Scotland. It is a fact that SMED is a key lean tool, hence it is also a part of the continuous improvement. Given the same, the SMED implementation can be made more effective and efficient, if the implementation environment should be seen in the context of lean interventions. The recommended prerequisites (SMED-ZERO) were categorized into four areas as mentioned in Table 6.1 (Moxham & Greatbanks, 2001).

On similar lines, Silva and Filho (2019) based on their study of 130 published research articles related to SMED (from 14 leading publications) of studies conducted in 24 years (from the year 1995 to the year 2018) across 28 countries, confirmed that the SMED technique should be combined with nine lean tools namely 5S, Standardized work, Kaizen, OEE, TPM, Poka-yoke, VSM, A3 methodology, and Visual management (Silva & Filho, 2019).

6.3 Style changeover in apparel manufacturing

Changeover: Henry (2013) defines changeover as *"the total process of converting a machine, line, or process from running one product to another."*

According to Henry (2013), *"Changeover time is the total elapsed time from the last unit of good production at normal speed and efficiency of the preceding run to the first unit of good production of the succeeding run at normal speed and efficiency."*

To understand rapid setup implementation, it is essential to understand different phases of the style changeover process. A changeover is the set of activities/ processes of changing between the manufacture of one style and the manufacture of an alternative style to the point of meeting specified production and quality rates. Primarily, it involves three phases as rundown, setup, and run-up as mentioned in Fig. 6.5 (McIntosh, Culley, Mileham, & Owen, 2001).

Rundown phase: This phase is the interval when production of old style is complete but set up for new style cannot be started as some of the pieces of the old product are still in line for alterations or waiting for some parts/trims. In this phase feeding is ceased, production of old style is finished on most of the workstations. However, few pieces of old style are in line for alterations and/or waiting for replacement of damaged/missing parts or trims. For example, damaged placket, missing collar, etc. Rundown typically stretches till all the pieces of old style are out from the line.

Rundown time (RDT): RDT is referred to the time when the feeding of the existing style is stopped in the line, and it only produces the remaining pieces in the line. A decline in the

Table 6.1 Prerequisites for SMED implementation (SMED-ZERO).

Area	Subcategory	Brief description
Teamwork approach to communication	Management commitment • Employee role • Employee meeting • Dedicated meeting-place	Promoting the active participation from all the concerned personnel right from the top management to the people going to carry out the improvement on the floor. This enhances the sense of belongingness as well as the commitment to the work.
Visual factory control	Visual factory control • Effective production systems	Promoting communication through visual management as it reduces the dependency on verbal communication.
Performance measurement	Performance measurement • Vision	It is vital to define the target and the means to achieve the same. To understand the scope of the intervention, it is critical to assess the current situation or stand. This also guides in setting up the targets for improvement.
Kaizen to simplify both assessment and measurement	Communication tools • Internal communication of SMED project progress • Continued management support	Promoting the variety of communication tools and techniques to enhance problem-solving and experimentation skills of the personnel involved in changeover (SMED implementation).

Source: Adapted from Moxham, C., & Greatbanks, R. (2001). Prerequisites for the implementation of the SMED methodology. *International Journal of Quality & Reliability Management, 18*(4), 404–414.

output is witnessed while run down and it makes the machines and workstations (the initial workstations) available for the setup for the new style.

Setup phase: It is the phase in which no manufacturing occurs. In this phase, machines and equipment are adjusted as per the requirements of the new style. In this phase, there is no output from the line.

Setup time: ST is the time calculated from the time of loading of the first piece of a new style on the very first workstation in the line till the time of output of the same piece from the last workstation of the line conforming required quality. Batch setting time (BST) also includes preparation time for a new style like process machine setting, operator setting, and initial line balancing. This is also referred to as BST and throughput time as well in some cases.

Style changeover time: It reflects the time difference between the time of output of the first piece of the new style and the time of output of the last piece of the previous style from the last workstation the line. Generally, COT directly depends on BST, but there may be instances when the COT is lesser in comparison to a higher BST (Makhija, 2012).

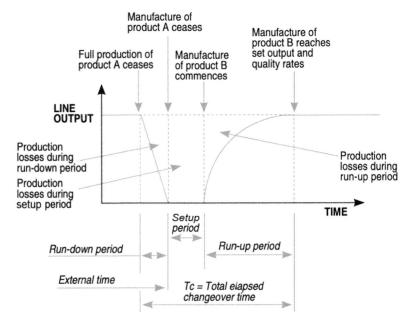

Figure 6.5 Phases of changeover (McIntosh et al., 2001).

Run-up phase: This phase starts when production for the new style is commenced and continues until consistent output at full capacity occurs. The run-up period extends until each operation (including new operations) reaches the specified production and quality rates consistently.

In apparel manufacturing, a production drop is frequently witnessed while there is a style changeover. Such a production drop is experienced in all the processes of manufacturing. For orders with smaller order quantities and diversified products, the impact of such product loss is significant on the lead time. This phenomenon is termed as start-up loss or changeover loss or shift-in/shift-out loss (Jana & Tiwari, 2018). A traditional style changeover process has been graphically depicted in Fig. 6.6.

6.4 Single Minute Exchange of Die in apparel manufacturing

SMED (along with 5S, VSM, and Visual management) is one of the most applied lean tools in apparel manufacturing. The shorter product life cycles, diversified products, and smaller order quantities have forced the apparel manufacturers for adopting QCO, and of course, SMED is the most appropriate tool to achieve the same.

Rapid setup is not an isolated process, where a specially trained workforce carries out the changeover activities. It requires s focus on Organizational Improvement and

Rapid setup

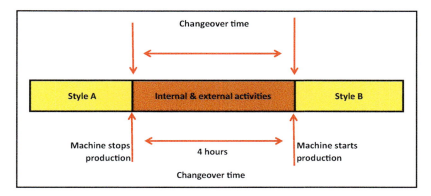

Figure 6.6 Traditional style changeover process.

Design Improvement. The following are the six techniques that should be considered in implementing rapid setup.

Separate internal from external setup operations (tasks)

1. Convert internal to external setup tasks
2. Standardize function
3. Adopt parallel operations
4. Use functional clamps or eliminate fasteners altogether
5. Eliminate adjustments

Refer to Fig. 6.7, step 1–6 focuses on organizational improvement, whereas 2 steps (related to improvement in design of machine/equipment and improvement in product design) focus on design improvement.

The limited scope of design improvement makes rapid setup implementation in the apparel industry to rely mainly on Organizational Improvement.

6.4.1 Organizational improvement

Following are a few examples of initiating Organizational Improvement in apparel Industry.

6.4.1.1 Convert internal task to external

- *Identifying internal and external activities*: Machine plan should be prepared by comparing the operation bulletin of the old and new styles. Machine plan should identify the following parameters (Refer to Table 6.2 for an example):
  1. Operation Type
  2. Machine Type (Segregated as internal/external)
  3. Machine Number
  4. Stitch per Inch (SPI)
  5. Required Settings

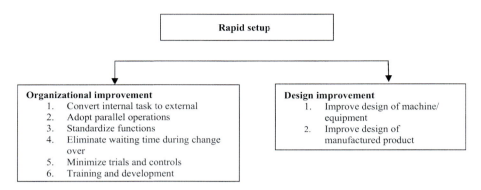

Figure 6.7 Organizational improvement and design improvement.

Table 6.2 Machine plan segregating internal and external tasks.

Sr.No.	Operation	Internal Elements					External Elements				
		Machine Type	Attachments	MC No.	SPI	Setting	Machine Type	Attachments	MC No.	SPI	Setting
	New Style	Old Style									
	Preparatory										
1	Make and cut beltloop	Flat lock 2N	F 339 Belt loop folder 1/4"	AMLFL18	14	No Change					
2	Overlock Zipper Fly & Facing	3THOL Pneumatic		AMLOL01	10	Change SPI Change Gauge					
3	Make button hole on Zip fly	Eyelet B/Hole		AMLBH02	10	Change SPI					
4	Sew Pintucks on Pocket Bag						SNLS w/UBT	T Guide	AMLSN06	14	
5	Hem Pocketbag mouth making Pleats	SNLS w/UBT	T Guide	AMLSN09	10	Change SPI					
6	Runstitch Pocket flap						SNLS w/UBT		AMLSN15	14	
7	Mark for Snap Buttons						Helper Table	FGP			
8	Attach snap button to pocket flap and pocket Bag						Snap Attaching M/C		AMLSA01		
9	Turn and top Stitch Pocket flap						SNLS w/UBT	CR 1/4"	AMLSN11	14	
10	Attach facing to front slant pocket's bag	SNLS w/UBT		AMLSN21	10	Change SPI					
11	Close pocket Bag	SNLS w/UBT		AMLSN03	10	Change SPI					
12	Turn Pocket Bag						Helper Table	Metal Frame			
13	Topstitch pocket bag	SNLS w/UBT	CR 1/4"	AMLSN28	10	change SPI CR1/8"					
14	Make dart @ Front Panel	SNLS w/UBT		AMLSN34	14	No Change					
15	Press Moon Patch						Iron Table				
16	Attach moon Patch to back with Runstitch @ Seat	SNLS w/UBT	C/L 1/4"	AMLSN76		Change SPI CR 1/8"					
17	Runstitch Pocket flap						SNLS w/UBT		AMLSN32	14	
18	Turn and T/S Pocket flap						SNLS w/UBT		AMLSN45	14	
19	O/L @ Pocket flap						3THOL Pneumatic		AMLOL03	14	
20	Thread Trimming						Helper Table				

Operations of old style and new style should be compared for the segregation of internal and external elements. If operation a for a new style is already being performed in the old-style then the element is internal, thus the machine type should be filled as an internal element. But if a machine has to be arranged from outside or a machine from the old style will become idle but can be utilized at some other operation in the new style, then that should also be filled as an external element. Along with machine type, attachment details should also be filled by referring to the operation bulletin.

- *Convert internal task to external*: Since reallocating tasks by converting internal activities to external activities significantly reduces set up time, opportunities for such conversions should be identified. Example, loading of parts before the end of old-style or template preparation in advance.

Rapid setup

- *Investigate the use of automates/centralized work stations*: Investigate opportunities to convert internal elements to external by use of automates in the form of centralized work stations. Depending upon the product mix, explore the use of automates like continuous bottom hemming machine and auto front placket stitching for knits Tee/Polo shirts (convert internal to external elements). The use of fusing machines in the cutting section rather than in the sewing line can help to convert the internal element to external. For trousers/ bottoms, explore opportunities to employ centralized automats for back pocket setting, auto belt loop setter, seam serging, j-stitch, loop/tape cutting, elastic cutting, pocket hemming, welt pocket making, etc. For Shirts/jacket, explore the use of auto sleeve setter, front pocket setting, pattern stitch machines, sewing unit for run stitching and centralized auto trimming of cuffs, etc.
- *External machine setting*: External machines should be set by the mechanic according to the machine layout plan in advance (in the run-down phase). Typically, machine setting should include machine parameters like gauge change, Folder/Presser foot change, Needle change/removal, SPI adjustment, Jigs/template preparation, Programming, etc as per the requirement of a new style. Machine setting should be approved for quality by Quality supervisor and machine marked with 'Setting OK' tag.

6.4.1.2   Adopt parallel operations

Set up parallel operations example, set-up of operations in different sections simultaneously (front and back sections). Note that the tasks must not be dependent upon each other. Usually, the creation of parallel tasks requires more manpower, so increased personnel must be considered.

6.4.1.3   Standardize functions

Formalize standardized operation instructions for change over process. Example develop standard operating procedures defining the accountable person and timelines.

6.4.1.4   Eliminate waiting time during changeover

Concentrate efforts to eliminate waiting time during the changeover process. A simple reallocation or reorganization of group tasks may help reduce this waiting time. For example, performing bottom hem operation before waistband attaches in denim for starting few bundles, as set up of waistband operation consumes more time.

6.4.1.5   Minimize trials and controls

Trials and controls can be minimized by formalizing procedures and utilizing standards. Checklists can be used, and technicians can be held accountable for key procedures by signing off on the checklists.

6.4.1.6   Training and development

Traditional changeover procedure can be modified by broadening the scope of a technician's abilities through training programs. This will allow a technician to perform tasks that he or she was not previously allowed to do—and thus remove the dependency on another technician for completion of a task.

6.4.2 Design improvement

6.4.2.1 Improve design of manufacturing machine & equipment

Investigate use of clamps instead of fasteners where ever feasible. Example use magnetic edge guides in place of screw fastener type.

6.4.2.2 Improve design of manufactured product

Exercise the scope of changes in product construction parameters in consultation with buyer to facilitate reduction in set up time. For example, margin for strip matching at sleeve joint for yard dyed knits, increasing/ decreasing SPI by a unit as the specified SPI cannot be produced on some machines or to increase the width of J stitch so that it can be easily performed on J Stitch machine. However, since such changes require buyer's consent, which often is a time-consuming process, they have limited application in apparel industry.

It is observed that in comparison to design improvement, organizational improvement is cheaper, easy to implement, and produces significant reductions in changeover time.

6.4.3 SMED examples in apparel manufacturing

If we compare the original SMED example (Shingo, 1985) with apparel manufacturing functions, jig changing in sewing automats will be the closest parallel. From the screw-fixed jigs in past to the quick-change, fasteners now made the changing of jig in sewing automats a perfect SMED example. Although the true SMED originated from a machine fixture changeover situation, necessitated by the change of specifications of the product from one style to another, in an apparel manufacturing scenario, this may be style changeover (cut pieces and accessories loading) in a sewing line, operation change in a sewing machine (change of attachments and jigs), starting of a new shift in a factory (allocation of sewing operators based on attendance), and so on. The following are some examples of SMED practices practiced in apparel manufacturing.

Presetting-up machine: It is observed that generally, around 15%−20% of operations in a line need different machines during style changeover (within similar product categories). In such a scenario the factory needs to keep 15%−20% extra machines. Before the style changeover, the extra sets of machines are readied with appropriate presser foot, attachments, and work aids based on the operation bulletin. During the style changeover, the respective old machines are pulled out from the line and new machines are pushed in and operators can start work with minimal COT. Here, the internal activity of the machine setup for the new style is "externalized" while the past style was going on.

Pretraining of the operator: Some critical or new operations (in a new style) may require some training for the operator. So, a couple of days before the style changeover, pull out the designated operator from the line and train her on the new operation a few hours a day. During this training period, the reduced output in the current operation is compensated by other available operators/floater or QCO team members. Once the style changeover takes place, these trained operators can achieve target output in the new operation without any learning curve. Here, the erstwhile internal activity, "the learning curve in new operation" is externalized.

Fast-changing: Bra cup molding machines in a lingerie manufacturing factory operate at $180°C-200°C$. Whenever there is a need to change the size or shape of the mold, first the mold has to cool down and then change the mold (by loosening and disengaging the old mold and engaging and tightening the new mold) and then heat it. This changeover of mold typically can take $40-50$ minutes. But by changing the mold fixing method, now mold can be removed when it is still hot and move specially designed trolley, and new mold also can be preheated and move back to the machine using the same method. So total changeover can be reduced to a maximum of $6-8$ minutes. Here, the cooling and heating of mold, which was internal activity, being converted to external activity.

Preattendance of workers: Every day in the morning when shift start, line balancing to be done due to absenteeism. Usually, this takes a minimum of $10-15$ minutes per day. If you are a 1000 machine factory, you lost 15,000 minutes per day. But if you can design a system (APP) for employees to confirm their attendance at least 40 minutes before shift start, line managers can do the allocation of operators for line balancing before operators reach and that can be communicated to operators also before they reach to the workplace. If your company has a transport system, then the attendance system can be installed in the bus and real-time data can be given to the production/IE team for line balancing. There are garment factories in India and Sri Lanka which records the worker's attendance once they boarded the bus, thus allowing the line manager to decide on the worker allocation beforehand. Here, the "daily operator allocation" activity, which usually takes place after the worker reaches the factory, is being made simultaneous with their bus journey duration.

Embroidery is a value-added process in the textile/apparel industry with a low-value-added ratio, due to considerably high feeding or COT. Even though the speed of the embroidery machines has been increased considerably by the manufacturer to increase productivity, the machine utilization is generally below 50%, adding to the cost and also bringing down productivity. Further, only one design can be stitched at a time, which limits flexibility. In a traditional embroidery machine, the frame on, which the fabric is mounted, can only run in a single batch. On completion, the machine has to be stopped, and the fabric panels taken off the frames, and the threads beneath the embroidery cut or finished; during this whole process, the embroidery machine is out of operation and it does not matter which machine is being used or at what speed it embroiders. The SMED intervention enables attaching an additional set of embroidery frames on to the $X-Y$ direction motor's shaft, that is, in place of a single row of frames, now two rows of embroidery frames are fixed at the same place. The simple intervention doubles the productivity by allowing the operator to make ready one set, whereas the other set is in process. The same innovation has successfully experimented at KOC Industries Ltd. (TJ Apparel) in Ludhiana (Apparel Resources, 2012). The target was a minimum 25% increase in production without compromising on quality standards. At the factory, the cut panels were being put into a frame, and then, the frame was fixed into the machine to do the embroidery. Before the SMED, the total cycle time was 5.35 minutes for one round of production for 1700 stitches design in 12 head machines with a production of 135 pieces per hour. After the kaizen, the total cycle time increased to 6.09 minutes for one round of production for 3250 stitches design in 12 head machine with increased production of 240 pieces which translate to a production increase of 77% (Apparel Resources, 2012).

The above examples are a mix of interventions that involves a change of style as well as a simple batch loading process. Other examples like bobbin prefilling while sewing operation in progress, automatic bobbin changing process in an embroidery machine, etc. are also often cited as examples of SMED applications in apparel manufacturing. While the technology to enable such fast changeover to include the use of clamps instead of fasteners, use magnetic edge guides or snap fasteners in place of screw fastener, etc. Other interventions include the scope of changes in product construction parameters in consultation with the buyer to facilitate a reduction in setup time. For example, the margin for stripe matching at sleeve joint for yard dyed knits, increasing/decreasing SPI by a unit as the specified SPI cannot be produced on some machines or to increase the width of J stitch so that it can be easily performed on J Stitch machine. However, since such changes require buyer's consent, which often is a time-consuming process, they have limited application in the apparel industry. It is observed that in comparison to design improvement, organizational improvement is cheaper, easy to implement, and produces significant reductions in COT.

Kordoghli and Moussa (2013) have studied COT and different types of wastes on the sewing floor. The study was conducted on two large apparel manufacturing setups (with more than 1000 employees and 10 sewing lines in each plant) producing jeans pants. The COT was calculated from the time of output the of last good piece of the previous style and the time of output of the first good piece of the new style. Here, the start-up time was not considered in the COT due to the practical difficulty in distinguishing start-up failures and other failures. The end of the COT was considered at the time of output of the fifth good piece from the line. As the study was conducted on the assembly line system, the COT was calculated for every workstation. The COT for the sewing line was derived by summing up the COT of all the workstations. The COT observed for the first work station at manufacturing setup 1 and manufacturing setup 2 was 50 and 25 minutes, respectively. The changeover values observed for the entire sewing lines at manufacturing setup 1 and manufacturing setup 2 were 3953 and 715 minutes, respectively (Kordoghli & Moussa, 2013). This gives an indication of huge changeover losses ranging between 1.49 and 8.23 days as average COT per line considering 480 minutes (8.00 hours) working per day.

Table 6.3 Categorization of internal and external activities.

Internal activities	External activities
• Machine overhauling • Machine setting like Hook set timing change, feed timing, etc. • Template making • Layout change (machine movement) • Operator waiting for pieces • Operator waiting for sitting instruction • Operator waiting for machine positioning • Operator waiting for the mechanic to come for initial setting • Operator waiting for sewing part accessories like button, sewing thread • Required machine not found at the correct time/extracted from a different line • Changing the attachment • Operator waiting for the previous style to clear off from the line • Supervisor searching a for required operator	• Operator training • Mock sample preparation • Online machine adjustment like tension, SPI (Stitch per Inch), etc. • Changing attachment (small setting) • Setting the required quality standard

Makhija (2012) discusses the changeover in apparel manufacturing with the batch setting process at 10 sewing lines at five different apparel manufacturing organizations handling small order quantities with frequent changeover. The study aimed to provide solutions for rapid changeover. The study was undertaken for 6 months to observe changeover for different types of products including woven products such as formal shirts, casual shirts, trousers, chinos, skirts, knitted products including ladies' top, polo-neck, and crew-neck T-shirts. It was observed that approximately 60% of the time was consumed in the line layout change (20%) and adjusting machine settings (40%). Remaining around 40% of the time was consumed in activities such as quality issues, template making, and nonavailability of resources. Following Shingo's SMED methodology, the identification of internal and external activities was done as indicated in Table 6.3.

After the categorization of Internal and External activities, efforts were put toward converting the maximum internal activities into external activity to achieve the best results, and the entire activities of the changeover process (refer Table 6.4) were suggested in the context of three phases as, (1) Preparation (before batch setting), (2) During batch setting, and (3) After batch setting.

While implementing SMED, it was observed that 10 activities out of a total of 13 internal activities (except for machine overhauling, machine setting like hook set timing change, feed timing, etc., and line layout change) could be converted into external activities. This resulted in COT reduced in the range of 4−48 hours depending on the product complexity (Makhija, 2012).

Table 6.4 Activities for an effective and efficient changeover in apparel manufacturing.

Preparation activities—before batch setting	Activities—during batch setting	Activities—after batch setting
• Allocation of the new style to the line where minimum changes/modifications are required product-wise lines categorization. • Preproduction meeting with all the concerned persons especially the batch Supervisor, Quality Control person, and the changeover team. • Strict adherence to the changeover checklist. • Changeover team, Production manager, and Floor in charge to discuss weekly line load plan. • Ensuring the availability of all the trims and accessories at the time of batch setting. • Ensuring the availability of a skilled workforce as per the skill matrix as well as the training of the operators for critical operations if required. • Ensuring the machine availability (along with folders and work-aids) and setting the machines for critical operations before the style changeover. • Ensuring all the templates to be ready and available for production people. • Ensuring operation Bulletin (with details of the machine, folders, work-aid, etc.) and line layout planning for the style to be loaded.	• Strict adherence to the standard work methods (as suggested by Industrial Engineer or the Work-study expert). • Availability of required trims & templates at respective workstations. • The spare machines should be allotted before the start of the process to stimulate offline batch setting • Ensuring that all the operators would sit at their assigned workstation (according to skill allocation) without losing time. • Ensuring quality output at each of the workstations.	• Effective monitoring of production at regular intervals until it reaches peak production • Preparation of the batch setting report with clear mention of reasons for the delay (if any) • Encourage skill development, sensitization of operators toward achieving Right First Time through training and retraining. • Proper documentation and control of the new process for future reference

Zerin, Hossain, and Zannat (2019) discussed a case of lead-time improvement by reducing COT through SMED at an apparel manufacturing setup of 130,000 units per month capacity. The study was conducted on a trouser manufacturing line with 60 sewing workstations, 5 in-line pressing, and 3 quality checking stations. A six-step methodology (inline of the Shingo's SMED methodology) was applied to implement SMED. Before changeover improvement, the internal and external activities consumed time as 9.83 and 3.03 hours, respectively, making total COT as 12.86 hours. Ten internal activities (out of total 23 internal activities) were converted to the external activities, and remaining internal activities were streamlined while SMED implementation. This has resulted in an improvement COT to 2.95 hours. Monitoring of the next five changeovers for the same line confirmed that the average COT was reduced to 3.0 hours from 12.86 hours as the earlier COT (Zerin et al., 2019).

Bajpai (2014) conducted SMED implementation in apparel manufacturing for knitted products (for two changeovers) at an apparel manufacturing setup in India. With the prime objective of reduction in style changeover. An observational study was conducted on the sewing floors to note down the list of activities/elements that are involved in the batch setting. The elements along with the timeline were noted and classified under the following headings, Machine time, BST, Demonstration time, and Time delay due to operator. The RDT and the run-up time was not under the scope of the study (Bajpai, 2014). The values of the observed percentage of time consumed in the batch setting have been mentioned in Table 6.5.

From Table 6.5, it can be observed that the in case 1, due to a high level of demonstration (high level of workmanship due to complexity of the garment), time consumed for demonstration was 49% followed by BST and machine time which is 25% and 22%, respectively, whereas in case 2, due to the product complexity and more number of machines required, the machine time consumed 59% of the time followed by BST as 24%. The activities were analyzed in terms of internal and

Table 6.5 Observation on batch setting activities.

Batch setting activity time(% time consumed)	Case 1Out-going style: Knitted Ladies polo shirt (Basic style) New style to be loaded: Knitted Ladies top with metal zippers at the back	Case 2Out-going style: Mandarin collar full sleeve woven shirt with three button placketA new style to be loaded: half sleeve woven classic shirt with two pockets with pocket flap
Machine time	22%	59%
Batch setting time	25%	24%
Demonstration time	49%	11%
Time delay due to operator	4%	6%

external, and it was observed that out of total 11 internal activities, 6 activities could be completely converted as external activities, whereas 4 activities could be partially converted as external activities. During SMED implementation, streamlining of activities under all heads were done. And as a result, the style COT was considerably reduced, as the COT in case 1 was observed reduced from 404 to 266 minutes cumulative in all four activities considered for observation (Bajpai, 2014).

Singla, Pal, and Prakash (2017) studied the style changeover at one of the leading apparel export houses in India. The study was conducted on knitted products and two different product styles, one with repeat order, and another as a new order were observed. As one of the research findings, it was observed that the average time in line setting up was up to 16 hours. The key reasons for such a high ST were unorganized machine setting (505 and 589 minutes for Style 1 and Style 2, respectively) and Operator demonstration and idle time (400 and 65 minutes for Style 1 and Style 2, respectively). The total COT observed was 20.4 and 13.45 hours for Style 1 and Style 2, respectively. SMED methodology was applied by segregating internal and external activities and focussing on maximizing the conversion of internal activities as external activities. The activities were streamlined further and standardized by developing relevant formats, the formation of the changeover team, enabling style preparation in advance, structuring preproduction meetings in a more organized manner, etc. The entire process and improvements were monitored closely and regularly. This resulted in improved line setup with 7.0 and 13.0 hours for the Style 1 and Style 2, respectively (Singla et al., 2017).

Hasan, Prasad, and Wadhwa (2008) studied the batch setting issues at a leading apparel manufacturing company dealing with frequent changeovers due to smaller order quantities and diversified products. The changeover process was carefully observed and analyzed. The key parameters affecting the batch setting were identified, and times were recorded for each of the parameters. The total time for each style as well as the time taken by each parameter per style was noted. The percentage timeshare of each parameter was calculated for each batch setup. The reasons for causing delay (due to that a particular factor is consuming more time in batch setting) were identified, and efforts were put to reduce the delays and inefficiencies. After the improvements, the times consumed in each parameter were again measured and calculated against the SAM (Standard Allowed Minutes) of respective styles. This gives the values of BST per parameter per SAM. As an outcome of the study, it was observed that the average BST was reduced to 12.59 minute per SAM from an earlier BST of 22.5 minute per SAM (Hasan et al., 2008).

6.5 Impact of rapid changeover

Lean manufacturing is a cross-functional discipline and would require full involvement from all other divisions of the organization. For standard work to be effective, enforcement is the key to change. Be it SMED or any other lean tool for a successful outcome, it is important to change the culture of change management and a sense of shared responsibility. Some key benefits of the rapid changeover may include:

- lot size can be reduced,
- help to reduce inventory,
- reduce the cost of setup labor,
- increase the capacity of bottleneck equipment,
- help to eliminate the setup scrap, and
- reduce potential quality problems and obsolescence.

With the advancement of technology, in these days, SMED is being implemented more effectively through real-time data capturing. Precise timings of all the activities related to machines, machine settings, maintenances, quality checks, outputs, etc. are captured on a real-time basis. This brings accuracy and effectiveness in the changeover process and makes the Muda visible.

6.6 Conclusion

Significant improvement in changeover performance can be achieved through rapid setup implementation. While implementing rapid setup, it should be guarded that the saving in COT is not negated by poor functional controls like nonavailability/shortage of fabric/trims after commencing production, nonadherence/compliance to PPM schedules, poor style clarity among production supervisors, etc.

SMED can prove to be a game-changer in this competitive and volatile manufacturing environment. To remain in the business (with shrinking profit margins), adapting to rapid changeover is a necessity, which can bring flexibility and responsiveness in the system along with reducing lead times. For a successful implementation of any lean tools including SMED, commitment from all the concerned personnel is vital. SMED can be made sustainable if it is applied holistically along with other lean tools especially 5S and Visual management. Another aspect is that improvement through SMED should be developed as a habit, not just a one-time activity, and this is only possible when there is are commitment and discipline from all the stakeholders involved in the improvement process.

References

Apparel Resources. (2012). *Improving productivity in embroidery with lean*. Retrieved from http://www.apparelresources.com. (Accessed 24 April 2020).

Bajpai, J.D. (2014). SMED (Single-Minute Exchange of Die) methodology in garment manufacturing Industry: Case study in reducing style change over time. In *5th international & 26th all India manufacturing technology, design and research conference (AIMTDR 2014)* (pp. 61−67). Guwahati, India: IIT Guwahati.

De la Vega-Rodríguez, M., Baez-Lopez, Y., Flores, D., Tlapa, D., & Alvarado-Iniesta, A. (2018). Lean manufacturing: A strategy for waste reduction. In J. García-Alcaraz, G. Alor-Hernández, A. Maldonado-Macías, & C. Sánchez-Ramírez (Eds.), *New perspectives on applied industrial tools and techniques* (pp. 153–174). Cham: Springer.

Feld, W. M. (2001). *Lean manufacturing tools, techniques, and how to use them*. St. Lucie Press/APICS Series on Resource Management.

Godina, R., Pimentel, C., Silva, F. J., & Matias, J. C. (2018). *A structural literature review of the single minute exchange of die: The latest trends. 28th international conference on flexible automation and intelligent manufacturing (FAIM2018) conference*. Columbus, OH: Elsevier B.V.

Hasan, Z., Prasad, S., & Wadhwa, S. (2008). *Reducing the batch set up time*. New Delhi: National Institute of Fashion Technology, Department of Fashion Technology.

Henry, J. R. (2013). *Achieving lean changeover: Putting SMED to work*. Boca Raton, FL: CRC Press.

Herrmann, J. W., Cooper, J., Gupta, S. K., Hayes, C. C., Ishii, K., Kazmer, D., & Wood, W. H. (2004). *New directions in design for manufacturing. International design engineering technical conferences and computers and information in engineering conferences* (pp. 853–961). Salt Lake City, Utah: The American Society of Mechanical Engineers.

Jana, P., & Tiwari, M. (2018). *Industrial engineering in apparel manufacturing*. New Delhi, India: Apparel Resources Pvt. Ltd.

King, P. L. (2009). SMED in the process industries: Improved flow through shorter product changeovers. *Industrial Engineer, 41*(9), 30.

Kiran, D. (2016). Total quality management: An overview. In D. Kiran (Ed.), *Total quality management* (pp. 1–14). Butterworth-Heinemann.

Kordoghli, B., & Moussa, A. (2013). *Effect of wastes on changeover time in the garment industry. 5th international conference on modeling, simulation, and applied optimization (ICMSAO)* (pp. 1–3). Hammamet, Tunisia: IEEE.

Makhija, A. (2012). Reduction of style changeover time is critical to success. *Stitch World, 10*(5), 36–41.

McIntosh, R., Culley, S., Gest, G., Mileham, T., & Owen, G. (1996). An assessment of the role of design in the improvement of changeover performance. *International Journal of Operations and Production Management, 16*(9), 5–22.

McIntosh, R., Culley, S., Mileham, T., & Owen, G. (2001). *Improving changeover performance: A strategy for becoming a lean, responsive manufacturer*. Butterworth-Heinemann.

McIntosh, R., Owen, G., Culley, S., & Mileham, T. (2007). Changeover improvement: Reinterpreting Shingo's "SMED" methodology. *IEEE Transactions on Engineering Management, 54*(1), 98–110.

Moxham, C., & Greatbanks, R. (2001). Prerequisites for the implementation of the SMED methodology. *International Journal of Quality & Reliability Management, 18*(4), 404–414.

Perkins, C., & Silvestro, B. (2019). *Watch the fastest tire change in the history of formula 1*. Retrieved from Rroadandtrack https://www.roadandtrack.com/motorsports/news/a29666/fastest-f1-pit-stop/. (Accessed May 2020).

Reik, M. P., McIntosh, R. I., Owen, G. W., Mileham, A., & Culley, S. J. (2005). *The development of a systematic design for changeover methodology. Proceedings ICED 05, the 15th international conference on engineering design*. Melbourne, Australia: The Design Society.

Reik, M. P., McIntosh, R. I., Owen, G. W., Mileham, A., & Culley, S. J. (2006). Design for changeover (DFC): Enabling the design of highly flexible, highly responsive manufacturing processes. In T. Blecker, & G. Friedrich (Eds.), *Mass customization: Challenges and solutions* (87, pp. 111−136). Boston, MA: Springer.

Shingo, S. (1985). *A revolution in manufacturing: The SMED system.* New York: Productivity Press.

Silva, I. B., & Filho, M. G. (2019). Single-minute exchange of die (SMED): a state-of-the-art literature review. *The International Journal of Advanced Manufacturing Technology, 102*, 4289−4307.

Singla, M., Pal, S., & Prakash, T. (2017). *Reduction of changeover time during style change.* Mumbai: National Institute of Fashion Technology, Department of Fashion Technology.

Spencer, M. S., & Guide, V. D. (1995). An exploration of the components of JIT: Case study and survey results. *International Journal of Operations and Production Management, 15* (5), 72−83.

Tu, Q., Vonderembse, M. A., & Ragu-Nathan, T. S. (2004). Manufacturing practices: Antecedents to mass customization. *Production Planning and Control: The Management of Operations, 15*(4), 373−380.

Womack, J. P., Jones, D. T., & Roos, D. (2007). *The machine that changed the world: The story of lean production.* Free Press.

Zerin, N. H., Hossain, M. L., & Zannat, M. (2019). Manufacturing lead time improvement by reducing changeover time with the application of SMED. *International Journal of Scientific & Engineering Research, 10*(7), 2088−2092.

Autonomation

7

Chandrark Karekatti[1] and Prabir Jana[2]
[1]Ananta Garments Ltd., Dhaka, Bangladesh, [2]Department of Fashion Technology, National Institute of Fashion Technology, New Delhi, India

7.1 Introduction

Jidoka is a Japanese word that is referred to as autonomation with the human element (or Human Automation). It aims to identify and eliminate the problem at the source itself. According to Imai (1986), Jidoka has two meanings in the Japanese language: (1) automation, which is referred to as changing a manual process into a machine process and (2) automatic control of defects, which is referred to as incorporated the insight or mind of a human to troubleshoot and correct failures (Imai, 1986). Jidoka ensures that the quality is in-built in the product by identifying and eliminating the defect at the point of its origin. In this context, Jidoka is a means of achieving the right first time; hence, it also helps in the elimination of Muda. At the same time, it is worth noting that the entire approach of building the culture of improving quality should not be referred to as Jidoka. Jidoka should be supported by an integrated exception management system called Andon (Technopak Advisors, 2011). Most often, automation is confused with autonomation, while automation is a function of autonomation. The term "autonomation" is a combination of autonomy and automation. It implies the independence of automation or allowing a process to be able to make its own decisions, thereby giving it a human touch (Adjei, Thamma, & Kirby, 2014). The difference between automation and autonomation in the context of people, machine, quality, and inspection is explained in Table 7.1.

Jidoka has proven to be a turning point in mass manufacturing and has been an integral part of the Toyota Production System as one of its pillars. There are instances when the machine or process is not functioning normally, in such cases it is necessary to stop the machine to avoid producing faults or faulty output. This requires manpower, which may be inefficient as well. In such situations, an autonomation in form of Jidoka may help immensely. It is a means of automatically stopping the machine or the process in the situation of a malfunction or a defect to avoid producing defective output. The lean tools such as Poka-Yoke (error proofing) and Andon (lamp/light) are considered as a means to achieve Jidoka and practically applied as an integral part of Jidoka implementation.

Lean Tools in Apparel Manufacturing. DOI: https://doi.org/10.1016/B978-0-12-819426-3.00002-3
© 2021 Elsevier Ltd. All rights reserved.

Table 7.1 Difference between automation and autonomation (Adjei et al., 2014).

Category	Automation	Autonomation
People	Manual processes become easier but still needs human supervision	Supervisors can multitask and productivity improves
Machines	Machines complete cycle until the stop button is activated	Machine detection of errors and correction is autonomous
Quality	Defects can be produced in mass quantities due to machine malfunction	Machine crashes are prevented by auto-stop, hence defects are avoided
Error and diagnosis	Errors are discovered later and root cause analysis is long term	Errors are discovered and corrected quicker

7.2 Jidoka

The roots of Jidoka lie in the early 1900s with the development of automatic loom stop motion by Sakichi Toyoda the founder of Toyota Industries Corporation (Lean Enterprise Institute, 2008). Sakichi Toyoda is also referred to as the father of the Japanese industrial revolution. Mr. Toyoda was facing a problem of broken weft (the width-wise yarns in the fabric that run perpendicular to the selvage) breakages in the fabric while weaving was going on, and that resulted in faulty fabrics (Toyota Industries Corporation, 2016; TCMIT, 2015). If the broken thread is not detected immediately, it may result in a fabric defect known as missing weft or broken weft. Among many inventions developed by Mr. Toyoda, one was an automatic shuttle changing mechanism in 1903 (Mass & Robertson, 1996). The other challenge was warp (the lengthwise yarns in the fabric that run parallel to the selvage) breakage in the looms causing fabric defect known as float. As another technological intervention at Toyoda Boshoku (now called Toyoda Loom Work Co., Ltd.) (Process Improvement Japan, 2010), Mr. Toyoda developed lightweight metallic pins (drop-wires) for each of the warp yarn, and the yarn was passed through a hole (called the eye of the pin) at the time of loom mounting. As soon as there was a warp breakage, the respective pin could fall (due to gravity) on a moving part of a slide over. This resulted in the activation of knock-off mechanisms and resulted in the stoppage of the loom (Kabir, 2017). Toyota's Model G was the first loom model that was equipped with this feature (Japan Patent Office, 2002). The operator could mend the broken warp and resumed loom working (Roser, 2019).

Such intervention did not only automate the work but also enhanced the ability to capture fault with ease and accuracy. This immensely improved the quality of the product, productivity as well as the efficiency. These are few examples referred to as the classic applications of Jidoka. According to Lean Lexicon, *"Jidoka highlights the causes of problems because work stops immediately when a problem first occurs. This leads to improvements in the processes that build-in quality by eliminating the root causes of defects"* (Lean Enterprise Institute, 2008).

Alex Warren, former Executive Vice President, Toyota Motor Corporation, Kentucky defined Jidoka in the iconic book on success stories of the Toyota Production System "The Toyota Way" (Liker, 2004) as:

> *"In the case of machines, we build devices into them, which detect abnormalities and automatically stop the machine upon such an occurrence. In the case of humans, we give them the power to push buttons or pull cords—called Andon cards —which can bring our entire assembly line to a halt. Every team member has the responsibility to stop the line every time they see something that is out of standard. That is how we put the responsibility for quality in the hands of our team members. They feel the responsibility—they feel the power. They know they count."*

7.2.1 Jidoka methodology

Hirano (2009) discusses different ways or approaches to make the same product. There may be different levels of human intervention involved in making a product from fully manual work to an automated process with minimal or no human involvement. There are four steps in developing Jidoka as suggested by Hirano, each of the steps involves a relation between machine and human (Hirano, 2009). Please refer to Table 7.2 to understand the steps of developing Jidoka in the context of apparel manufacturing.

Jidoka as a quality control philosophy is a four-step process (refer Fig. 7.1) that engages whenever an abnormality occurs (Kachru, 2009).

Step 1 and Step 2 are related to automation, means machine and equipment can be equipped with such features where it can indicate the error or fault or any abnormality. Application of limit switches, sensors, indicators, buzzers, and so on may be employed. Lean tools such as Andon and Poka-Yoke may also be employed for quick detection of abnormality. Step 3 and Step 4 require human intervention where the machine operator or in-charge or supervisor needs to attend the machine and rectify the issue. There are several techniques for root-cause identification, which have been discussed in a separate chapter in this book.

Hirano (2009) discusses the importance of Jidoka through an interesting case of machine moving and machine working. Sometimes machine works while that adds value to the product, in other situations machine just performs the task with its movement (this generally happens when there is no or minimal automation). To reduce the labor in a process, people opt for technological upgradation of the plant and install new automatic machines, though this requires a different skill set to operate these advanced machines, as the normal operator (who used to handle the earlier machine may not be able to competent enough to operate an advanced machine). Furthermore, such advanced equipment may require different supervisory skill sets as well. This all is an added cost to the process. This may be a further costly affair if the machine is still moving (not working to add value), and keep creating defects or defective output. In such a case, automation may not yield in the expected results, as it still requires people to constantly monitor and supervise the process.

Table 7.2 Levels of developing Jidoka.

Level	Level of automation	Brief description	Example (in the context of apparel manufacturing)
1	Manual labor	Work is completely done by hand. Suitable when the labor cost is cheap or the work is simple, which can be done quickly	Sewing a buttonhole by hand or manual needle work
2	Mechanization	Some part of human work is done by machine. The majority of the work is done manually only	Making a buttonhole by machine but the cutting is done manually
3	Automation	All the manual work is performed by the machine. The worker loads the piece at the machine and operates the machine by pressing the ON/OFF switch. After loading the piece operator only have to watch the process (need to be alert) and in case of a thread breakage the operator has to manually stop the machine	Making a buttonhole by industrial buttonhole machine where sewing is followed by automatic cutting
4	Jidoka (human automation)	All the work is performed by the machine itself, though the operators need to operate the machine by pressing the ON/OFF switch. After loading the piece to the machine, the operator need not be alert and can concentrate on another operation. The machine shall perform the task and stop automatically. In the case of thread breakage, the machine will automatically deactivate the cutter to avoid irreparable damage to the piece	Making a buttonhole by industrial buttonhole machine with the auto-stop mechanism of cutting in case of thread breakage

Application of Jidoka is the next level intervention, where we are enabling machine (may be through upgradation) to reduce (without or minimal) the requirement of constant supervision or human intervention by applying human wisdom to change the machines. In this process, we aim to enhance machine abilities with

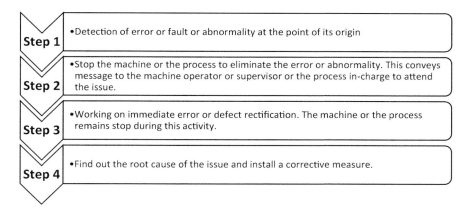

Figure 7.1 Jidoka steps.

human automation. The machine and equipment with Jidoka intervention can work continuously without disrupting the flow. Such machines are referred to as *"flow-oriented human automated machines."*

Hirano advocates for the step-by-step application of Jidoka in the process rather than opting for complete automation, as such automation in one go may lead involve not only financial implications but process disruption as well (Hirano, 2009).

7.2.2 Jidoka functions

The automation should aim for improving efficiency, quality enhancement in a cost-effective manner. There is practically no meaning of automation, where we are still producing defective pieces, and systems still require supervision. Given the same, the application of Jidoka can be initiated to separate human work from the machine. As obvious, human work can be performed either by hands (one hand or both) and/or by the legs. The start point may be critical observation and analysis of what and how the left hand, right hand, and the legs are performing the work. The second step may be critical thinking in the directions as:

- How the hands (first the left hand and then the right hand) can be relieved from the work?
- Is there any possibility of some intervention where this work can be automated through some intervention?

Such an intervention should not affect the functioning of the machine but should aim to improve productivity and quality. Also, there is a need to work on making this intervention reliable, through constant monitoring and amendments; it should be ensured that the machine is capable of continuously producing defect-free output. With this, we ensure the machine automation, which is efficient and productive. According to Hirano (2009), *"the development of defect-prevention devices for automated equipment is the heart and soul of Jidoka."* The machines with Jidoka intervention must be able to detect the errors and stop the process by themselves. Such detected issues or abnormalities should also be communicated to the concerned people may be through a light or sound an alarm so that the machine can be attended immediately for corrective action.

Three main functions of Jidoka may be summarized as (Hirano, 2009):

Function 1: Separation of manual work from the machine work—Enabling machine to perform a major share of the work and the human is relieved. Human skills may be utilized somewhere else for value addition.

Function 2: Defect prevention—With Jidoka interventions, machines are modified in such a manner that it no longer keeps producing defective output. Here the machine is having the ability to sense if anything goes wrong. This relieves the supervisor or in-charge of constant monitoring. This makes the machine working, not just moving.

Function 3: Application of Jidoka in assembly operations—As soon as the defect or abnormality is identified, the work of the assembly line should be stopped. This ensures that the defect or defective pieces are not moving forward in the system. By doing this we can prevent the cost of poor quality and can control several Muda in the manufacturing environment.

Salinas-Coronado, Aguilar-Duque, Tlapa-Mendoza, and Amaya-Parra (2014) find Jidoka a common-sense philosophy with serveal benefits, as (Salinas-Coronado et al., 2014):

Increases trust—Jidoka boosts the people's empowerment, as it encourages people to report defects and problems without fear of blame.

Improves communication—Effective and efficient notification of the error, defect, abnormality, or a problem to the concerned people.

Creates urgency—Communication through signals (may be a sound or light or by any other means) creates an urgency to attend and rectify the issue.

Contains the problem—Prevents the defect or defective piece to move further and reduces the cost of poor quality.

Involves others—Encourages the involvement of all the concerned people through effective and immediate communication of the problem.

Drives prevention—Motivates people to investigate the root cause of the problem and makes technical intervention sustainable.

Cultural change—Encourages the culture of collective efforts with coordination and cooperation.

7.2.3   Jidoka in apparel manufacturing machinery and equipment

While the genesis of Jidoka is an auto-stopping mechanism of loom during warp breakage, the machinery and equipment used in apparel manufacturing also have incorporated several such mechanisms. Here are some prominent ones (Table 7.3).

Although the autonomation process has been essentially incorporated into machine design, allowing the machine to stop during an unusual process, modern applications of autonomation do not limit it to just the stopping of machines when they malfunction. Real autonomation includes a complete manufacturing process, with series of steps to automate fabrication and assembly operations, as well as an

Table 7.3 Autonomation in apparel manufacturing equipment

Machine	Autonomation
The auto-stop mechanism in a fusing machine	• The machine automatically switched off in case of overheating • The machine automatically shuts down in case of belt misalignment to avoid costly damage to the belt
Auto-protection of the belt in a fusing machine	• Automatic cooling system is switched on which cools the machine before completely stopping to prevent the belts from burning or getting damaged • If the temperature is abnormally high, then the machine belt speed will be increased automatically and keep running with the alarm is on (to protect the belt from getting damaged)
Knife Intelligence mechanism in CNC cutting machine from Gerber Technology	The Knife Intelligence feature senses the deflection of the knife that occurs when cutting difficult materials or high-ply spreads and automatically corrects the knife angle (by using a powerful algorithm) to compensate for this deflection. The result is more accurately cut parts
The auto-stop mechanism in flat bonding press or bra cup molding machine	In a flat press bonding or bra cup, the molding machine temperature is pre-set based on fabric and operation. As the temperature is a very crucial parameter for the quality output, the machine automatically locks itself from functioning until the pre-set temperature is achieved. Only once the press or mold is heated up to pre-set temperature, the machine can be operated
Auto-stop mechanism of multihead embroidery machine	When any thread of any embroidery head breaks in a multihead embroidery machine, the machine stops immediately, ensuring no faulty merchandise is produced
The auto-stop mechanism in a sewing automat during wrong thread tension	In case the thread tension in a certain sewing automat varies from the set standard, the sewing automat will stop automatically, ensuring no faulty merchandise is produced
The auto-stop mechanism in a sewing automat during thread breakage	In case of thread breakage, all sewing automat will stop automatically, ensuring no faulty merchandise is produced, ensuring ease in repair

(Continued)

Table 7.3 (Continued)

Machine	Autonomation
The auto-stop mechanism in a hot air seam sealing machine	In case of any nonconformity of air pressure, the heater will be turned off and the machine will not work, ensuring no faulty merchandise is produced In case of any nonconformity related to tape tension, the heater will be on, but Mc will stop functioning, ensuring no faulty merchandise is produced
The auto-stop mechanism in an automatic button sewing machine	In a button sewing machine with hopper feeder if the feeder is not able to supply the button to the machine clamp three times, automatically it will try to feed the machine but after that, it will stop

approach to managing the daily interactions between humans and machines on a manufacturing shop floor (Adjei et al., 2014).

7.2.4 Jidoka system in the apparel manufacturing process

7.2.4.1 Autonomation for sewing defect (RFT) control

A homogeneous Jidoka system capable of detecting and triggering a corrective mechanism is developed by modifying the conventional in-line and end-line inspection systems. What is interesting is this system is developed by using the existing quality checkers with no requirement of additional manpower. This inspection system now comprises of a digital display board at end-line inspection with a data posting interface. The in-line and end-line inspection report is modified by adding operation and operator name, this enables to identify operator- and operation-wise hourly defect status (refer to Fig. 7.2). The quality inspector marks the defect code against the corresponding operation/operator name. Hourly inspection data at the front, back, waist, and assembly (end line) inspection points are combined and posted in the system. Hourly RFT is displayed on the digital display screen. For a sewing line running say at 90% RFT target, any downward deviation of RFT for any section trigger's the alarm. The alarm is ON until time the abnormal condition is rectified and the RFT is within control range in the next hour RFT data. The moment line RFT or section RFT falls below 90%, the control mechanism is set on whereby the line chief, supervisor, IE, quality controller scroll the system for stratified data to identify the operation/operator generating high defects (refer to Fig. 7.3) Once the root cause for deviation is identified, corrective action is taken which may be related to the mechanical condition, skill improvement, monitoring, work aid, and so on.

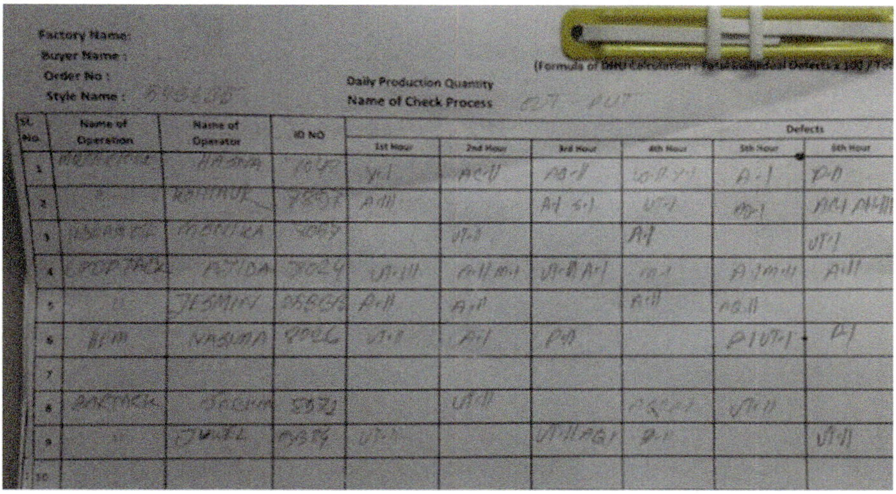

Autonomation

Figure 7.2 Operator/operation-wise Defects per Hundred Units (DHU) record.

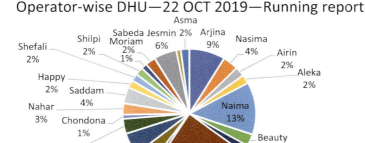

Figure 7.3 Running operator-wise Defects per Hundred Units (DHU) data.

7.2.4.2 Jidoka system for automatic fabric roll allocation to marker

Entreprise Resource Planning (ERP) system or cut planning software is interfaced with actual fabric warehouse stock. The fabric stock is posted with key quality parameters like shrinkage percent, shade group, fabric width, running length, and so on. The ERP/cut planning system allocates fabric rolls having shrinkage/shade/width corresponding to the marker selected. For any deviation, an alarm is sounded and the roll allocation is only permitted after approval from the responsible person.

Figure 7.4 Direct data posting in cutting section.

7.2.4.3 Online data posting at fabric spreading

During lay spreading and lay quality checking, lay sheet data are directly posted via tabs into the ERP systems (refer Fig. 7.4). These data are electronically transferred for bundle card preparation. This eliminates human errors and multiple handwritten reports. The quality-related data collected at spreading are made available at cut panel inspection for defective panel replacement.

7.3 Poka-Yoke

Shigeo Shingo invented Poka-Yoke (pronounced "POH-Kah YOH-Kay") in the 1960s. Because they serve to prevent (or "proof," in Japanese, *yoke*) the sort of inadvertent mistakes (*Poka* in Japanese) that anyone can make (Shingo, 1986). Poka-Yoke, meaning error proofing in Japanese (also called *Baka-yoke*), is a process improvement tool, which has seen wide applications in manufacturing industries like automobiles. Poka-Yoke is a tool under the gambit of Jidoka. In other words, Poka-Yoke is a means to achieve Jidoka (Autonomation) (Roser, 2018). However, its applications in the Apparel industry are very limited. Many people presume Poka-Yoke as limit switches, optical inspection systems, guide pins, or automatic shutoffs that should be implemented by the engineering department. This is a very narrow view of Poka-Yoke. In practice, Poka-Yoke mechanisms can be electrical, mechanical, procedural, visual, human, or any other form that prevents incorrect execution of a process step. It gives machines and operators the ability to stop producing nonconforming products or detect when an abnormal condition has occurred and immediately generate an alert and stop work or rectify the process. This enables operations to build-in quality at each process stage.

7.3.1 Poka-Yoke categories

Poka-Yoke devices fall into two major categories: control method and warning method. When abnormalities occur, the control method shuts down the machines or lock clamps to halt operations, thereby preventing the occurrence of serial defects. The warning method calls abnormalities to worker's attention by activating a buzzer or a light. Since defects will continue to occur if workers do not notice these signals, this approach provides a less powerful regulatory action than control methods. Thus control method has a more powerful regulatory function in achieving zero defects (Shingo, 1986). A good example of control method Poka-Yoke is a buttonhole sewing machine shuts down once the thread breaks, while a buzzer sound or blinking light in a pattern-taking machine once, the predefined amount of thread remains in the bobbin is an example of warning method Poka-Yoke. The setting functions of Poka-Yoke systems can be divided into three categories: contact method, fixed-value method, and motion-step method.

7.3.1.1 Contact method

Methods in which sensing devices detect abnormalities in product shape or dimension by whether or not contact is made between the products and the sensing devices are called contact methods. These can be as simple as guide pins or blocks that do not allow parts to be seated in the wrong position before processing. Contact type helps by physically guiding the manufactured products with little or no human intervention to produce goods that conform to the requirements. This method filters out the part right away that is not in the correct position. The automatic button feeder machine uses this method of Poka-Yoke.

A button sewing machine with an automatic hopper feeder mechanism (refer to Fig. 7.7) uses a vibrating circular tray (refer to Fig. 7.8) to feed the button to the clamp. Refer to Fig. 7.5, the button cross-section generally has a chiseled out circle on the face side distinguishing the face of the button from the backside. This cross-section feature is used as a guideline for the contact method of Poka-Yoke. The tray path (refer to Fig. 7.6) has a portion zigzag cut out pathway depending on the

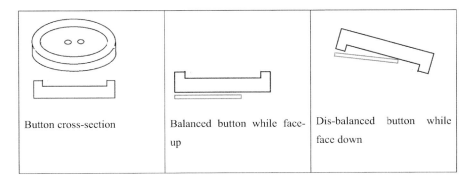

Figure 7.5 Button cross-section and movement.

Figure 7.6 The zigzag pathway in the vibration tray.

Figure 7.7 The button sewing with auto feeder.

diameter of the button. This zigzag cut out pathway acts as a sensing device for this Poka-Yoke method. Due to the chiseled out cross-section those buttons when face up are stable while crossing the zigzag pathway. However, those buttons which are upside down got dis-balanced while crossing the zigzag pathway and ultimately fallen. Therefore finally only those buttons faced up will reach the clamp to be attached. As the buttons to be attached in a garment face side up, this method filters out those buttons that are not face up (Figs. 7.7 and 7.8).

Another example of Contact Method Poka-Yoke is by physically guiding the manufactured products with little or no human intervention to produce goods that conform to the requirements. Binding rib attaches by binder attachment (refer to Fig. 7.9) ensures the width is maintained.

7.3.1.2 Fixed-value method

With these methods, abnormalities are detected by checking for the specified number of motions in cases where operations must be repeated a predetermined number

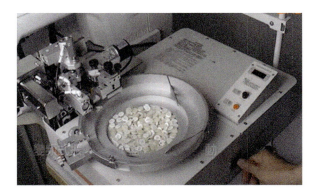

Figure 7.8 The vibration tray in an auto feeder.

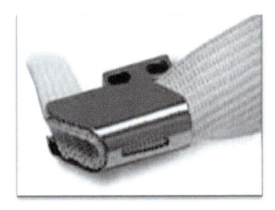

Figure 7.9 Detection type Poka-Yoke-Contact Type (rib binder).

of times. This method can be used when a fixed number of operations are required within a processor when a product has a fixed number of parts that are attached to it. Suppose a total of seven buttons to be attached to a shirt front and button sewing operator sometimes forgets to attach one or two, the following Poka-Yoke can be followed:

In a traditional set up (refer to Fig. 7.10), the tray has countless buttons and the operator simply picks up one by one, even if he/she attaches one button lesser, the tray still has enough buttons, and no anomalies can be detected. Rather if the operator first takes seven buttons from the first tray and places in the second tray and then starts attaching from the second tray, then if all buttons from the second tray need to be attached for every shirt front. Any number of buttons left in the second tray means such several buttons are not attached to the shirt front by mistake. This logic is very similar to the originating example of Poka-Yoke where two springs were to be attached to the limit switch.

Figure 7.10 Traditional set up with the countless buttons in the tray.

Figure 7.11 Cycle sewing machines: pattern tacking and pocket attach.

Other examples of fixed value method employ automatic counters or optical devices and control the number of moves, rate, and length of movement as well as other critical operating parameters. Typically, these are programmable systems (refer to Fig. 7.11) that work as per values punched in microprocessors. Shop floor trials have demonstrated improvement in product quality through the use of machines working on similar principles. Examples are programmable profile stitching machines that eliminate human intervention in sewing operation.

7.3.1.3 Motion-step method

Abnormalities are detected by checking for errors in standard motions in cases where operations must be carried out with predetermined motions in these methods. This method ensures the processors or operator does not mistakenly perform a step that is not part of the normal process. A simple example of this is the color coding of storage bins, oil containers, components, and so on to prevent using the mixed or incorrect component. Color coding of bins holding hangtags, labels in the finishing section, or at stores can help in improving visual controls (refer to Fig. 7.12).

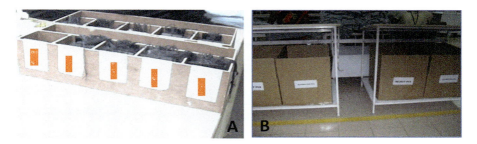

Figure 7.12 Motion step method—color coding in finishing section.

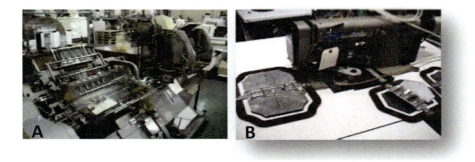

Figure 7.13 Motion step method—fully automatic robotics systems.

Other examples of Motion Step type are fully automatic robotics systems that work as complete workstations (refer Fig. 7.13). These workstations carry out material handling, sewing, and disposal functions thereby eliminating any chances of human errors resulting from the handling, placement, fatigue, and so on.

7.3.2 Poka-Yoke implementation in the apparel industry

The most common Poka-Yoke mechanism in-built in the sewing machine is bobbin case fixing in the hook set. The prong in the bobbin case and groove in the hook set need to match and only then the bobbin case will fit correctly in the hook set (refer to Fig. 7.14). The mechanism is done in such a way that the bobbin case cannot be fixed the wrong way, if the prong is not aligned with the groove, the bobbin case cannot be fixed at all.

The practitioners and consultants always lookout for designing new Poka-Yoke applications based on abnormality or mistakes that commonly happen, frequency of mistakes, the severity of the mistake, and downstream consequences. The following are a few examples demonstrating the use of the control method for error detection as used in the apparel industry.

Figure 7.14 Bobbin case fixing in hook set.

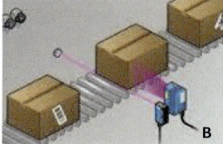

Figure 7.15 Barcode/RFID scanning demonstrating control method of Poka-Yoke application.

1. Barcode/RFID scanning (refer to Fig. 7.15) to ensure that the production progress on subsequent operation/process cannot be made unless the preceding operations/processes are complete. Thus ensuring that no production steps/operations are skipped/incomplete, before moving the garments/panels to the next process step.
2. ERP systems can be configured to block the warehouse receipt or stock transactions until production progress quality audits/inspections (refer to Fig. 7.16) in all preceding steps are complete.
3. ERP systems can be configured with controls such that incomplete sales order cannot be released for production until a true manufacturable configuration is posted. For example, a sewing order can not be released for production unless the respective trims and accessories are inhouse and updated in ERP.
4. We have witnessed a drastic improvement in cutting quality through Poka-Yoke implementation for auto cutting/sewing maintenance schedules. Modern automatic cutting machines (CAM) can be programmed to disable cutting when preventive maintenance schedules have lapsed. Similarly, programmable sewing machines can be programmed to stop if periodic maintenance/process routines are not maintained. This can help in controlling errors resulting from poor machine conditions.

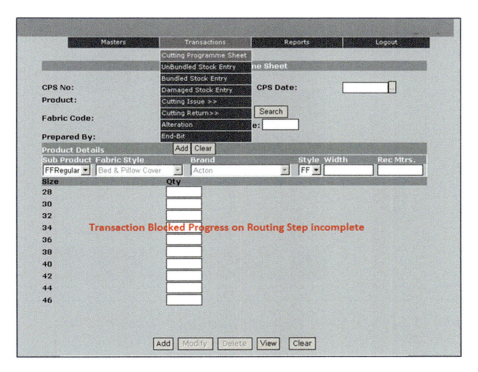

Figure 7.16 ERP system configured to block warehouse receipt if the quality check is incomplete.

5. Modern fusing machines that discontinue operation if the set process parameters are not attainable can be used to ensure conformance to set process parameters. These machines have the "Diagnostic system" for heating elements (refer to Fig. 7.17), compressed air supply, and belt tracking that ensure that the actual working parameters conform to set process parameters.
6. Another important element in the control method is the identification and elimination of Human Work Error Mode. Various Human Error Modes are classified into 16 categories as shown in Table 7.4. Standards operating procedure for continuous fusing operation (refer to Table 7.5) is demonstrated as an example for developing these controls. For each of the process steps as indicated in SOP, the potential Human Work Error Mode is identified (refer to Table 7.6). The last step encompasses developing controls for preventing the occurrence of identified human Error Mode. Typically, these controls are like checklists, flip boards, and so on.

7.3.2.1 Monitoring/warning method

The warning method of implementing Poka-Yoke signals the occurrence of a deviation or trend of deviations through an escalating series of buzzers, lights, or other warning devices. However, unlike the control method, the warning method does not

250　　Lean Tools in Apparel Manufacturing

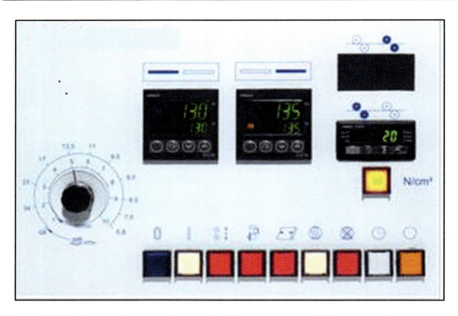

Figure 7.17 Fusing machines equipped with "Diagnostic system" for heating elements, compressed air supply, and belt tracking.

Table 7.4 Sixteen human error modes.

S. No.	Failure mode	Failure mode No.
1.	Omission	Human Error Mode 1
2.	Excessive/insufficient repetition	Human Error Mode 2
3.	Wrong order	Human Error Mode 3
4.	Early/late execution	Human Error Mode 4
5.	Execution of restricted work	Human Error Mode 5
6.	Incorrect selection	Human Error Mode 6
7.	Incorrect counting	Human Error Mode 7
8.	Misrecognition	Human Error Mode 8
9.	Failing to sense danger	Human Error Mode 9
10.	Incorrect holding	Human Error Mode 10
11.	Incorrect positioning	Human Error Mode 11
12.	Incorrect orientation	Human Error Mode 12
13.	Incorrect motion	Human Error Mode 13
14.	Improper holding	Human Error Mode 14
15.	Inaccurate motion	Human Error Mode 15
16.	Insufficient avoidance	Human Error Mode 16

Autonomation

Table 7.5 Standard operating procedures for fusing operation.

	Department: Cutting Dept	Quality Target:	Quality Target:Rej < 0.25 %
	Process Step:Fusing	Productivity Target:	Productivity Target:2200 Pcs /Shift/Person
	Machine: Fusing Machine		
Sr. no	Set Up Instructions	Operating Instructions (with check points)	Reaction Plan
1	Ensure avaibility of Fusing Rolls from Stores WRT to work Order and Product Pack. If not available Inform Sup'	Cut the rolls as per width mentioned in KNP sheet and qty' required for the shift by taking the roll to straight knife cutting machine.	Inform Cutting In charge
2	Clean Machine top, belt and table with cloth	Load the cut roll on Machine stand	Inform cutting In charge
3	Switch ON main switch using Green switch. Inform Maintenance if not working	Place the cut panel on the machine bed	
4	Set Temp in temp controller and time by using the speed regulator as given below Fusing Material Temp Time(Sec) Kufner 150c 5 Freudenberg 150c 5 7070 150c 5 This Spec can be modified as per Fabric requirement. Approval to be taken from TSD dept.	Aligne the fusing on the fabric and feed to conveyor	Realign
5	Wait till desired temp is achieved	Trim the Fused parts using scissors as soon as they exit the rear end	
		Check at least one Pcs for gum marks by visual inspection every Hour	Inform Cutting In charge 2 Hrly Audit of time and Temp by auditor. Stop prod & adjust the setting till found OK
		Make bundles of fused parts and hand over to part mixing	
		Update production data in the register	
	Consequences of wrong work: 1 Could cause loss due to scrap/rework 2. Delay in supply of lay could cause capacity loss at cutting machine 3. Delay in meeting customer requirements		Safety Instructions:

shut down the process on every occurrence. This method is used when a bandwidth of acceptance exists, for a process. For example, drop in steam pressure in ducts, drop in suction pressure, Defects per Hundred Units (DHU) level at sewing section, and so on. Fig. 7.18 demonstrates the traffic light system indicating the current DHU level. Traffic light indicates the DHU level with red, yellow, and green colors. The occurrence of red light calls for halting the production line until the situation is addressed.

Table 7.6 Identifying potential Human Work Error Mode.

Work instruction	Potential human error mode
Check availability of fusing rolls from stores WRT to work for order and product pack	1. Forgetting to read sheet (mode 1: omission) 2. Reading the wrong sheet (mode 6: incorrect selection) 3. Misreading the sheet (mode 8: misrecognition)
Clean machine top, belt, and a table with cloth and cleaning liquid	1. Forgetting to clean (mode 1: omission) 2. Using unspecified cleaning liquid cloth for cleaning (mode 6: incorrect selection) 3. Cleaning not complete (mode 8: misrecognition)
Set temperature in temperature controller and time by using the speed regulator as given below | Fusing material Time (s)	Temperature (°C)
Kufner 5	150
Freudenberg 5	150
7070 5	150 | Inform maintenance department if the temperature does not reach 150c within 20 min	1. Forgetting to set the temperature and time controller (mode 1: omission) 2. Incorrect setting of temperature and time controller (mode 6: incorrect selection)
Fusing operations	1. Forgetting to do fusing (mode 1: omission) 2. Wrong fusing with substrate (mode 6: incorrect selection) 3. Holding a damageable panel of the subassembly product (mode 10: incorrect holding) 4. Fusing the panels in the wrong position (mode 11: incorrect positioning) 5. Fusing the parts in the wrong way around (mode 12: incorrect orientation) 6. Inaccurately fusing of the part (mode 15: inaccurate motion)

Mistake (Poka) happens due to various reasons; the relationship between different types of Poka and order of causes are explained beautifully by the following Table 7.7. While such a matrix helps to design of Poka-Yoke system, this also helps in predicting the mistakes (Trout, 2020).

Autonomation

Figure 7.18 Traffic light system in apparel production indicating Defects per Hundred Units (DHU) level.

Table 7.7 Poka types and order of causes (Trout, 2020).

CAUSES OF DEFECT	Intentional Error	Misunderstanding	Forgetfulness	Misidentification	Inexperience	Willful Error	Inadvertent Error	Slowness	Non-supervision	Surprise
Not Following Procedures	■	□	■	□	□	□	■	□	□	
Processing Errors	■	■	□	□	■	■	■	■	■	
Errors in Part Setup	□	□	■	□	□		■	□	□	
Missing Parts	■	□	□		□	□	■		□	
Wrong Parts	■	■	■	■	■	■	■		■	
Processing Wrong Workpiece	□	■	■	□	□	■	■		□	
Misoperation			□				□		□	■
Adjustment Error	□	□	□	■	□	■	□	□	□	□
Improper Equipment Setup							■			■
Improper Tools and Jigs							■			□

■ Strongly Connected   □ Connected   © Noria Corporation

7.4 Conclusion

Jidoka, in association with Poka-Yoke, plays a vital role in quality enhancement through elimination or reduction of errors at the source itself. These tools bring much-needed transparency (through effective and immediate communication) and better control in the process. The most important aspect is that the application of these tools relieves the operators, supervisors, and other concerned personnel from nonvalue-added continuous supervision and monitoring, and their skills can be utilized in some more meaningful ways.

The best Poka-Yoke in the world is a robust design foolproof of generating errors. The second best Poka-Yoke is "Training" and "Awareness." The automotive industry is a leader in this area because they invest in training and retraining. Given the benefits Poka-Yoke brings, it should not be long before the apparel industry adapts to the "Designed to Conform" philosophy.

References

Adjei, K.B., Thamma, R., & Kirby, E.D. (2014). Autonomation: The future of manufacturing. *IAJC-ISAM International Conference.*

Hirano, H. (2009). *JIT implementation manual: The complete guide to just-in-time manufacturing (Vols. V (Standardized Operations—Jidoka and Maintenance/Safety)).* Boca Raton, FL: CRC Press.

Imai, M. (1986). *Kaizen: The key to Japan's competitive success.* New York: McGraw-Hill Education.

Japan Patent Office. (2002). *Sakichi Toyoda Wooden Hand Loom, Automatic Loom. (J. P. Office, Producer)* Retrieved from www.jpo.go.jp:https://www.jpo.go.jp/e/introduction/rekishi/10hatsumeika/sakichi_toyoda.html. (Accessed April 2020).

Kabir, R. B. (2017). Exploration of the advancement in warp & weft stop motion: Primitive to electronic system. *Engineering and Technology, 4*(1), 1–8.

Kachru, U. (2009). *Production & operations management.* New Delhi, India: Excel Books India.

Lean Enterprise Institute. (2008). *Lean Lexicon: A graphical glossary for Lean Thinkers.* One Cambridge Center, Cambridge, USA: Lean Enterprise Institute, Co.

Liker, J. K. (2004). *The Toyota way.* New Delhi, India: Tata McGraw-Hill Publishing Company Limited.

Mass, W., & Robertson, A. (1996). From textiles to automobiles: Mechanical and organizational innovation in the Toyoda enterprises, 1895-1933. *Business and Economic History, 25*(2).

Process Improvement Japan. (2010). *Toyota History and the Origins of Jidoka.* Retrieved from http://www.process-improvement-japan.com/toyota-history.html. Process Improvement Japan & Australia (Accessed June 2020).

Roser, C. (2018). What exactly is Jidoka? Retrieved from https://www.allaboutlean.com/jidoka-1/. (Accessed April 2020).

Roser, C. (2019). *The Toyoda Model G Loom.* Retrieved from https://www.allaboutlean.com/toyoda-model-g/. (Accessed May 2020).

Salinas-Coronado, J., Aguilar-Duque, J. I., Tlapa-Mendoza, D. A., & Amaya-Parra, G. (2014). Lean manufacturing in production process in the automotive industry. In J. L. García-Alcaraz, A. A. Maldonado-Macías, & G. C.-R. Cortes-Robles (Eds.), *Lean manufacturing in the developing world methodology: Case studies and trends from Latin America*. Switzerland: Springer International Publishing.

Shingo, S. (1986). *Zero quality control: Source inspection and the Poka-Yoke system*. CRC Press.

Technopak Advisors. (2011). *Lean manufacturing—The way to manufacturing excellence*. Gurgaon, NCR: Technopak Advisors Pvt. Ltd.

Toyota Industries Corporation. (2016). *The story of Sakichi Toyoda. (Toyota Industries Corporation)*. Retrieved from https://www.toyota-industries.com/company/history/toyoda_sakichi/. (Accessed May 2020).

TCMIT. (2015). *Development of loom technology*. Retrieved from http://www.tcmit.org/english/exhibition/textile/fiber03/. Toyota Commemorative Museum of Industry and Technology (Accessed April 2020).

Trout, J. (2020, April 3). Poka-Yoke Explained. (Noria Corporation) Retrieved from Reliable Plant: https://www.reliableplant.com/poka-yoke-31862. (Accessed June 2020).

Process balancing

Chandrark Karekatti[1] and Manoj Tiwari[2]
[1]Ananta Garments Ltd., Dhaka, Bangladesh, [2]Department of Fashion Technology, National Institute of Fashion Technology, Jodhpur, India

> *Balance flow, not capacity: you should not balance capacity with demand. What you need to do instead is balance the flow of product through the plant with demand from the market.* — Dr. Eliyahu M. Goldratt

8.1 Introduction

The manufacturing of apparel has evolved to include the application of various production systems depending on the nature of business and type of product. In the pursuit of running excellent operations, organizations have implemented various production systems and adapted different philosophies with varying levels of success and failure. Most of such efforts have been driven by a primary focus on reducing costs through an emphasis on increasing local improvements. However, a global focus on improving the overall system is required based on a logical operating principle with supporting mechanisms to govern local improvements. Before local improvements can be effectively made, the system must be stabilized on a reliable operating mechanism (Dixit, Iyer, & Tiwari, 2010). According to Akturk and Erhun (1999), Just-in-Time (JIT) has become popular in repetitive manufacturing, which encourages change and improvement through inventory reduction.

Garment manufacturing factories are often characterized by longer supply chain lead times and excessive in-process inventory. The problem is further compounded by the seasonal nature of the Apparel trade, causing peak and slack demand cycles. This often results in escalating inventory levels across the supply chain. Also, scarce use of demand forecasting tools (such as expert, exponential smoothing, multiple regression), further makes production scheduling in the apparel industry critical. Careful observation of the shop floor can reveal that most of the work sits idle in the queue without any progress. Technically, a serially linked mass manufacturing process (as garment manufacturing process) is balanced for rated efficiency. However, despite theoretically balanced manufacturing capacities/line plans, garment factories often struggle with huge quantities of work in progress (WIP), thereby indicating unbalanced capacities.

This results in poor resource utilization, higher throughput time, and poor delivery adherence OTIF% (on time in full). As a normal practice, planner in the garment factory is busy scheduling and rescheduling work everywhere and every time to meet local contingency. However, in doing so, he tries to fix part of the problem and not the whole problem. It is observed that responding to fluctuating production demand results in oscillating demand magnification upstream a supply chain, a phenomenon termed as "Bullwhip Effect."

In such a scenario, certain lean process balancing tools like continuous flow manufacturing, Heijunka, Kanban, and the drum−buffer−rope (DBR) system can be effectively utilized for capacity balancing, production smoothing, and product mix leveling. The goal is to produce goods at a constant rate, so that upstream and downstream processes can also operate at a constant and predictable rate, thereby reducing inventory. This chapter aims to discuss the application of such a process balancing lean tools in the practical apparel manufacturing environment.

8.2 Challenge of high inventory levels in apparel manufacturing

Inventory is considered as a necessary evil, as it is used to prevent the workstations and processes from running out of material to work upon. In typical apparel manufacturing (at least in Asian and African continents), traditional practices with large inventories are still very much prevalent at most of the apparel manufacturing plants. People feel safe and secure with large inventories, and just want to stay in their safe cocoon without any botheration about the hidden losses of keeping high inventory levels. In the apparel industry, despite being so dynamic, with shorter product life cycles and diversified products, the fear of handling small orders with diversity is always there. Even though, the era of large order quantities has passed long back, there are many professionals who feel comfortable working with high inventory levels. Overloading and underloading of processes, mismatch of demand and supply between the processes (sometimes sewing lines are flooded with cut panels, and at some other times sewing lines are just starving for feeding), creation of bottlenecks, piling up of huge inventories in form of unpacked garments, or packed garments but waiting to be shipped (orders completed much ahead of time), delayed shipments are some of quite frequently visible scenarios. Most of the time, the reason for all such issues is lack of processes balancing and interruptions in the manufacturing flow.

A higher level of inventories is nothing but concealing the problems (process imbalance, shortage or surplus of materials, machine issues, quality issues, nonadherence to schedules and time lines, and capacity underbooked or overbooked to name a few) with a tape just line band-aid. So, in the presence of high inventory levels, all the problems are covered up (or not visible) though these all are very much live behind the thick layers of inventory. Hirano (2009) cites several probable losses due to high inventory levels, as a burden on factory's operating capital (as money is blocked in terms of inventory), increased cost due to storage of inventory and its maintenance, cost of obsolesce and deterioration, consuming additional space (which could be utilized in some other meaningful activities), creating wasteful activities (such as counting, recording, picking, transporting, and storing consuming available infrastructure), consuming extra management and extra energy (efforts) to handle inventories, and advance procurement of materials, which may not match the buyer requirements. In today's competitive environment, wasting money and efforts on concealing the problems (with higher levels of inventories) is nothing but inviting bigger problems, which may be fatal to an organization's survival.

Just in contrast to this, for an effective and efficient redressal, the lean approach firmly advocates for transparency in the process, and identification of issues and problems. This can only happen when such thick layers of inventory are removed maybe through reduction or elimination of excess inventory. This leads to the real lean transformation where efforts are put to achieve JIT condition supported by Heijunka and Kanban.

8.3 Continuous flow manufacturing

A smooth and uninterrupted flow of information and materials indicates the good health of the organization. It not only brings the transparency (in terms of knowing the demand) but a balance between the demand and the supply also. The traditional approach focused on increasing the production (with keeping higher inventory levels), while the flow-manufacturing approach emphasizes improving the flow (with keeping optimum or minimum required inventory levels) with optimization of resource utilization.

The traditional approach of mass manufacturing of apparel products using assembly lines with progressive bundle systems is prone to create bottlenecks or interruptions in the manufacturing flow. Here, in a progressive bundle system, the work can pass to the next operation only when all the work of a bundle is completed. As the pieces can move only with the bundle, the in-process inventories are generally kept higher. This leads to an increased throughput time which is contributed by a significant amount of waiting time associated with the work in the bundles. The work in the form of the bundle is moved from one workstation to the next with operators stationed at a designated place performing individual skill-based tasks or operation. Here, the focus is on increasing the speed of work (to get more output from an individual workstation or the process). In apparel manufacturing setups the most visible approach is to invest heavily in automation and installing high-tech machines. Many times, such decisions are taken without considering process balance (within and between) of demand and supply. This makes the entire manufacturing setup working at a different pace and different rhythm resulting in-process imbalances.

Ohno (1988) in his famous book *Toyota Production System: Beyond Large-Scale Production* recommends focusing on two key points while seeing complete elimination of waste from the process, as:

1. Efficiency improvement should be linked with the cost. The only products should be produced which are required using minimum manpower.
2. Efficiency improvement should be seen at every level (each operator, each line, and the entire plant). Efficiency must be improved at each step as well as for the plant as a whole (Appell, 2014).

This can be understood with a simple example, assuming one production line has 10 workers and makes 100 products per day. This means the line capacity is 100 pieces per day and the productivity per person is 10 pieces per day. Observing the line and workers in further detail, however, we notice overproduction, workers waiting, and other unnecessary movements depending on the time of day.

Suppose we improved the situation and reduced manpower by two workers. The fact that 8 workers could produce 100 pieces daily suggests that we can make 125 pieces a day with 10 workers. This is possible by increasing efficiency through controlling or eliminating wastes in the process. Though the capacity to make 125 pieces a day existed before, it was being wasted in the form of unnecessary work.

The concept of flow manufacturing focuses on achieving overall rhythm in the entire manufacturing process (while the traditional approach, as discussed in the previous paragraphs, the progressive bundle approach focuses on rhythm at individual workstation aiming to increase the speed). This over rhythm is maintained right from the very first step to the final stage of getting the finished product. In the case of apparel manufacturing, it can be treated from cutting of the garment panels to the final finishing & packing of the end product. The tempo of this rhythm may be set by the customer demand (takt), and each of the processes or the workstation needs to just work enough to meet the takt. Interestingly, there is no requirement of hurry or pressure to increase the speed to get more output. According to Hirano (2009), it should be "slow enough to remain in the overall flow" (Hirano, 2009, p. 332).

In recent years, flow manufacturing has evolved as a solution to produce products with small order quantities and diversification. The consumers' demand for customization (mass customization) has further established the necessity of continuous flow manufacturing. The traditional way of manufacturing products may not be suitable in such circumstances; there is a need to produce well in a batch flow system, such as single-piece flow manufacturing (Miltenburg, 2001).

The flow production may be categorized into two categories: (1) flow within the factory and (2) the flow between the factories. The flow within the factory primarily covers the flow at the manufacturing site right from the processing of raw materials to the final finishing & packing of the end product. An example of flow production within a factory may be cut to pack processing of apparels maintaining single-piece flow throughout the manufacturing. The flow between the factories includes transactions of a factory to its raw material suppliers, subcontractors, distributors, warehouses (outside the factory premises), etc. It primarily includes, demand-driven supplies of raw materials (such as fabrics, sewing & packing trims) from the suppliers, regular supplies of cut panels from embroidery subcontractor,

Process balancing

regular transactions between factory and washing contractor, movement of packed garments to the warehouse.

8.3.1 Key traits of flow manufacturing

As the name indicates, flow manufacturing is regarding producing the goods through maintaining a flow of materials throughout the process. As the word "flow" is used it's obvious that the flow should be uninterrupted. At the same time, it is important to note that the rate of flow should be regulated by the customer demand also called takt. In this context an example of a flowing river may be taken, we can maintain the river clean when the water is freely flowing. Any kind of interruptions may result in overflow or no flow of the water at some point. In the case of manufacturing, such disturbances create bottlenecks and process imbalances. Hirano (2009) discusses some conditions of flow manufacturing as indicated in Table 8.1.

8.3.2 Key time-related definitions in flow manufacturing

Throughput time: The term "throughput" in apparel manufacturing is originated from "thruput".

Throughput terminology in the supply chain context and apparel manufacturing context is different. Throughput time in a sewing line is defined as the time that elapses between the point at which material enters the sewing line to the point at which it exits; ironically, it is the same as the definition of material flow time (Jana & Tiwari, 2018, p. 27).

Throughput time for a garment in a sewing line can be calculated by the formula:

$$\text{Thruput time} = \text{Standard time of the garment}$$
$$\times [1 + (\text{Average WIP between each sewing operation})]$$

The relation between throughput, material flow time, and inventory in a supply chain is explained by Little's Law as:

$$\text{Inventory} = \text{Throughput} \times \text{material flow time}$$

where throughput is defined as the rate at which goods are produced and material flow time is the time that elapses between the point at which material enters the supply chain to the point at which it exits (Chopra & Meindl, 2013).

If the standard time (for stitching) of a garment is 25 minutes and the garment is produced in the unit production system (UPS), with on average 5 hangers of WIP before each operation, the throughput time for one garment would be 150 minutes [$= 25 (1 + 5)$]. If the same garment is being produced in the progressive bundle system and there is only one bundle (of five pieces) before each operation, the throughput time should not be calculated on Standard Allowed Minutes (SAM) of 25 minutes, rather SAM in the critical path (which will be lower than 25 minutes).

Cycle time: Cycle time is the average time consumed in performing one cycle of a repetitive task or activity. The unit of measurement of cycle time is in seconds or

Table 8.1 Conditions of a flow-manufacturing system.

Sr. number	Condition	Brief	Objective	Example in the context of apparel manufacturing
1.	One-piece flow	The single piece at a time and the same piece moves to the subsequent workstations for further processing. Focusing on different hidden wastes is the start point, and then devising the ways to eliminate this waste.	Total uncovering of all the hidden/ concealed wastes in the process	Achieving a one-piece flow in apparel manufacturing requires precise balancing, standardization, and synchronization. It can be implemented in relatively simple products such as crew-neck T-shirts; however, a "near" one-piece flow situation is achieved with 1−2 pieces as WIP between workstations.
2.	Layout arrangement	Rearrangement of the layout to minimize the conveyance (material movement). With single-piece flow, there may be a heavy movement where each piece needs to move forward as soon as work is done at a workstation.	Minimizing the conveyance (material movement) through arrangement of workstations in a sequence	The machines are arranged according to the sequence of operations. Operators sitting face-to-face with zigzag flow of material (sometimes in brick-arrangement) are used by the factories working with single piece (or near single-piece flow)
3.	Synchronization	This is related to setting-up the tempo or rhythm in the system. Each of the workstation to work in a synchronized manner to meet the demand, else the entire single-piece flow may get disturbed.	Maintaining an uninterrupted flow in the setup	Most probably, this is the most critical and difficult thing to achieve in apparel manufacturing. All the operators should work on the same pitch. Most of the time, this is practically not possible to maintain the same pitch; however, efforts are put by clubbing or splitting the operations, and making cycle time near to the pitch time or takt time.

				Operator skill, machine capabilities, operation similarity, and operation sequence (of the operations which are clubbed and split) play a vital role in achieving synchronization.
4.	Multiprocess operations	The operator performing more than one process on different workstations.	The objective is manpower reduction using the multiskilling capabilities of the workers	Multiskilling in apparel manufacturing is very much practiced. Here one operator performs work(s) using different machines arranged in the sequence. By doing so, he carries the piece to the next process. Machine-to-man ratio is kept more than 1 so that the machines to work are available for operators.
5.	Training of multiprocess workers	Preparing the workers with multiskilling abilities so that they can handle more than one process at different workstations.	Achieving standardized working on different workstations to achieve the required pace	In apparel manufacturing, multiskilling is quite popular. However, in the context of single-piece flow special focus areas are • thorough standardization of machines • thorough standardization of processes
6.	Standing work-posture	Bringing flexibility in operators' movement using standing posture.	To enhance the flexibility in a movement when switching over from one work to another, helping out other operators, quickly addressing the issues (such as quick movement to resolve imbalances), ease of material transportation (where operator) carrying his work with him	It is recommended to have standing work-postures, but in apparel manufacturing (in the Indian subcontinent) a great resistance is observed (by the workers as well as supervisors and managers) for standing work-posture. Standing work-postures are quite common in the Latin American and European countries in apparel manufacturing.

(Continued)

Table 8.1 (Continued)

Sr. number	Condition	Brief	Objective	Example in the context of apparel manufacturing
7.	Compact equipment	Keeping the workstations as compact as possible and compactly arranging them.	To save time and energy in transportation and motion. To achieve a lesser distance traveled Enhancing space utilization	Arrangement of workstations adjacent or in brick arrangement.
8.	U-shaped layout	The workstations are achieved in the U-shaped layout. Though single-piece flow can also be achieved in straight layout, though in the context of multioperation (multiskilling) U-shape layout is recommended.	Ease of the operator's movement from one station to another with lesser distance traveled Ease of supervision	Loading point and output points are near and on the same side. This enables easy feeding to the system. Also, it helps operators switching over from one workstation to another with ease and minimum movement. As the layout is compact in U-shape, supervision, and noticing & attending issues/problems becomes easy.

WIP, Work in progress.

minutes. A task is considered repetitive if the cycle time is less than 30 seconds or the task is performed by more than 50% of the work hour (Jana & Tiwari, 2018, p. 186).

In simple words, cycle time is the elapsed time between the start time of nth cycle to the start time of $(n + 1)$th cycle. In a practical environment the reference point can also be taken as end time or any other point in a cycle which can be recognized. But it is important to have clear reference points to calculate the cycle time. Necessary allowances (such as personal, fatigue and delay, and machine allowances) should be incorporated in the cycle time to make it realistic. There should be efforts to reduce the lead time (maybe using method improvement, improvised workplace arrangement, improving on the technology, providing work-aids or attachments, training & skill development, etc.). A combination of reduced cycle time and uninterrupted flow (controlled WIP) may do wonders toward achieving the overall efficiency of the plant.

Pitch time: Pitch time is the average time an operator is going to contribute while making a product. Pitch time is the theoretical operation time; each operator should opt for a planned balanced line (Jana & Tiwari, 2018). The unit of measurement of pitch time is time (in seconds or minutes or hours)/operator.

Assuming, the standard time to stitch a garment is 25 minutes and there are 10 operators in the line. Then, on an average, 2.5 minutes will be contributed in sewing by an individual operator. However, practically depending on the complexity, individual operations of that garment may take different times (that may be more or less than the pitch time). Here to improve the line balance, the operations may be split or clubbed making operation time value near to the pitch time.

Takt time: The word Takt is from the German language that is referred to as the beats or rhythm of a drum. In the manufacturing environment, takt refers to the rate of output required to meet the demand. This indicates the number of products should come out from the last workstation as final output after a particular time interval (called takt). Takt time is a simple concept, yet counter-intuitive, and often confused with cycle time or machine speed. Takt time is the pace of production needed to meet customer demand or production targets (Jana & Tiwari, 2018, p. 26). The unit of measurement of takt time is time (in seconds or minutes or hours)/unit.

Assuming, the buyer requirement is 24,000 units per month from a plant and the plant operates for 24 days in a month (including weekly break and other holidays). It means, on an average 1000 units should be produced per day. If the working hours of plant are 8 hours (480 minutes), it means that the output rate should be 0.48 minute per unit, means one final product should be out from the line after every 0.48 minute (or 28.8 seconds).

According to Roser (2015a, 2015b, 2015c), takt time gives an important indication about the customer demand, in the context of speed of producing the goods, shorter takt time means faster output rate, and accordingly, the resources need to be arranged and balanced. Also, in pull-based systems, takt time may be used in determining the number of Kanban. The line takt should include all wastes, while the cycle time should be as much as possible excluding any kind of wastes. The ratio between the cycle time and the takt time is the overall equipment effectiveness (Roser, 2015a, 2015b, 2015c). In apparel manufacturing, typical line balancing is done based on two aspects: (1) target- and (2) resource-based line balance (Jana & Tiwari, 2018, p. 143). The target-based line balance is similar to the takt time—based balancing, as it is based on the output rate to meet the delivery deadlines. However, it is important to note that there may be instances where the demand rate is higher than the line output (it may be limitations due to line setup, layout, and workstation availability). In such a scenario the industrial engineers and planners opt for increasing the number of lines to meet the demand.

Lead time: Lead time is referred to as the elapsed time between placing an order (demand) by the customer and the fulfillment of the demand in terms of product or service supply. It can be also said as the time between initiation of a request and the time of delivery.

In industry environment, lead time is referred to in different contexts, such as delivery lead time, manufacturing lead time or production lead time, and process lead time. For example, in apparel manufacturing production, lead time is a popular time, which is considered from the date of receiving raw materials (fabric and sewing trims) to the date of shipment when the order is ready for dispatch at the manufacturing site. The other commonly used term is order lead time, which is considered from the date of placing the order (by the buyer with purchase order—PO) to the receiving of the finished goods at the buyer's warehouse. The order lead time includes transit time as well.

8.3.3 Implementing single-piece flow in apparel manufacturing

Single-piece flow is the core idea of flow manufacturing. This is referred to as one-piece flow and continuous flow as well. A single piece can be referred to with the movement of all the panels of one garment at a time. It can also be treated as a movement of the bundle with a single piece.

An operator may operate a workstation, and the same bundle may be passed on to the next workstation for further processing. After passing on from one workstation to the next workstation, the work is being done on the pieces, and eventually, it results in a stitched garment. Such a bundle of (single component) may be moved in the sewing line by hand-to-hand or in a tray. Such bundles are not usually moved using an overhead conveyor. While in UPS, single-piece flow means all components of a single garment move in a tray/or hanger, here single-piece flow means a single component of a garment moving from one operation to another. As an extension, the flow is further maintained to finishing & packing sections as well, where the sewn pieces (one piece at a time) are moved further to finish and pack. In such scenario, finishing & packing workstations are arranged after the sewing workstations. Here, it is important to note that sometimes (as a variant to flow manufacturing) instead of making bundles of a single piece, the cut panels (as feeding panels directly from the lay, the number of such panels may depend on the layers in that cut) are made available at the first workstation. Similarly, the other panels are placed at respective workstations, and operators attach the panels one by one to the semifinished work as soon as it has arrived. This makes such arrangements a combination of pull and push, though the pace or tempo of the flow is still regulated by pull approach only.

Given the same, flow-manufacturing setup requires work centers and processes to be arranged in a different configuration as compared to traditional assembly line configuration. The pace of the flow is regulated by the demand (customer demand as reference) which makes it a pull-based system. In demand-driven scheduling the production schedule is triggered by a pull signal from the customer order. A pull signal links the upstream and downstream operations across the value stream (processes).

Karekatti (2014) discusses the development of a flow-manufacturing system for a full sleeve shirt at a vertically integrated setup. A lean-based flow-manufacturing system was developed to achieve an efficient flow with improved efficiency and reduced waste. The steps followed are mentioned in Table 8.2.

Table 8.2 Developing a flow-manufacturing system.

Step	Activity	Activities brief	Key objectives
1.	Identify and schedule pacemaker process	Identification of the pacemaker process based on the master schedule of shipments As the flow shall be regulated by the output of the pacemaker process, hence it is ensured to provide sufficient input (feeding) to the pacemaker process	To set the pace or tempo of the manufacturing process Pacemaker process to pull the raw material (inputs, here fabric) from the previous process (here fabric store)
2.	Develop work cells	Configuration of manufacturing setup (machines & equipment) for single-piece flow The design of work cells should ensure producing only what is needed to maintain the flow without creating bottlenecks The availability of multiskilled operators is an important aspect, as it helps in improving the rhythm of work at different workstations by clubbing or split of the operations	Ensuring uninterrupted flow Due to technical issues and limitations, it may not be always possible to achieve single-piece flow, but the efforts should be there to reduce the bundle size (if not the bundle of a single piece)
3.	Develop intermittent supermarkets	Creation of intermediate supermarkets with a minimum buffer to tackle the contingencies and fluctuations	To ensure flow throughout the system

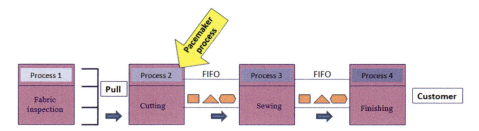

Figure 8.1 Pacemaker process.

8.3.3.1 Identifying the pacemaker process

As indicated in Table 8.2, identification and scheduling the pacemaker process is the very first step. It is an important factor to set the pace for upstream (processes before the pacemaker process) and downstream (processes after the pacemaker process) processes in a manufacturing environment. The pacemaker process is the process that controls or regulates the pace of the entire value stream. Ideally, the pacemaker is the most downstream process and it impacts all the upstream processes. It can also be considered where the customer order enters the apparel manufacturing value chain. Given the same, cutting process was treated as the pacemaker process (refer Fig. 8.1), while raw material sourcing, fabric manufacturing, and fabric processing were considered as the upstream process. As the cutting process is treated as the pacemaker, it sets the pace for all the upstream and downstream processes. Based on the delivery date of the finished garments (buyer's shipment schedule), the cutting schedules are drawn with planned cut dates (PCDs) for each of the orders. The PCD triggers the pull for sourcing of raw material (fabric, sewing trims, packing trims, etc.) and scheduling activities upstream. The downstream processes such as sewing, finishing, and packaging work on first-in-first-out principle (FIFO) (Karekatti, 2014).

8.3.3.2 Setting up work cells

The next step is the development of setup aiming to achieve single-piece flow. The work cells (line layouts) facilitating single-piece flow should be developed by the industrial engineer. Considering the practical implications and balance between efforts and cost, an ideal configuration need not provide a single-piece flow on all workstation. In such situations, it is recommended that a combination of bundle system (at sewing preparatory and subassembly sections) and single-piece flow (at the final assembly section) may be opted for. To improve the rhythm at different workstations, it is important to utilize the multiskilling abilities of the operators. The clubbing or split of operations (to make operation cycle times near to pitch time or the takt time) improves resource utilization as well by eliminating the nonvalue-adding activities/movements.

The conventional assembly line and modified work cell configuration for full sleeve shirt have been shown in Figs. 8.2 and 8.3 respectively. The work centers

Operation Bulletin: Long Sleeve Shirt			
Parameter			
Projected Output per shift	452	Minutes Per Day	480	No Of Operators	30 (Inc Absm)
Man to Machine Ratio	1.3:1	Total Number of Machines	27	Total SAM	18.8
Out Put at 100%	894	Actual Efficiency	51%	No Of workplaces	35
Absenteeism	10%	Pcs Per Operator Per Day	16.00		

Sl No.	OPERATION	SAM	M/C Opr	Opr. Grade	ACT TIME	MACHINE TYPE CLASS	OPERs CAL	OPERs REQ	M/Cs CAL	Projected Output Per Shift (8 hrs)
	COLLAR									
1	Runstitch collar	0.40	Y	C	0.67	Autojig	0.8	1	1	720
2	Clip and turn collar	0.35	N	N	0.44	Collar Turning Machine	0.5	1	0	1097
3	Topstitch collar	0.45	Y	B	0.69	SNLS w/UBT	0.8	1	1	693
4	Hem collarband	0.35	Y	C	0.58	SNLS w/UBT	0.7	1	1	823
5	Attach collar band to collar	0.50	Y	A	0.67	SNLS w/UBT	0.8	1	1	720
6	Turn and Iron collar pick	0.30	N	N	0.38	Iron Table	0.4	0	0	1280
		2.35			**3.42**					
	CUFF PREPARATION									
7	Hem cuff	0.35	Y	C	0.58	SNLS w/UBT Prog				
8	Run Stitch Cuff	0.35	Y	B	0.54	SNLS w/UBT Prog			1	891
9	Clip & Turn Cuff	0.35	N	N	0.44	Collar Turning Machine	0.5	1	0	1097
		1.05			**1.56**					
	SLEEVE PREPARATION									
10	Attach under sleeve placket	0.50	Y	A	0.67	SNLS w/UBT Prog	0.8	1	1	720
11	Attach and Finish Upper Sleeve Placket	1.40	Y	A	1.87	SNLS w/UBT Prog	2.1	2	2	514
12	Inline Checking	0.85	N	N	1.06	Checker Table	1.2	1	0	452
		2.75			**3.60**					
	FRONT PREPARATION									
13	Sew buttonhole placket	0.35	Y	A	0.47	Kansai	0.5	1	1	1029
14	Sew button placket	0.60	Y	B	0.92	SNLS w/UBT Prog	1.1	1	1	520
15	Hem pocket	0.40	Y	B	0.62	SNLS w/UBT Prog	0.7	1	1	780
16	Iron Pocket	0.35	N	N	0.44	Helper Table	0.5	1	0	1097
17	Clip Pocket and placket	0.40	N	N	0.50	Helper Table	0.6	1	0	960
18	Attach pocket to front	0.70	Y	A+	0.88	SNLS w/UBT Prog	1.0	1	1	549
19	Inline Checking	0.35	N	N	0.44	Checker Table	0.5	1	0	1097
		3.15			**4.26**					
	BACK PREPARATION									
20	Attach size label to main label	0.20	Y	C	0.33	SNLS w/UBT Prog	0.4	1	1	1440
21	Attach label to yoke	0.30	Y	C	0.50	SNLS w/UBT Prog	0.6	1	1	960
22	Attach yoke to back	0.70	Y	A	0.93	SNLS w/UBT Prog	1.1	1	1	514
23	Topstitch back-yoke join	0.40	Y	B	0.62	SNLS w/UBT Prog	0.7	1	1	780
		1.60			**2.38**					
	ASSEMBLY									
24	Join shoulder	0.70	Y	A	0.93	SNLS w/UBT Prog	1.1	1	1	514
25	Attach collar to body and finish	1.70	Y	A+	2.13	SNLS w/UBT Prog	2.4	2	2	452
26	Attach sleeve	0.75	Y	A+	0.94	SNLS w/UBT Prog	1.1	1	1	512
27	Topstitch armhole	0.70	Y	A	0.93	DNLS F/B	1.1	1	1	514
28	Join side seam	0.90	Y	B	1.38	FOA 2N	1.6	2	2	693
29	Attach cuff to body	1.20	Y	A+	1.50	SNLS w/UBT Prog	1.7	2	2	640
30	Bottom Hem	0.75	Y	A+	0.94	SNLS w/UBT Prog	1.1	1	1	512
31	End Line inspection	1.20	N	N	1.50	Checker Table	1.7	2	0	640
		7.90			**10.25**					
	Total SAM's	**18.80**			**25.13**		**28.4**	**35**	**27**	

Annotations in figure: Operation breakdown; 0.4/60%; Standard allowed minutes; ACT' time X target per day / minutes available per day (0.67 X 550/480); (Minutes available per day /ACT' time) X no. of operators

Figure 8.2 Assembly line plan—conventional.

are divided into three sections (work cells): sewing preparatory, sewing preassembly, and sewing final assembly (refer to Fig. 8.4). Some operations (refer to Fig. 8.3) are clubbed to improve resource utilization and single-piece flow in the final assembly that are highlighted in yellow. The preparatory stage comprises a collar, cuff, and sleeve section. The two preassembly sections comprise back and front preparatory or subassembly sections, respectively.

A rack is installed to receive the cut panels from the cutting sections, with a maximum permissible WIP equal to hourly consumption. Back, collar, cuff, and sleeve sections are fed simultaneously in bundle form, while the material movement at the subassembly section (front and back) and the final assembly section is maintained in single-piece form. The maximum of one bundle (with bundle size as 15 pieces) is allowed between two workstations. As soon as the back section is

			Operation Bulletin: Long Sleeve Shirt								
						Parameter					
	Projected Output per shift		452			Minutes Per Day	480		No Of Operators		28 (Inc Absm)
	Man to Machine Ratio		1.4:1			Total Number of Machines	25		Total SAM		18.6
	Out Put at 100%		825			Actual Efficiency	55%		No Of Workplaces		32
	Absenteeism		10%			Pcs Per Operator Per Day	18				

Clubbed Oper'	Sl No.	OPERATION	SAM	M/C Opr	Opr. Grade	ACT TIME	MACHINE TYPE CLASS	OPERs CAL	Man Power Req.	M/Cs CAL	Projected Output Per Shift (8 hrs)
		COLLAR									
	1	Runstitch collar	0.40	Y	C	0.67	Autojig	0.8	1	1	720
	2	Clip and turn collar and cuff 0.70		N	N	0.88	Collar Turning Machine	1.0	1	0	549
	3	Topstitch collar	0.45	Y	B	0.69	SMLS w/UBT	0.8	1	1	693
	4	Hem collarband	0.35	Y	C	0.58	SMLS w/UBT	0.7	1	1	823
	5	Attach collar band to collar and top stitch	0.80	Y	A+	1.00	SMLS w/UBT	1.1	1	1	480
			2.70			**3.82**					
		CUFF PREPARATION									
	6	Hem and Run Stitch Cuff 0.60		Y	B	0.92	SNLS w/UBT Prog	1.1	1	1	520
7,6	7	Run Stitch Cuff 0.00		Y	B	0.00	SNLS w/UBT Prog	0.0	0	0	
8,2	8	Clip & Turn Cuff 0.00		N	N	0.00	Collar Turning Machine	0.0	0	0	
			0.60			**0.92**					
		SLEEVE PREPARATION									
	9	...under sleeve placket	0.50	Y	A	0.67	SNLS w/UBT Prog	0.8	1	1	720
	10	Attach and finish Upper Sleeve Placket	1.40	Y	A+	1.75	SNLS w/UBT Prog	2.0	2	2	549
	11	Inline Checking	0.85	N	N	1.06	Checker Table	1.2	1	0	452
			2.75			**3.48**					
		FRONT PREPARATION									
	12	Sew buttonhole placket	0.35	Y	A	0.47	Kansai	0.5	1	1	1029
	13	Sew button placket	0.60	Y	B	0.92	SNLS w/UBT Prog	1.1	1	1	520
	14	Hem pocket and attach pocket pkt to front	1.30	Y	B	2.00	SNLS w/UBT Prog	2.3	2	2	480
	15	Iron Pocket	0.35	N	N	0.44	Helper Table	0.5	1	0	1097
	16	Clip Pocket and placket	0.40	N	N	0.50	Helper Table	0.6	1	0	960
17,14	17	Attach pocket to front	0.00	Y	A+	0.00	SNLS w/UBT Prog	0.0	0	0	
	18	In Line checking and matching front and back	0.00	N	N	0.00	Helper Table	0.0	0	0	
			3.00			**4.33**					
		BACK PREPARATION									
	19	Attach size label to main label	0.50	Y	C	0.83	SNLS w/UBT Prog	1.0	1	1	576
20,19	20	Attach label to yoke	0.00	Y	C	0.00	SNLS w/UBT Prog	0.0	0	0	
	21	Attach yoke to back and top stitch	1.00	Y	A	1.33	SNLS w/UBT Prog	1.5	2	2	720
22,21	22	Topstitch back-yoke join	0.00	Y	B	0.00	SNLS w/UBT Prog	0.0	0	0	
	23	Inline Checking	0.35	N	N	0.44	Checker Table	0.5	1	0	1097
			1.50			**2.17**					
		ASSEMBLY									
	24	Join shoulder and Top Stitch	0.70	Y	A	0.93	SNLS w/UBT Prog	1.1	1	1	514
	25	Attach collar to body and finish	1.70	Y	A+	2.13	SNLS w/UBT Prog	2.4	2	2	457
	26	Attach sleeve	0.75	Y	A+	0.94	SNLS w/UBT Prog	1.1	1	1	512
	27	Topstitch armhole	0.70	Y	A	0.93	DNLS F/B	1.1	1	1	514
	28	Join side seam	0.90	Y	B	1.38	FOA 2N	1.6	2	2	693
	29	Attach cuff to body (attach and finish)	1.20	Y	A+	1.50	SNLS w/UBT Prog	1.7	2	2	640
	30	Bottom Hem	0.75	Y	A+	0.94	SNLS w/UBT Prog	1.1	1	1	512
	31	End Line inspection	1.20	N	N	1.50	Checker Table	1.7	2	0	640
			7.90			**10.25**					
		Total SAM's	**18.63**			**25.18**		**29**	**32**	**25**	

(Annotation: Clubbed operations (7 clubbed with 6))

Figure 8.3 Assembly line plan (modified).

completed (at operation no. 11), the pieces are transferred to matching with the front (at operation no. 18).

8.3.3.3 Creating intermediate supermarkets

It is a challenge to design a continuous flow system from the cutting section to the final assembly section, intermediate supermarkets are designed to ensure uninterrupted supply at different work cells (refer to Fig. 8.5). Given the same, a provision of intermediate buffer (supermarket) is provided to control the flow of production at the workstation numbers 11, 18, and 22.

The intermediate supermarkets ensure maintaining minimum buffer inventory between sections to tackle any contingency or fluctuations in the flow (in the case of bottleneck may be due to underproduction, overproduction, or any breakdown).

Process balancing 271

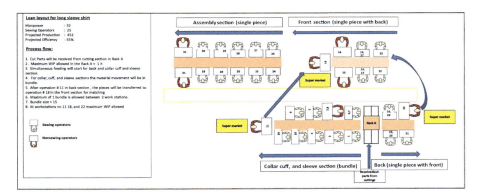

Figure 8.4 Flow-manufacturing setup layout.

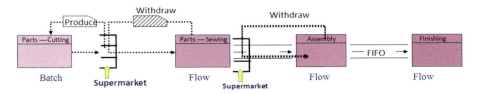

Figure 8.5 Intermediate supermarkets.

To facilitate smooth flow, it is important to define and maintain the buffer quantity at the supermarket stages and the supermarkets should not be treated as the inventory stock points.

This flow-manufacturing model is projected to achieve 12% higher manpower productivity per day and a 1% improvement in the standard allowed minute as compared to conventional assembly line manufacturing. As lean philosophy is all about working on achieving continuous improvement, to make the flow-manufacturing setup sustainable constant efforts should be put through Kaizen interventions.

8.3.4 Challenges in flow manufacturing

Achieving single-piece flow is challenging when it comes to its practical implications. Several factors work as a hindrance in achieving single-piece flow. Apparel manufacturing environments are full of uncertainties. We witness a day-to-day issues (that disturb the plan and of course flow) about the styles which are to be loaded for manufacturing, these issues may be from any area such as nonavailability, shortage or delayed supply of raw materials, defective raw materials, machine breakdowns, machine & equipment settings—related issues, delays in the process setting-up, fabric & trims—related approval delays, pressure from the buyers, and sudden fluctuations in the demand (order quantity changes at the last moment). In addition to this, there are some major hurdles to adopt flow manufacturing, to name a few are the mental blocks to adopt single-piece flow, fear of process balance with lesser WIP, and preference to stay in the comfort zone by the supervisors and managers.

Apart from it, there are some other reasons as well which are related to the nature of apparel manufacturing. For example, practical difficulties with changeover, generally the style changeover times, are high which leads to larger lot sizes. Many times, it is observed that the manufacturing time of the style (with relatively small order quantity) is lesser than that of the changeover time. In such a situation, people prefer going for larger lot sizes. In such scenarios, it is recommended to first focus on reducing the changeover time, and then working on gradually reducing the lot size and the bundle size with close monitoring of manufacturing flow.

In apparel manufacturing, it has been witnessed that most of the time the other traits of flow manufacturing are missed out. Considering the complexity of nature and high level of human intervention, apparel manufacturing setups do not follow the flow manufacturing same as it is followed in other industries especially automobile manufacturing. Generally, in apparel manufacturing, the continuous flow installations are limited to the U-shaped process layout only.

Another common practice observed is a grouping of machines that perform processing steps in a sequence. Such grouping often termed as a "cell," however, achieving a real continuous flow is rarely seen. Technically, an ideal continuous flow system should cover the entire manufacturing process right from cutting the panels to final finishing & packing. Achieving this state of uninterrupted flow may be a difficult proposition in complex manufacturing environments. As a start point, it is recommended to install the flow system at sewing assembly operations. Such a system should be aided with a pacemaker process that regulates the pace (rate) of the output of the preceding and succeeding process and intermittent supermarkets to facilitate the flow of production from the upstream operation. Another thought (on reality grounds) is that considering the nature of apparel manufacturing, where garments are cut in bulk (as it provides an opportunity to save fabric through improved marker utilization), and the high level of human intervention especially for complex products, it may not be practically possible to keep a bundle size of a single piece. Here, we need to see the balance between the cost and the benefit. It may not be a financially viable option to keep the bundle size as small as one. Of course, there should be efforts on reducing the bundle size, and we should determine the optimum bundle size ensuring uninterrupted stable manufacturing flow.

It is important to understand that the lot size reduction (to achieve single-piece flow) cannot be achieved in one go, but it has to be carefully implemented in a step-by-step manner with regular monitoring. Plan−do−check−act (PDCA) may be the right approach for such implementations. Roser (2015a, 2015b, 2015c) suggests that as a start point, the lot size reduction should be implemented only in one part of the plant (where there is a high probability of success) in a controlled manner. Otherwise, such implementations and changes may be taken as a failure and waste (Roser, 2015a, 2015b, 2015c). Since the issues mentioned are present with most of the apparel factories, it is suggested to implement process enhancement tools such as 5S, Kanban, visual management, Heijunka (product and process leveling), Total Productive Maintenance

Process balancing 273

(TPM) (machine reliability and capacity), and rapid setup (style changeover performance) while implementing flow-manufacturing system (Karekatti, 2014).

8.4 Constraint management

Apparel manufacturing is often associated with longer lead times and a high level of WIP. The apparel manufacturing environment is full of inefficiencies that are fueled by internal factors as well the external factors such as the seasonal nature of apparel trade, causing peak and slack demand cycles, and delays in raw material supplies. This leads to the accumulation of huge inventories across the supply chain that is nothing but a Muda where a significant operational capital is blocked without any favorable outcome (Karekatti, 2013a, 2013b). This accumulation of inventories results in unbalanced capacities affecting operational performance. Given the same, apparel companies continuously explore different philosophies to improve their operations. Amongst others, Theory of Constraints (ToCs) provides the simplest solution for production in the form of DBR scheduling, an application that does not require large sets of data, extensive worker training, or lower-level buy-in (Dixit et al., 2010). ToCs, popularly referred to as ToC was introduced by Dr. Eliyahu M. Goldratt in his book *The Goal* in 1984, are a management philosophy that advocates a systemic view of the business. It considers the system as a chain whose strength is governed by its weakest link, the constraint. It implores the elimination of decision making based on local efficiencies and encourages improving the global optimum by exploiting the constraint. DBR is the logistical application of ToCs—a scheduling system providing planning and controlling methods (Goldratt & Cox, 1984).

8.4.1 Drum—buffer—rope system

DBR is an important element of ToC. The performance of the entire system is regulated or controlled by the weakest link. This weakest link is referred to as the drum, rate-limiting process that sets the pace of the system. Other resources, irrespective of their capacity and efficiency, have to function according to the rate-limiting resource (weakest link or drum). To understand it better, let's take an example of an apparel manufacturing setup with spreading & cutting, stitching, dry process, washing, and finishing & packing processes. Assuming the daily output capacity of finishing & packing section is 10,000 pieces in a shift of 8 hours. The other spreading & cutting, stitching, and washing except the dry process are capable of producing 10,000 pieces to meet the output requirement of the system. However, it is noticed that the dry process is heavily loaded due to the demand for dry process effects (such as handstand, whiskers, chevrons, destructions, and laser effects) which are relatively high time-consuming activities. Hence, the maximum output from the dry process is only 8000 pieces. In such a situation the dry process works as a limiting resource (drum) and results in the daily output of washing and finishing & packing process as 8000 pieces only despite having the capacity to produce more.

In such a case, to maximize the output of the system, efforts should be directed toward protecting the dry process (drum) from any kind of disruption. According to ToC, the overall improvement of the system can be done by strengthening the weakest link (here dry process). The strengthening of the dry process may be achieved by providing "time buffers" and synchronizing or subordinating all other resources and decisions to this process. Here the synchronizing or subordinating mechanism is termed as "rope."

To achieve a stainable overall improvement the process of seeking and strengthening the weakest link in a system is a continuous activity. As soon as the weakest link is strengthened (or improved), the focus should be directed to find out the next weakest link.

8.4.2 Implementing drum−buffer−rope system

There is at least one constraint present at any point of time in a system. In fact, in an apparel manufacturing environment, they are generally more than one constraint or limiting resources available. ToCs recommends focusing on the weakest link (which is the biggest constraint at a particular time), and strengthening of same for the overall improvement of the system. Steps of implementing the DBR system are mentioned in Fig. 8.6. It can be seen from the figure that the loop to be repeated constantly, as soon as a constraint is resolved.

As step 1, identifying and exploiting the constraint (drum) can be achieved by preparing and posting a detailed production schedule of this process (here dry

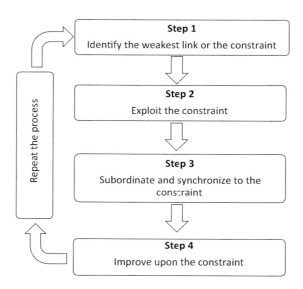

Figure 8.6 Implementing drum−buffer−rope system.

Process balancing

process). A production schedule is a primary document that reveals the overloading of the processes against available capacity, delays, accumulation of inventories and formation of bottlenecks, and other related issues. To strengthen the constraint process, it is important to first focus on improving the utilization of existing resources. For this, it is vital to identify the potentially problematic areas that are yet to be utilized to the fullest. The exploitation of the process (strengthening the process) can be initiated by applying performance enhancement measures such as appropriate work allocation, capacity leveling or balancing of subprocesses, improving upon the quality management system, working on reducing machine idle times and breakdowns, and setup reduction. This will further help in enhancing sewing output.

As soon as the constraint process is scheduled and exploited, it is important to synchronize and subordinate all the other processes to the constraint process. This can be understood in a way that all the supporting processes shall work toward fulfilling the requirements of the constraint process (drum), hence it cannot be kept idle or waiting for the feeding (input) from the previous process, or all the output of the constraint process shall be moved forward (or accepted) by the next process to avoid creating any bottleneck. The buffers (in the form of safe inventories) should be planned to tackle the fluctuation and preventing the constraint process from getting idle and unproductive. The management of this buffer, which is also referred to as rope in ToC, can be done effectively by precise planning (either planning or backward planning depending on the requirement).

Such a well-tuned rope mechanism provides better control over the entire process. Precise functioning of rope can help in reducing WIP as well as the lead time. This control can be selectively modified and utilized according to the demand patterns of finished goods. In the case of high demand, the improved (newly exploited) dry process capacity can be utilized to pump out excess WIP, while in normal times with static demand, the feeding can be controlled and the drum can produce just what is required to meet the shipment schedule.

8.5   Production leveling

The typical apparel manufacturing environment presents real day-to-day challenges. Most of the time of a production planner (the department is called process planning & control) is consumed in firefighting instead of actual planning and controlling. There are several instances when planners are busy adjusting and readjusting the plans, this creates chaos in the entire manufacturing process. The reason for the same is the inability to understand the demand patterns in advance; hence, the process scheduling is adjusted or now and then. Further, in the traditional approach of planning, it is preferred to run longer runs of same/similar styles to keep minimum start-up losses and avoiding frequent changeovers. Up to some extent, the wish for working with longer runs is good as well, as start-up and changeover are never value-added activities. But the core issue is a slower start-up and longer changeover times, which are generally not addressed properly. Provided, there is a rapid start-up and quick changeover, there should not any issue with frequent changes.

In such scenarios the traditional approach of planning brings uncertainties within and outside the organization. Due to sudden changes in the plan, there is pressure on the upstream processes. For example, if the stitching plan changes suddenly, it may lead to a sudden and urgent need for cut panels for style X, while the style Y, which was there in an earlier schedule, is ready with cut panels but is now not required due to the changed priorities. Hence, the cut panels of style Y maybe just lying in the cutting section. Also, sewing and packing trims for style X may be required immediately, and there are chances that the trims store is not ready with these trims or the trims may be in transit. Such situations are very common in most of the apparel manufacturing setups, and it results in higher inventory levels greater opportunities for defects, idle time, and/or overtime. Moreover, facilitating longer runs of similar style can lead to the overproduction of garments not in immediate demand. A relatively nonfluctuating demand and an even mix of styles during production are critical to avoid these impacts.

8.5.1 Heijunka approach for production leveling

Heijunka is load leveling which is based on running small batches of different styles (mix production) in the intermediate process; in this way, it brings an evenness of load in the entire manufacturing process. According to Matzka, Mascolo, and Furmans (2012), Heijunka is a key element of the TPS (Toyota Production System) that levels the release of production Kanban to achieve a uniform flow. In the case of apparel manufacturing, suppose there are three key processes—cutting, sewing, and finishing & packing, then products of different styles in small lots may be produced in sewing (intermediate process) (Karekatti, 2013a, 2013b). The production rate (or flow) at the intermediate process is kept as constant in such a manner that other processes (cutting, finishing & packing, etc.) can be planned accordingly with more predictable output. As the flow rate is predictable with lesser fluctuations (due to lead leveling), the common facilities (centralized setups with specialized operations such as dry process effects and special washing effects) can be utilized effectively and efficiently.

Production leveling (demand smoothing) and product leveling are two key aspects of Heijunka. Production leveling (demand smoothing) is based on the long-term prediction of demand for the products and aims to achieve a production run with constant output meeting the predicted demand. In such situations, production scheduling is done based on the predicted long-term demand only and the short-term fluctuations in the demand are ignored. Based on the demand pattern, the operations and inventories can be managed (refer to Table 8.3) by demand smoothing on a four weekly basis. The demand smoothing can be done even for a weekly basis, fortnightly basis as well depending on the specific nature of manufacturing.

While doing production scheduling through demand leveling, there may be instances when the weekly demand is more than the leveled production (refer weeks 1, 3, 6, 8, and 12 in Table 8.3). In such situations, it is recommended to maintain

Process balancing 277

Table 8.3 Production leveling with smoothened demand.

Week	Weekly demand	Weekly leveled production (4 week cycle)	Difference
1	11,055	10,952	−103
2	10,097	10,952	855
3	12,100	10,952	−1148
4	10,556	10,952	396
5	9020	9555	535
6	9700	9555	−145
7	8920	9555	635
8	10,580	9555	−1025
9	10,800	10,865	65
10	9752	10,865	1113
11	10,124	10,865	741
12	12,784	10,865	−1919
Total	125,488	Total difference	0

Table 8.4 Buyer-wise order quantities and delivery schedule.

Buyer	Style	PO	Order quantity	Delivery date
A	@	A-1	12,000	15-Mar
	@	A-2	3000	30-Mar
	@	A-3	3000	07-Apr
	@	A-4	3000	14-Apr
B	#	B-1	5000	15-Mar
	#	B-2	5000	21-Mar
C	$	C-1	6000	07-Apr
D	X	D-1	6000	14-Apr
Total demand (pieces)			43,000	

a standard inventory of finished products at the end of the production process as a safety buffer. The size of this inventory can be determined based on the fluctuations in customer demand, production process stability, and the shipment schedule.

8.5.2 Heijunka approach for product leveling

The demand leveling can be understood with other examples, assuming there are four orders (refer to Table 8.4) from buyers A (Style @), B (Style #), C (Style $), and D (Style X). The order quantities for buyers "A," "B," "C," and "D" are 21,000, 10,000, 6000, and 6000 pieces respective as per the delivery dates mentioned in Table 8.4. All the orders are to be delivered in 1 month from 15 March to 14 April. The factory can produce an average of 3000 pieces per day. Style A has four POs—A-1, A-2, A-3, and A-4—with different delivery dates, where PO A-1 has the largest order size of 12,000 pieces. There are two POs—B-1 and B-2—each

278 Lean Tools in Apparel Manufacturing

with 5000 pieces for buyer B, while buyers C and D having a single PO of 6000 pieces each.

Following the traditional approach, a planner may plan the orders style-wise and delivery schedule-wise (refer to Fig. 8.7), where the entire quantity of 21,000 pieces (for buyer A) is planned first and then orders of B, C, and D. Here, longer runs are preferred to avoid style changeover and reduce start-up losses. Here, it can be observed that three POs of buyer A are ready much before the delivery schedule (deliveries on 30 March and in mid-April) and maybe lying in the warehouse in waiting for shipment, while the POs of buyer B (which are delivered in mid-March) are a loaded bit late. It may result in a difficult situation for B in case there is any major issue that occurs. This may also impact the deliveries of buyers C and D as well where there is a threat of delayed shipment due to late loading of the orders while production scheduling.

As indicated in Fig. 8.7, an alternate approach of load leveling for option 1 may be applied, where quantities are planned as per the delivery schedule in a balanced mix manner. This starts with the loading of the largest order (PO A-1 with 12,000 pieces to be shipped at the earliest in the schedule), followed by orders (B-1 and B-2) of buyer B. Then, after completion of orders of buyer B, PO A-2 is loaded

Traditional approach		Demand leveling approach — Option 1		Demand leveling approach — Option 2	
Buyer A	500	Buyer A	500	Buyer A	500
Style @	800	Style @	800	Style @	800
Quantity 21,000 pieces	1000	Quantity 12,000 pieces	1000	Quantity 12,000 pieces	1000
PO A-1, A-2, A-3, A-4	1500	PO A-1	1500	PO A-1	1500
	2000		2000		2000
	2500		2500		2500
	3000		3000		3000
	3000		700		700
	3000	Buyer B	1000	Buyer B	1000
	3000	Style #	1500	Style #	1500
	700	Quantity 10,000 pieces	2000	Quantity 10,000 pieces	2000
Buyer B	500	PO B-1, B-2	2500	PO B-1, B-2	2500
Style #	800		3000		3000
Quantity 10,000 pieces	1000	Buyer A	1000	Buyer A	1000
PO B-1, B-2	1500	Style @	1500	Style @	1500
	2000	Quantity 3000 pieces	500	Quantity 3000 pieces	2000
	2500	PO A-2	1000	PO A-2, A-3, A-4	2500
	1700	Buyer C	1500		2000
Buyer C	500	Style $	2000	Buyer C	1000
Style $	800	Quantity 6000 pieces	1500	Style $	1500
Quantity 6000 pieces	1000	PO C-1	1000	Quantity 6000 pieces	2000
PO C-1	1500	Buyer A	1500	PO C-1	1500
	2000	Style @	2000	Buyer D	1000
	200	Quantity 6000 pieces	1500	Style X	1500
Buyer D	500	PO A-3, A-4	1000	Quantity 6000 pieces	2000
Style X	800	Buyer D	1500	PO D-1	1500
Quantity 6000 pieces	1000	Style X	2000		
PO D-1	1500	Quantity 6000 pieces	1500		
	2000	PO D-1			
	200				

Figure 8.7 Production scheduling (traditional approach vs load-leveling approach).

followed by C-1, and at the last PO A-3 and A-4 (of buyer A considering deliveries scheduled at the later stage). As another option (refer Option 2 of Fig. 8.7), to avoid changeover and setup losses, quantities of PO A-2, A-3, and A-4 may be clubbed and can be loaded after the orders of buyer B (PO B-1 and B-2). Product leveling will ensure that demand for parts upstream is also leveled thereby reducing the amplitude of bullwhip effect across the supply chain. Here, it is to be noted that to get the best results of load leveling, it is important to improve upon the changeover and setup, otherwise frequent style change may lead to inefficiencies, cost, as well as disturbances in the flow. The popular lean tool Single Minute Exchange of Die (SMED) may be immensely useful in achieving quick changeovers and rapid setups.

8.5.3 Heijunka box

Heijunka box is a visual scheduling board where Kanban cards are placed. A Heijunka box is generally a wall-mounted board with pockets or slots to keep the Kanban cards. Heijunka box plays an important role in load leveling and flow management, as this is the place from where necessary instructions (to replenish or reproduce) are given to the concerned process to meet the demand. Hence, the Heijunka box is referred to as a leveling box as well (Lean Enterprise Institute, Marchwinski, Shook, & Schroeder, 2008, pp. 29–30).

The rows in a Heijunka box indicate products or product families, while the columns represent the period. The Kanban cards are placed in respective slots according to the product and its quantity required. The visual representation of the box makes it easy to understand the status of the job sequence at a glance. By issuing the Kanban cards (according to product leveling or production leveling) at definite intervals, Heijunka the box enables load leveling and ensures uninterrupted flow without the creation of bottlenecks.

8.6 Kanban

Kanban is a key lean tool that plays a vital role in practical implementations of flow manufacturing and Heijunka. Kanban is so important that it is considered as the operating system of the TPS (Ohno, 1988, p. 93). Kanban is a Japanese word that is referred to as "sign" or "signboard." A Kanban may be a card, cart, container, instruction card, visible record, doorplate, or a poster as well. It can also be in the form of a marked square (referred to as Kanban Square) using tape or color. Usually, Kanban signals are physical (such as card, cart, label, marked sign), but it can also be communicated digitally (through e-mail, EDI, ERP, Barcode, QR code, RFID, etc.), such Kanban is revised as e-Kanban (Roser, 2019), Kanban communicated through fax are also termed as fax ban (Hill, 2012, p. 133). According to Ohno (1988), the idea of Kanban was first practiced in Japan (in the 1940s at the Toyota's machine shop); however, the same concept was applied in the US supermarkets (the first US-style supermarket appeared in Japan in the 1950s), where the customers used to get what was needed (the right product), at the time needed (the right time) and in the amount needed (the right quantity). Kanban is a pull-based (demand-driven) signaling device that is used to control (through authorization and instructions) the production or withdrawal of items (Lean Enterprise Institute, 2008). Kanban works as a key element of a controlling mechanism, where instructions related to initiating or halting an activity (maybe a service or producing goods) are communicated to achieve "JIT" (Monden, 2012).

8.6.1 Kanban card

Kanban cards are pull-based visual signals that trigger the requirement of material as demanded by the downstream process. The Kanban cards also play a role in controlling (and authorization) the flow of materials (inventories) in the process; hence, these cards are critical to the productivity and WIP of the entire system. A Kanban card works as a token to initiate the production or replenish the items to maintain the decided inventory level. Table 8.5 and Fig. 8.8 indicate the necessary information a Kanban card should have on it. These days Kanban is also referred to as information (where earlier it was referred to as a "sign") that is attached to the process, which initiates the replenishment after the part is used (Roser, 2017).

8.6.2 Types of Kanban

Monden (2012) classifies the Kanban into two classes: production and withdrawal. The production Kanban can be further classified into ordinary production Kanban and triangular Kanban, while withdrawal Kanban can be classified interprocess Kanban and supplier Kanban (Monden, 2012, p. 41). Withdrawal Kanban (also called conveyance Kanban and transportation Kanban) specifies the type and quantity of the product to be withdrawal by the downstream process from an upstream process (Akturk & Erhun, 1999). The production Kanban specifies the type and quantity of the product which is to be produced by an upstream process to meet the demand as mentioned in the Kanban card.

Ordinary production Kanban is generally used for producing the items which are not produced in lots, while the triangular Kanban is used for lot productions. Interprocess Kanban is used for parts or item withdrawals within a factory. For example, replenishment of packed garments by the finished goods warehouse from the supermarket between the finishing & packing section and the finished good warehouse can be done using interprocess withdrawal Kanban. If the goods are consumed by an external agency (such as the distribution center of the buyer) from the finished good warehouse of a vendor, then supplier withdrawal can be used.

Apart from these basic classifications, there are many other types of Kanban as well that are in practice by different industries. Such additional Kanban types include express Kanban (issued in extraordinary situations when there is a shortage of parts or items and collected just after its use), emergency Kanban (issued temporarily in emergencies when some parts or items are required to rectify defective

Process balancing 281

Table 8.5 Information on Kanban card.

Information	Details	Example in the context of apparel manufacturing
Card serial number	The serial number of the card (e.g., if there are 20 cards for button replenishment, then each card can be numbered as 1–20), in case of a card missing, tracking can be done using card number	Card no. ST-B-9
The item number or item code	The item number of the product, which can also be in the form of Alpha-numeric code like AXP3256	For example, if it is a sewing trim (button) for a particular orderST-B-L14-BL-4358 (sewing trim–button–size–color–order reference)
Item name	**Name of the item**	**Button**
Quantity	How many units are required (number of items needed)	1700 (assuming a shirt having 10 buttons, and it is required for 170 shirts)
Unit	Unit of measurement (meter, gram, kilogram, liter, gross, etc.), sometimes it may be like how many packets, boxes, rolls, pallets, etc. as well	12 Gross (12 × 12 dozen) = 1728 buttons (1 gross = 12 dozens) Or1 Packet (if there are 12 gross buttons in a packet)
Item image	Illustration or image of the item may be placed on the card for better understanding	The image of the button desired to indicate the correct color and size or an image of the desired button attached to the garment so that the people can also connect with the purpose in the end product
Packaging type	Type of package, like a bundle, roll, crate, palate, box, bottle, carton, trolley, tray, bin, etc.	Box (in context of buttons)
Customer information	The name of the consuming process where the item is being used	Sewing process (floor number, line number, etc. also may be mentioned)
Supplier information	The name of the internal and external supplier	Internal supplier—Trims StoreExternal supplier—Maybe the brand name of the button or the supplier company name
Supermarket detail	Name or code of the supermarket	S-ST-1
Other information	Barcode, QR code	In case there is any requirement for digital transfer of information

S-ST-1, Supermarket-Sewing trims-1; ST-B-9, Stitching-Button-9.

pieces, machine maintenance–related issues, sudden emergency of making some extra production, etc., such Kanban should be just collected after its use), job-order Kanban (issued for specific job order), through Kanban (when two or more

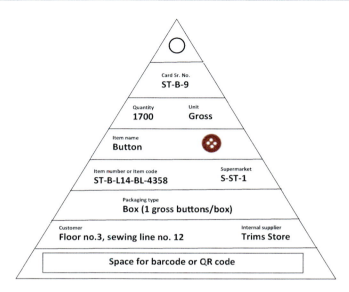

Figure 8.8 Triangular Kanban card.

workstations of processes are so closely linked that it can be treated as one process only, and there is no need of issuing separate Kanban. Such Kanban is also referred to as Tunnel Kanban.), common Kanban (a single Kanban is used as withdrawal Kanban as well as production Kanban, this is generally done when two workstations or processes are situated very near to each other). Apart from it, a trolley, cart, crate, bin, truck, etc. can also be used as Kanban. These types of Kanban are especially helpful when the number of transactions is relatively lesser. In such situations an empty trolley, cart, truck, etc. communicate as a Kanban which needs to be filled in.

8.6.3 Supermarket

Kanban and supermarkets are the backbones of pull systems. Supermarkets play a vital role in the successful functioning of Kanban systems. Supermarkets work as a bridge between the consuming process and the supplier (producing process). A buffer inventory is maintained at the supermarkets to the ensure feeding of items and materials in the right amount at the right time. Adequate replenishment and producing parts at the right time in the right quantity become more vital in complex working environments such as apparel manufacturing environment. Supermarkets are not just the stock place of inventories but much beyond this. Supermarkets are logically designed and installed facilities that work according to the laid rules only. Roser (2015a, 2015b, 2015c) discusses three conditions for a supermarket:

1. *The products are organized as per their type:* The items or parts are stored physically in the group of their types. In a supermarket, there may be several different parts, but this

has to be placed in designated cells in an organized manner. It helps in a quick assessment of the available and issued quantities.

2. *FIFO is maintained:* The FIFO sequence should be followed which means that the part of an item of a type that entered in the supermarket as first (oldest) should be issued at the earliest.

3. *Triggering the demand to replenish or reproduce:* As soon as a part of an item leaves the supermarket, the demand signal should be triggered for replenishment or reproducing the item. It is an important requirement of flow management to maintain the WIP levels as per specific requirements. Also, the fluctuations in the demand (based on the consumption pattern of the downstream process) should be addressed in a balanced manner (it should neither starve nor block the bottleneck) while signaling for replenishment or reproduction. Such signals for replenishment or reproduction can also be communicated using an indicator light or a color flap. For example, a green light may indicate that inventory at the supermarket is full, yellow light can indicate to upstream process to prepare for replenishing or reproduce, while red light can be used as a signal for immediate requirement. The installation and positioning of the supermarket can be done according to the consuming and supplying processes (in terms of the traffic of parts or items movement), size, volume, and weight of the parts or items, type of packages, mode of transport (used for replenishment), the distance traveled, availability of the space.

8.6.4 *Implementing Kanban*

Kanban plays a critical role in achieving the desired objectives of lean transformation. The successful implementation of Kanban can bring a new wave of improvements in the overall manufacturing environment of an organization. As it controls the flow, the impact on inventories can be significantly visible if applied effectively and efficiently. It is imperative to do some preparation (as groundwork before implementing Kanban). Such preparation provides a valuable opportunity to become better equipped to implement a sustainable Kanban system. Ortiz (2015) discusses different planning aspects as preliminary work for an effective and user-friendly Kanban implementation (Ortiz, 2015). The preliminary work involves some key activities (refer Table 8.6) such as preparing a Kanban sizing report, cycle counting, identification of parts, vendors, inventory quantities, on-hand inventory, and calculating the cost of on-hand inventory.

The report prepared based on the preliminary activities performed can be utilized in determining the requirements related to the storage area and work area. The information related to the new inventory levels (maximum and minimum inventories, reorder levels) can be utilized to determine the size of bins, shelves, supermarket size, and floor space.

Kanban is a pull-based material replenishment mechanism, where the replenishment fulfillment of the item consumed takes place as per the information of the Kanban card. The functioning of a Kanban is illustrated in Fig. 8.9.

The functioning of the Kanban can be understood with a simple example of replenishment of stitched pieces in the finishing & packing process. As illustrated in Fig. 8.10, a supermarket is installed between the sewing process and the finishing & packing process. The withdrawal of stitched pieces for finishing & packing is

Table 8.6 Preliminary activities to implement Kanban.

Activity	Key elements	Objective
Creating a Kanban sizing report	• Department-wise categorization of inventory • Part description • Part number • Vendor details • Cost of the part • Number of parts available at the moment • Cost of the parts available at the moment	To perform cycle counting and to get the baseline information
Cycle counting, part identification, vendors, and on-hand inventory	• Tagging of the parts with necessary information (as indicated in previous activity) • Conducting cycle counts	To improve recording of parts to match it with Kanban cards

Identification of inventory quantities	• Careful going through of each part, its use, lead times, and consumption pattern to decide
	  ◦ Max. level of new inventory
	  ◦ Min. level of new inventory
	  ◦ Reorder level (maintaining safety stock)
		To assess the appropriate inventory levels of parts

Calculating the on-hand cost of new inventory	• Cost calculation of the new inventory (new on-hand inventory) • Comparing the new on-hold cost of inventory with the previous cost	To identify the financial benefits

done from the supermarket. Assuming a bundle of 50 pieces along with the withdrawal Kanban is taken place. The withdrawal Kanban card is removed in the finishing section, and the same card is placed in the Kanban postinstalled nearby. Then the withdrawal Kanban card is taken from the Kanban post and put in the supermarket. This is the trigger point for the demand for new stitched pieces to be replenished in the supermarket. Then, a production Kanban card is issued and moved to the Kanban board (which works as Heijunka Board as well for production leveling). From the Kanban board, the production instructions are released to the sewing process in FIFO manner. After receiving the production Kanban card (issued from the Kanban Board) the pieces are produced according to the demand triggered. Keeping a buffer of 50 pieces, assuming stitching to produce 100 pieces, then two production Kanban cards are issued to the stitching process. As soon as the pieces are stitched, these are moved to the supermarket as a replenishment to the consumed items by the finishing & packing process.

Process balancing

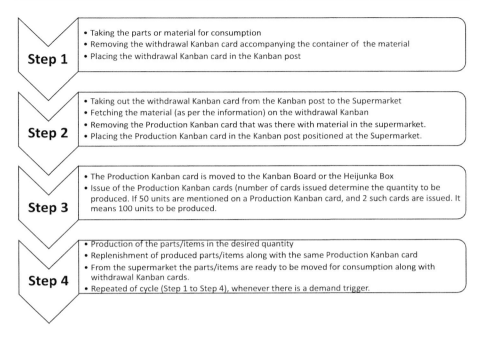

Figure 8.9 Steps in Kanban implementation.

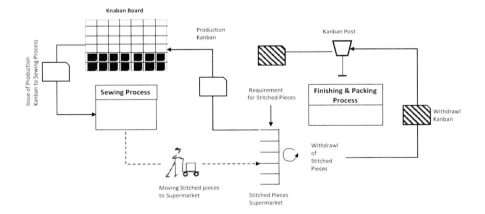

Figure 8.10 Functioning of Kanban.

In the entire mechanism of Kanban functioning the discipline has to be maintained and some rules should be followed. Monden (2012) and Deshpande (2016) discuss such rules in the context of apparel manufacturing:

1. *Predetermined consumption:* The downstream processes to consume the materials or pieces in a limited predetermined manner to maintain the flow and WIP. Consumption more than the decided limit is not allowed.

2. *Predetermined production:* The upstream processes that are responsible to supply the material or items shall be produced that are mentioned in the Kanban card instructions. Producing more than the decided limit is not allowed. This is avoiding flooding the downstream process, which is an indication of a push approach, while Kanban is a pull-based approach.
3. *Quality pass pieced to be moved ahead:* Working on a defective piece is adding cost to the process, and it is considered as a nonvalue activity that increases the burden on the upstream processes. Hence, defective parts, pieces, or items must not move forward to the downstream processes.
4. *Adjustments as per demand pattern:* Adjusting according to the demand pattern is an important factor for the success of flow manufacturing. There should be effective and timely communications (through Kanban cards) in case there is a fluctuation in the demand. It enables upstream processes to respond appropriately and avoid the creation of imbalances and bottlenecks.
5. *Limited numbers of Kanban:* As soon as the process is settled, the number of Kanban should be reduced, as a greater number of Kanban are a wasteful activity.
6. *Follow the rules:* To make the Kanban system effective, efficient, and sustainable, it is important to follow the rules laid down, hence the abovementioned rules (1−5) are nonnegotiable (Deshpande, 2016; Monden, 2012).

8.6.5 Determining the number of Kanban cards

Determining the number of Kanban cards is an important aspect, as it directly affects the inventory levels as well as the replenishment cycles. Vatalaro and Taylor (2005) suggested a Kanban equation to determine the number of Kanban required.

$$\text{Number of Kanban} = \frac{\text{Average daily demand} \times (\text{order frequency} + \text{lead time} + \text{safety time})}{\text{Container quantity}}$$

This can be understood with a simple example of apparel manufacturing. Assuming the factory is working for 8 hours a day with a daily requirement of stitched garments for a finishing & packing section is 5000 pieces. Order frequency (from finishing & packing section to stitching section) is twice a day. The lead time to produce the pieces required and make it available in the supermarket is 4 hours. Safety time to tackle any abnormality is 2 hours. The stitched pieces are transported to the finishing & packing section in a caged trolley with a capacity of 250 garments. Then, using the equation mentioned above, the number of Kanban can be determined as:

Average daily demand: 5000 pieces
Order frequency: 2
Lead time: 4 hours (or 4/8 = 0.5)
Safety time: 2 hours (or 2/8 = 0.25)
Container capacity: 250 pieces

$$\text{Number of Kanban} = \frac{5000 \times (2 + 0.5 + 0.25)}{250} = 55 \text{ Kanban}$$

8.7 Alternatives to Kanban

In the manufacturing environment apart from Kanban, there are few alternative techniques used to control the WIP. Constant WIP (CONWIP) and paired-cell overlapping loops of cards with authorization (POLCA) are two such techniques that are widely used in industries.

CONWIP:It is one of the alternatives to Kanban (Ip, Huang, Yung, Wang, & Wang, 2007; Spearman, Woodruff, & Hopp, 1990). CONWIP is a planning and control technique that focuses on maintaining a constant WIP in the process. This works on constantly issuing the feeding (in fix quantity) at the first workstation as soon as the last step produces the same quantity as output. It can also be referred as to a single Kanban used for the entire manufacturing process. Given the same, it is also referred to as single-stage Kanban and extended Kanban as well (Dallery & Liberopoulos, 2000).

Let's take an example of a sewing line producing 1200 pieces in a shift of 8 hours at an apparel manufacturing plant. The average hourly output is 150 pieces from the last workstation of the line. As soon as the end-line quality checker clears the pieces after quality check, the same may be flagged by using a light or any other visual control, as finally out from the line. This may communicate the signal to line feeders, and cut panels for 150 pieces may be loaded at the first workstation of the line. Here the buffers at individual workstations in the line are not taken care, also there are no signals followed whenever there is a bottleneck or interruption in the flow. This makes CONWIP easy to implement and operate. Many of the apparel manufacturing setups follow CONWIP as a WIP control mechanism. Simplicity, ease of managing, ability to handle made-to-order (small order quantities with complex customized products) are some of the key benefits of CONWIP.

POLCA: This is a hybrid push—pull production technique that is considered along with Kanban and CONWIP in the context of production management. According to the Encyclopedia of Operations Management, POLCA was developed by Professor Rajan Suri at the University of Wisconsin (Hill, 2012, p. 265). POLCA is most suited for low-volume manufacturing that combines pull-based Kanban systems with push-based MRP systems. According to Roser (2018), POLCA is designed for job-work-centers or cellular manufacturing with smaller order quantities with diversified or customized products. In such POLCA-based systems the jobs move differently from one workstation to another workstation or from one cell to another, and between each possible pair of workstations or cells, POLCA establishes a loop (Roser, 2018). A similar kind of approach is followed by few of the apparel manufacturers in NCR (National Capital Region) of India, where high priced, complex garments (with a lot of handwork and embellishment) in small order quantities (even 30 pieces per order) are manufactured by a set of multiskilled operators (generally 3—4 in number), and they complete the operations by switching over from one workstation to another workstation as and when there is work available. However, the authors are not sure if any POLCA cards are issued in such systems or not, but the basic functioning of the systems looks similar to POLCA.

8.8 Conclusion

Effective and efficient process balancing plays a vital role in transforming an organization into a lean enterprise. It provides much needs responsiveness and flexibility to tackle the dynamic demand patterns without overloading or underloading of the resources. By doing so, several Muda are avoided, including Muda of overproduction, excess inventory, waiting, defects, and unused human potential. Effective flow management and load leveling not only reduce the stress level (by minimizing or eliminating instances of day-to-day firefighting) but also bring stability to the entire supply chain.

To achieve the best results, it is imperative to see the process of balancing holistically, as it is not that can be performed in a stand-alone manner. The processes are very strongly linked with each other within the manufacturing setup as well as outside the factory (with its customers, vendors, suppliers, etc.). Also, the lean tools are interlinked as well. The success of one tool affects the performance of other tools, hence it can be said that the lean tools work collectively. So, given the same, other lean tools such as 5S, visual management, SMED, TPM should be applied to achieve the best results of JIT, flow management, Heijunka, and Kanban implementation.

References

Akturk, M. S., & Erhun, F. (1999). An overview of design and operational issues of Kanban systems. *International Journal of Production Research, 37*(17), 3859–3881.

Appell, K. (2014). The lean post: cost reduction, waste, and purpose. Lean Enterprise Institute. Retrieved from: https://www.lean.org/LeanPost/Posting.cfm?LeanPostId = 239#: ~ :text = Ohno%20says%2C%20%E2%80%9CWhen%20thinking%20about,we%20need%20using%20minimum%20manpower. (Accessed March 2020).

Chopra, S., & Meindl, P. (2013). *Supply chain management: Strategy, planning, and operation.* Pearson Education, Inc.

Dallery, Y., & Liberopoulos, G. (2000). Extended Kanban control system: Combining Kanban and base stock. *IIE Transactions, 32,* 369–386.

Deshpande, A. (2016). Reflexive production system: Use of Kanban. A.R. Ltd., Producer. Retrieved from: https://in.apparelresources.com/business-news/manufacturing/reflexive-production-system-use-of-kanban/. (Accessed March 2020).

Dixit, A., Iyer, V., & Tiwari, M. (2010). *Regaining control: Drum-buffer-rope in made-to-order apparel manufacturing.* Gandhinagar: Department of Fashion Technology, National Institute of Fashion Technology.

Goldratt, E. M., & Cox, J. (1984). *The goal: Excellence in manufacturing.* NY: North River Press.

Hill, A. V. (2012). *The encyclopedia of operations management: A field manual and glossary of operations management terms and concepts.* NJ: Pearson Education, Inc.

Hirano, H. (2009). JIT implementation manual: The complete guide to just-in-time manufacturing . (Vol. III). Boca Raton, FL: CRC Press.

Ip, W. H., Huang, M., Yung, K. L., Wang, D., & Wang, X. (2007). CONWIP based control of a lamp assembly production line. *Journal of Intelligent Manufacturing, 18,* 261−271.

Jana, P., & Tiwari, M. (2018). *Industrial engineering in apparel manufacturing.* New Delhi: Apparel Resources Pvt. Ltd.

Karekatti, C. (2013a). Constraint management the drum buffer rope system. *Stitch World, 11* (10), 28−32.

Karekatti, C. (2013b). Heijunka in production scheduling for garment industry. *Indian Textile Journal, 124*(1), 69−72.

Karekatti, C. (2014). Developing flow manufacturing in apparel factory (A. R. Ltd., Producer). Retrieved May 2019, from Apparel Resources: https://apparelresources.com/business-news/manufacturing/developing-flow-manufacturing-apparel-factory/.

Lean Enterprise Institute (2008). Marchwinski, C., Shook, J., & Schroeder, A. (Eds.), (2008). *Lean lexicon: A graphical glossary for lean thinkers.* Cambridge, MA; Brookline, MA: Lean Enterprise Institute.

Matzka, J., Mascolo, M. D., & Furmans, K. (2012). Buffer sizing of a Heijunka Kanban system. *Journal of Intelligent Manufacturing, 23,* 49−60.

Miltenburg, J. (2001). One-piece flow manufacturing on U-shaped production lines: A tutorial. *IIE Transactions, 33,* 303−321.

Monden, Y. (2012). *Toyota production system: An integrated approach to just-in-time.* Boca Raton, FL: CRC Press.

Ohno, T. (1988). *Toyota production system: Beyond large-scale production.* Portland, OR: Productivity Press.

Ortiz, C. A. (2015). *The Kanban playbook: A step-by-step guideline for the lean practitioner.* Boca Raton, FL: CRC Press.

Roser, C. (2015a). Introduction to one-piece flow leveling − Part 2 Implementation. AllAboutLean.com. Retrieved from: https://www.allaboutlean.com/one-piece-flow-leveling2/. (Accessed April 2020).

Roser, C. (2015b). Theory and practice of supermarkets − Part 1. AllAboutLean.com. Retrieved from: https://www.allaboutlean.com/supermarket-basic/. (Accessed May 2020).

Roser, C. (2015c). How to determine takt times. AllAboutLean.com. Retrieved from: https://www.allaboutlean.com/takt-times/. (Accessed June 2019).

Roser, C. (2017). Anatomy of the Toyota Kanban. AllAboutLean.com. Retrieved from: https://www.allaboutlean.com/toyota-kanban/#more-9512. (Accessed January 2020).

Roser, C. (2018). What is POLCA? AllAboutLean.com. Retrieved from: https://www.allaboutlean.com/what-is-polca/. (Accessed June 2020).

Roser, C. (2019). Should you use physical or digital Kanban cards? AllAboutLean.com. Retrieved from: https://www.allaboutlean.com/digital-kanban/#more-16823. (Accessed November 2019).

Spearman, M. L., Woodruff, D. L., & Hopp, W. J. (1990). CONWIP: A pull alternative to kanban. *International Journal of Production Research, 28*(5), 879−894.

Vatalaro, J. C., & Taylor, R. E. (2005). *Implementing a mixed model Kanban system: The lean replenishment technique for pull production.* Boca Raton, FL: CRC Press.

Apparel manufacturing systems

Manoj Tiwari[1] and Prabir Jana[2]
[1]Department of Fashion Technology, National Institute of Fashion Technology, Jodhpur, India, [2]Department of Fashion Technology, National Institute of Fashion Technology, New Delhi, India

9.1 Introduction

Manufacturing plays a critical role in industrial production irrespective of the type of industry. It has become even more important in the present era of cutthroat competition, where speed and accuracy have become vital for survival. The manufacturing of present times is much more dynamic and flexible than ever and expected to touch newer levels in times to come, though this trend is varying from industry to industry depending on the nature of the industry. This has forced organizations to go beyond lean manufacturing and adopting agility in all aspects (Choudri, 2002). According to the definition given by the Merriam-Webster dictionary, the term manufacturing is used for something made manually or by using a machine (Merriam-Webster, 2004). Manufacturing is also used as a "process" of making something as well. The word manufacturing is also used in the context of the "industry" where productive work is achieved using mechanical power or machinery. Etymologically, manufacturing has roots in a French word, manufacture which is a compound of "manus" and "factura," and "manus" are referred to hand, while "factura" is referred to as making of something (Online Etymology Dictionary, 2011).

The demand and supply dynamics are rapidly changing, and technological interventions making it even more dynamic. Traditional manufacturing practices are no longer relevant, as longer lead times are becoming shorter and order quantities or batches are also becoming smaller. This has put pressure on the manufacturers from all sides and they are left with no choice but to adopt flexible manufacturing systems. In the current times, one cannot compete with keeping huge inventories, and static organizational style as the concept of "Economy of scale" has become a concept of yesteryears.

Beaumont, Thibert, and Tilley (2017) in the *Same Lean Song, Different Transformation Tempo* emphasizes the tempo that has been changing very rapidly. According to them, new technologies, new customer expectations, and new sources of competition of recent times are forcing to adopt new business models. To survive, change is inevitable but it has to be quicker, faster, deeper, and successful (Beaumont et al., 2017). In the context of apparel manufacturing, in the past when

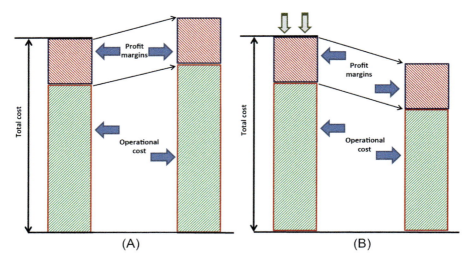

Figure 9.1 Competition and survival (Jana & Tiwari, 2018): (A) no competition and (B) cutthroat competition.

there was lesser competition in the market and manufacturers were able to increase the price of the goods and services to maintain the profit margins, whenever there was an increase in operational costs (refer to Fig. 9.1A). But in the current scenario (refer to Fig. 9.1B), there is tough competition among the vendors and service providers as there is immense pressure on each of them to offer better prices, improved quality, and shorter lead times (Jana & Tiwari, 2018).

At the same time, they have to combat higher costs in terms of higher labor costs, raw material, and other overhead costs that create tremendous pressure on resources and prevent the vendors from increasing the total costs. To survive in the business, one has to continuously work toward minimizing the operational costs by developing better ways of work (Somers, 2015). Achieving operational excellence that is agile but strongly built on the foundation of lean manufacturing principles is the key. Some key manufacturing philosophies and systems in the context of apparel manufacturing shall be discussed in this chapter.

9.2 Manufacturing systems

As indicated in the definition, manufacturing is also a process of making. We can consider manufacturing as a subset of operations where several activities are executed to accomplish the task of getting the work done. Input and output are essential in any kind of manufacturing (refer Fig. 9.2), where a transformation takes place to convert inputs into outputs. Inputs may include raw materials, machinery/equipment, labor/workforce, information/knowledge, space, and time, etc., while

Apparel manufacturing systems

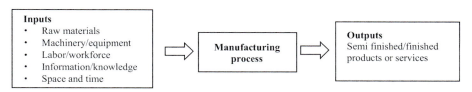

Figure 9.2 Manufacturing process.

the output is generally a semifinished/finished product or services that may be tangible or intangible.

9.3 Evolution of apparel manufacturing systems

Apparel manufacturing is a predominantly make-to-order scenario. Though apparel manufacturing involves three main operation categories—presewing, sewing, and postsewing, the focus is on managing the sewing operations. Apparel manufacturing has started with custom tailoring called whole garment system (Solinger, 1988), and with the increase in volume per style, it was graduated to *make-through* system of manufacturing, where a whole garment is still stitched by one tailor, only nonsewing operations such as cutting, fusing, pressing were done by different persons. The success of assembly line manufacturing in the automobile industry inspired the application of industrial engineering techniques in garment manufacturing and the *assembly line* manufacturing system evolved. The sewing and peripheral machine makers worked in tandem with industrial engineers to develop machines that sew faster, easier, and require less operator intervention once set. The sole objective was to improve productivity thus reduce labor costs. This may be termed as the *fluid stage* of system evolution (refer to Fig. 9.3). Also post–World War, "Baby Boomer" phenomena fueled the growth of department stores and branded apparel in Europe and the United States, which supported the growth of the assembly line production system in apparel manufacturing. Unlike the automobile industry where the maximum of assembly operations are robotized, apparel assembly largely depended

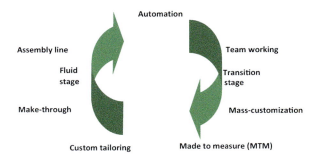

Figure 9.3 Evolution of apparel manufacturing systems (Jana, 2004).

on manual labor, resulting in humane nuisances such as absenteeism, inconsistent work method, and pace, error leading to the quality problem (Jana, 2004).

To realize low labor cost advantage, gradually but slowly apparel manufacturing facilities started migrating from developed to developing countries resulting in huge job losses in the United States, the EU, and Japan. Initially, it was thought that by automating the entire sewing operation, it will be possible to eliminate (or minimize) dependency on human operators; thus migration of apparel manufacturing to low-wage countries can be arrested. Many research initiatives in the United States, the EU, Scandinavia, and Japan spent millions of money and man-hour to develop such laboratory prototypes. Meanwhile, market characteristics also started changing. Flexible specialization thesis (Piore & Sabel, 1986) describes the crisis in manufacturing for mass (or Fordist) markets. Individualization sparked the breakdown of high volume to small volume merchandise; fashion-led merchandise requires quick response (QR) manufacturing, whereas assembly sewing has the slowest response time (due to higher throughput time). All these factors necessitated a desperate search for an alternative production management system that can handle low volume orders, flexible enough to handle frequent style change, faster response time, and ergonomically safer. This stage of development was driven by *time reduction* parameters ably supported by JIT (just in time), QR strategies, and can be termed as *transition stage*.

9.4 Apparel manufacturing systems

According to Solinger (1988), the apparel manufacturing system can be classified into two broad categories: continuous and intermittent. A manufacturing system is considered as continuous when the product being produced is not temporarily stored in the process while manufacturing. A system is treated as an intermittent system when the product is stored temporarily in the manufacturing process between successive processing operations (Solinger, 1988). The discrete category of manufacturing is also referred to as the systems, where discrete products (which can be counted as an individual unit product) are manufactured. Continuous manufacturing is referred to the manufacturing of the products which are measured or metered rather counted. Manufacturing of products such as automobiles, apparel, furniture, computers is some of the examples of discrete manufacturing, while oil and paints, textile fabrics and yarns, and wires are considered products from continuous manufacturing.

From work done or division of labor, manufacturing systems can be classified into three broad categories: make-through, assembly line, and modular system. In the make-through approach of manufacturing an individual operator performs all the sewing process by using different machines exhibiting multiskill capabilities. This is one of the earliest manufacturing approaches. The assembly line system maximum professes division of labor that aims to reduce the work content of any operation by two ways: repeatability of operation resulting rhythm and improved dexterity and use of special-purpose sewing machine and use of attachments and work aids resulting simpler work method with easier hand movements (Jana, 2004). Modular manufacturing or teamwork is a midway approach between make-through and assembly line system. It focuses on the maximum utilization of operators using their multiskilling and team working capabilities. In this approach the division of work is done to a small team of multiskilled operators, who work in a balanced, synchronized, and QR mode to achieve the desired output (Chandler, 1994). There are many synonyms to this teamwork-based approach of manufacturing, some popularly used interchangeable terms for teamwork system include flexible manufacturing, self-managed work teams, QR manufacturing, agile manufacturing, cellular manufacturing, single-piece flow, and Toyota Sewing Systems (popularly known as TSSs).

9.4.1 Make-through

This is a traditional approach and is generally not considered an industrial way of manufacturing. It is a multiskill-based approach because the entire work while manufacturing a product is executed by a single worker. It yields lesser machine utilization as well as inconsistency in the output. Such systems are not suitable for mass manufacturing, as they cannot handle the larger volumes due to the limitation of usage of specialized machinery and de-skilling devices. The setup costs are relatively less, at the same time, cost of work content and labor costs are higher.

Despite all these limitations, make-through systems are unique in terms of some key strengths: easy and economical manufacturing setup, faster throughput, ease of supervision, negligible impact of absenteeism, easily managed work in progress (WIP), lesser chances of musculoskeletal disorders [repetitive strain injuries (RSIs)] as the nature of work is not repetitive, availability of skilled workers, and their high self-esteem (Jana, 2013).

9.4.2 Assembly line

The assembly line way of manufacturing was first introduced by Henry Ford in 1913 for automobile manufacturing. The concept is based on the maximum division of work where operators with unique skill sets were allotted to execute-specific tasks throughout the product manufacturing.

This approach is different from the make-through, and the product is not made by a single operator but several operators contribute by performing their part. Once an operator completes the task, the semifinished product is moved forward to the next activity. This movement of work goes on continuously until the manufacturing of the product is completed. The entire process is divided into different operations or activities that need to be performed sequentially. Special-purpose machines and equipment are arranged accordingly as the setup of the assembly line, and the operators with exclusive skills are deployed to perform predefined tasks on the designated machines. The assembly line may be balanced according to the work content of individual operation and skill availability to achieve optimum output. This manufacturing approach results in improved operator and machine utilization, faster learning curve, improved dexterity, and reduced labor cost. Such systems can handle the larger volumes efficiently which make it a natural choice of manufacturing system suitable for mass manufacturing.

The material (on which manufacturing process to take place) may be in the form of bundles or may be in the form of a single piece. A bundle may consist of different cut panels that are to be assembled to make a complete garment or cut panels of the same single job/task together. The bundle that consists of cut panels of a single task which is to be executed on one or more workstations is referred to as job bundles (Solinger, 1988) or component bundles (Jana, 2013). In case the bundle contains all the parts/panels of a single garment, it is referred to as Garment bundle (Solinger, 1988). The same bundle moves forward in a single-piece flow system by passing through each workstation. Individual operators execute their assigned tasks on the same bundle and eventually, it results in a stitched garment. This system is also called a unit production system or unit flow production system. As an alternative to the bundles, the cut panels may be moved forward using a tray, basket, polybag, or a hanger as well depending on the logistic suitability.

As there is a bundle movement where all the components are tied in a single bundle, this system offers the advantage of lesser material handling and requirement of lesser WIP creating a possibility of achieving lesser throughput time. However, there are also chances of an increased throughput where simultaneous activities are converted into sequential ones as due to bundle limitation, one activity can be performed at a time on a given bundle. To achieve the best results, synchronization among the operations is recommended, which requires accurate line balancing with maintaining equal/nearly equal cycle time for each of the operations. In the absence of precise line balancing, there is always a threat of bottlenecks due to work imbalance. Such systems are also called single-piece flow systems and are generally based on kanban which is a pull-type system. The systems which work on JIT concepts are also inspired by such systems. Progressive bundle systems (PBSs), where bundle has the same component or panels that are to be completed by the operator at the same time, offer flexibility in terms of process control. In such arrangements, simultaneous operations of the same garment can be executed at the same time, which may result in faster throughput. However, such systems lead to an increased level of WIP which is against lean thinking. The most commonly observed layout arrangement of PBS-based assembly line systems has dual parallel lines of sewing machines; however, the sitting arrangement of the operators varies (facing in one direction, facing each other, zig-zag, U-shaped, etc.) as per product-specific requirements. There are some further variants to the assembly line system that also observed in the industrial apparel manufacturing where large order quantities are handled. The entire manufacturing process is divided into parts or sections and accordingly machine layouts are arranged. Such systems are called sectional assembly lines, and parts of the same garment (following PBS approach) are prepared simultaneously in different sections and then these parts are assembled in a separate section to make a complete garment. Such systems result in improved resource utilization and are generally easy to manage the operations. One of the limitations of

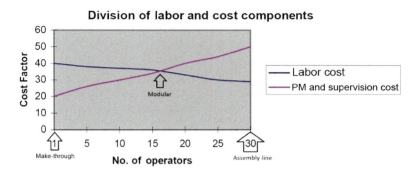

Figure 9.4 Division of labor and cost components (Jana, 2004).

such systems is the requirement of keeping separate section-wise WIP, which results in increased throughput time. A high level of synchronization and coordination is required for the smooth functioning of such systems, otherwise it may result in process imbalance by creating bottlenecks affecting the overall performance of the system. Fig. 9.4 depicts the relation between making through, modular, and assembly line system in the context of the use of workers and cost factor.

9.4.3 Modular manufacturing

Modular manufacturing is considered a supporting system toward achieving the benefits of lean management. According to the American Apparel Manufacturing Association, modular manufacturing is defined as "a contained manageable work unit of 5 to 17 people performing a measurable task. The operators are interchangeable among tasks within the group to the extent practical and incentive compensation is based on the team's output of first quality output" (Wang, Schroer, & Ziemke, 1991). Several advantages make this system most suitable to be used in lean practices; also, it proves this kind of manufacturing as an ideal solution that is extensively used in recent years. Modular manufacturing is a type of flexible manufacturing systems, which provides faster and efficient throughput and better balancing of work by keeping lower WIP and needing less space. The modular system enhances the multiskill development of operators, which helps in QR to style change with achieving quality standards. Modular manufacturing is a teamwork-based system where multiskilling, operator utilization, and team spirit are utilized at the best. This indeed results in achieving higher productivity and quality standards without compromising the ability to handle larger production quantities. At the same time, lesser levels of WIP ensure quick throughput, which makes this manufacturing system most preferred in current times. The concept has gained popularity with different interchangeable terms such as cellular manufacturing, flexible manufacturing, QR manufacturing, TSS, and agile manufacturing. Though these systems are referred to as variants to modular manufacturing the basic philosophy remains the same which is nothing but "Teamwork."

There are multiple drivers behind the rise of modular manufacturing in apparel; and interestingly those are complementary. First, the setback of automation; while by the early 1990s the researchers, consultants, and industry realized that full automation of sewing line was not possible (Jana, 2004), triggering efforts toward worker empowering. Second, the birth of QR system, fashion-led retailers started demanding frequent style change, small quantity, and quicker lead time, prompting manufacturers to dump assembly line manufacturing, which is volume-led, in-flexible to style change, and require long lead time. Third, the rising concern over the ergonomic vulnerability of sewing workers (Textile World, 1991), as it was established that sewing workers in assembly line manufacturing are vulnerable to RSI due to repetitive motions and awkward posture. And, there was need felt to address the ergonomic issues of the apparel workers in the developed world. And modular addressed all; supported worker empowerment through self-managed work teams supported QR manufacturing due to low WIP, addressed the ergonomic vulnerability by job rotation, and stand-up work.

9.4.3.1 Cellular manufacturing

Cellular manufacturing is one of the most popular forms of manufacturing in lean setups worldwide. A cell capable to accommodate product variations is arranged with specialized machinery along with multiskilled operators who can perform as a team. The machine and equipment arrangement ensures a smooth, uninterrupted flow of work resulting in a seamless transition of raw materials to the semifinished or finished products. Teamwork is the key to the smooth functioning of such a system, as the issues and challenges are addressed by the team, and responsibility is shared to achieve the targets. An important key characteristic is a flexibility or ability to adapt according to customer requirements. Smaller quantities with significant variations can be handled efficiently by maintaining lesser WIP levels. Such systems are known for quick changeover ensuring no or minimum start-up losses which is an essential requirement in a lean work environment.

As cellular manufacturing is quite popular in lean setups, it is widely believed that there has to be a single piece of movement in such systems. But this is not always true, as in apparel manufacturing due to several factors affecting the process performance, maintaining single-piece flow may be a challenging task. The single-piece flow in sewing means lean implementation, but the reverse is not true, as lean implementation does not necessarily require a single-piece flow (Jana, 2012). Single-piece flow means there is no WIP between any two operators in the sewing line. Here the pieces are moved from hand to hand in a stand-up work format, and the pieces are never kept in bin/trolley. To make this system work, ideally, the work content of all sequential operations required to be the same. However, as there will be some variation of work content between operation to operation, operators need to work in one or more fractions of any operation, and change of guard may happen in between any operation (to maximize operator utilization).

9.4.3.2 Toyota Sewing System

TSS is a modular manufacturing system. The concept is widely used in lean setups worldwide. The system was first introduced by Toyota Corporation, Japan in 1978 as one of the lean initiatives (as part of JIT) used for manufacturing car seat covers which were recognized as TSS globally. To ensure smooth functioning of the system by interchanging the operators, the U-shaped machine layout arrangement was suggested by Monden, where operators need lesser movements while switching over from one workstation to the other (Bischak, 1996). According to Monden, the U-shaped arrangement (refer to Fig. 9.5) imbibes the flexibility to change the throughput by adding/removing (adjusting) the operators as per requirement ensuring improved production leveling, such flexibility is not easily available in the traditional seated-work arrangements (Bischak, 1996).

Although it is widely believed that the source of TSS is the Toyota Production System (TPS), there is no explicit mention of the same in literature. In their seminal work *Decoding the DNA of the Toyota Production System*, 40 plants were studied in the United States, Europe, and Japan from different industries such as autoparts and assembly, prefabricated housing, cellphone, computer printers, injection molded plastics, but no apparel (Spear & Bowen, 1999). The tacit knowledge that underlies the TPS can be captured in four basic rules. First, all work should be highly specified as to the content, sequence, timing, and outcome. Second, every customer−supplier connection must be direct, and there must be an unambiguous yes-or-no way to send requests and receive responses. Third, the pathway for every product and service must be simple and direct. Fourth, any improvement must be made following the scientific method under the guidance of a teacher, at the lowest possible level in the organization.

During the 1980s some people called this modular system TSS because it is one of a family of so-called Lean Production system, which cuts the cost of about 14%, reduces the WIP inventory from weeks to days, needs 25%−30% less space, while workers enjoy more variety in their job, evinces higher satisfaction, resulting in lowering turnover (Stewart, 2007). Instead of breaking and sewing and assembly into a long series of small steps, modular production entails grouping tasks, such as the entire assembly of a collar, and assigning that task to members of a module or a

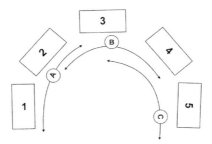

Figure 9.5 Layout diagram of TSS in sewing (Jana, 2004). *TSS*, Toyota Sewing Systems.

Figure 9.6 TSS in sewing (Jana, 2004). *TSS*, Toyota Sewing Systems.

team of workers (Stewart, 2007). It is a combination of make through and assembly line system (refer to Fig. 9.6), where a small group of multiskilled workers divides work among them, and assemble apparels in QR mode. Modular enjoys the best of both: optimum operation breakdown to enjoy the benefit of using special machines and work aids while still retaining low WIP, and accountability and empowerment of multiskilled workers (Jana, 2004).

Modular was touted as *the* production system in needle trades and advocated by an economist (Eileen Appelbaum, Economic Policy Institute, Washington, DC), consultants (Charles Gilbert, United States; Alan Chandler, United Kingdom), associations (American Apparel Manufacturers Association), research organizations (Textile Clothing Technology Corporation, United States; Work & Technology Group, United Kingdom), institutions (National Institute of Fashion Technology, India; The Nottingham Trent University, United Kingdom) and alike. There were conflicting reports regarding the popularity of TSS in developed countries, while some report says from 1980 to 1996, TSS could capture the imagination of only 10% of all clothing assembly line in the United States (Stewart, 2007); another report (The Economist, 1994) suggested that at least 56% of all manufacturing plants (not specifically apparel) have adopted some form of cellular manufacturing. While the comparative figure for Europe is not available, the adoption of teamwork is lower than the United States (Hague, 1995). TSS did never overtook Progressive Bundle Unit (PBU) due to multitude of reasons; the undue strain on operators due to zero WIP (Hague, 1995), and companies intent on outsourcing production could not have cared less about TSS (Stewart, 2007). Notably, the "fast fashion" giants such as Zara and H&M, Gap, Primark are still dependent on outsourcing.

9.4.3.3 Bucket brigade, Rabbit Chase, and Chaku-Chaku system

The TSS is also known as bucket brigade or bump-back production system due to some of its inherent properties. The term bucket brigade originally comes from

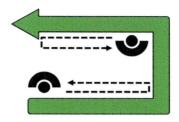

Figure 9.7 Bucket brigade (Roser, 2017a).
Source: By Christoph Roser at AllAboutLean.com under the free CC-BY-SA 4.0 license.

firefighting and is in general used for transportation using a human chain. People create a human chain by standing approximately 1 m apart (refer to Fig. 9.7). Goods are transported by handing them from person to person. This could be, for example, buckets of water, boxes for loading and unloading, or any other type of goods (Roser, 2017a). The advantage of a bucket brigade is to avoid waiting times (one type of waste) for operators in the case of an uneven workload. To even out the fluctuations (or mura for unevenness) in an assembly line, there is particularly a neat trick for manual lines that self-organizes fluctuations in the workload. The TSS conforms to some of the prerequisites of bucket brigade like the operations should have short cycle times, there must be a clear and identical sequence of steps for all parts passing through the system, there should be more process steps than workers, workers to be multiskilled, and there should be no buffer inventories. The TSS also does not conform to bucket brigade characteristics in few aspects like TSS does involve walking of operators (whereas no walking in bucket brigade), or keeping the bottleneck at the beginning of the line, which is not always possible due to style constraint. For a TSS or bucket brigade, there are three rules for an individual worker. The worker gets the part from the preceding worker and then moves with the part along the line until he can give the part to the next worker. After handing over the part (to the succeeding worker) the worker moves backward empty-handed until the worker bumps into the preceding worker and gets the part from the preceding worker. Workers move forward in the line always with a part and move backward always empty-handed.

Another technique used in some manufacturing cells to organize the production is called "Rabbit Chase" manufacturing. The worker moves along the line with his part (refer to Fig. 9.8). When the worker reaches the end of the line, he picks up a new part and starts again. Naturally, this works best for U-shaped lines, or in general lines where the end is close to the beginning of the line. In this cell configuration the faster operator "chases" the slower one (that is why the name "Rabbit Chase"), this also leads to a reduced system speed where the faster operator has to wait for the slower worker.

It consists of several workers following the product flow and processing in sequence all the operations. It has the advantage of low inventory, fast lead time, and clear responsibility leading to better quality. The main disadvantage is the total travel time for workers, as the workers are always in movement. The other

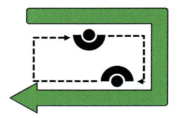

Figure 9.8 Rabbit Chase (Roser, 2017a).
Source: By Christoph Roser at AllAboutLean.com under the free CC-BY-SA 4.0 license.

disadvantages are the lowest worker sets the speed of the system (unlike the bucket brigade), possible mental pressure on slower worker due to the worker hindering everybody else, significant walking distances, may cause a traffic jam behind the slowest worker, every worker needs to be multiskilled (precisely should know all processes), and not very scalable.

Although bucket brigade or TSS has successfully experimented in apparel sewing, due to the above disadvantages the Rabbit Chase system is not reported in any sewing line. The only reported application of Rabbit Chase U-shaped line was for hanging garments order assembly, based in manual sortation from an aerial conveyor, in Inditex. The length of the line was in the range of 100 m and was occupied by several operators with enough space between, not to create delays. The same principle was followed for flatpacked garments, where the units were transported on trolleys at ground level, with an exception of fast operators that could overtake slow ones (Roser, 2017b).

Chaku-Chaku can be compared with more than one sewing automat being attended by one operator, which is common practice in sewing. A good example can be a decorative stitch in a pocket and back pocket attaching in a Jeans line. Chaku-Chaku (refer to Fig. 9.9) is a way to operate a semiautomated manufacturing line. One or more workers walk around the line, add parts to the processes, and then start the process. While the process works on the part automatically, the worker adds the next part to the next process, and so on. The word "Chaku-Chaku" in Japanese means either "Load, Load," or it can simply be the sound the machine makes while unloading similar to "Clack-Clack." Usually, the loading of the machine at the beginning is manual, while the ejection of the part after the process is completed is automatic. The sewing operator simply loads the component and presses the start button, once the sewing is complete the part is automatically disposed of by stacker.

Differing from the conventional setup of assembly lines where one worker is assigned a task to be performed on a single machine, TSS causes the enhancement of worker utilization. To get the best results of U-shaped modular arrangement, and to enhance the manpower utilization and ensuring the multiskilling capabilities of the operators, generally, the number of machines is kept more than the number of workers. In apparel manufacturing the average number of machines per worker maybe around 2.5. One interesting observation of TSS as done by Wang et al. (1991) that

Apparel manufacturing systems

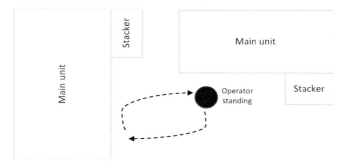

Figure 9.9 Chaku-Chaku system: one operator attending multiple sewing automats.

the work always moves forward in such manufacturing systems, and the workers have forward and backward movements as and when required to carry the work to the workstation and moving backward for additional work. In this way of movement, there are possibilities that the operator cannot find any work waiting to be completed; this may result in an interruption in the process. Despite this, in such manufacturing, arrangements exhibit high performance, even though the work content of operations (to be performed at different workstations) varied significantly. It was also observed that after some time the process settles down and the team members develop their exclusive working patterns (Wang et al., 1991).

The availability of extra machines/workstations motivates the team members to deliver faster and better results. As teamwork is emphasized, the operators are compensated (collectively as a team) based on the quality output of the team, this brings togetherness and zeal to perform better for collective benefits.

Some key advantages of TSS in the context of a lean work environment may be stated as the improved production leveling, effective control on the output which can be increased or reduced just by modifying the number of workers without much change in the machine layout setup. This also results in the ability to handle volumes (in terms of order quantities) with the same effectiveness. Another major advantage of using such systems is a significant reduction in the WIP in comparison to the traditional assembly line systems. According to a survey conducted with apparel manufacturers, a good 60% reduction in inventory levels was reported while working on modular systems (Gilbert, 1989); this also improves the quality levels as due to lesser WIP, it is easy to trace and rectify the problems in lesser time. Adopting such systems also results in a significant reduction in the labor cost, and improvement in productivity (in terms of output/total employee) was reported over 20% (Gilbert, 1989). The other significant advantages of such systems have increased the morale of the workers and reduced absenteeism and turnover (Kulers & DeWitt, 1990). Despite all these reported advantages which make these systems as the natural choice of manufacturing systems in the lean work environments, there are a few disadvantages as well, such as an increased level of training requirements, difficulty in replacement of skilled workers in case of absenteeism, and satisfying wage compensation to all the operators working as a team.

While the layout of machinery and equipment in TSS draws special mention being "U" or "C" shape resulting minimum movement of operators (removal of waste), another form of sewing machine layout is very popular in lean-implemented garment factories called "LEGO layout." In such a layout the sewing machines are face to face leaving no space in between and operators sit on either side. The origin of this layout is believed to be MAS Holding in Sri Lanka, the company credited with popularizing lean in apparel manufacturing. During the early year 2000, a team of executives led by Mahesh Amalean visited Toyota Motor Corporation in Japan for a knowledge-sharing program to understand the lean concept, and in return the executives experimented and implemented their learning in several plants of MAS Holding, Sri Lanka. Now several such practices are followed in apparel manufacturing across the globe and "brick-laying layout" is one of them (Wickramasinghe, 2020).

In the traditional layout, it is assumed that sewing machines are to be kept in regular rows with a center table in between, while one grid in the row takes up a space of 4 ft.[2]. All the lean effort of minimizing material flow assumes that the machines are to be placed in such regular orientation with uniform spacing. However, in lean layout, no such preassumed grid, rows orientation is followed. From the operation bulletin the sequence of operations is followed and machines are placed with free orientation ensuring the minimum length of material movement. One example of the same will explain the concept better (refer to Figs. 9.10 and 9.11); suppose the first operation is label attach in a yoke which is done by one operator and the next operation is yoke join with back which is done by two operators. Therefore one machine output will become input for two machines. In traditional options (options A, B, C), we try to place the machine in grids, while in the lean layout, we try to place the machine changing the orientation by 90 or 180 degrees to achieve the least possible path of material movement (options X, Y, Z). And such blocks are repeated to complete the line.

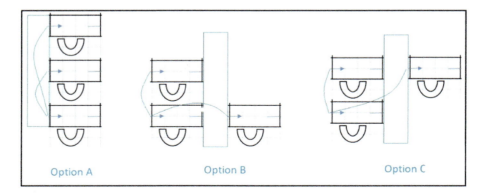

Figure 9.10 Traditional layout grid.

Apparel manufacturing systems 305

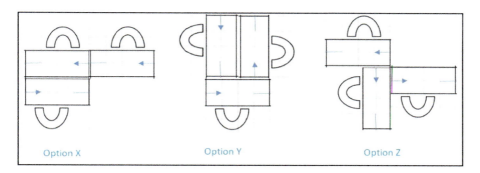

Figure 9.11 Lean layout grid.

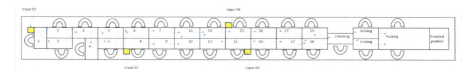

Figure 9.12 Schematic diagram "LEGO layout": operations 1–7 are the preparation of the front panel; operations 8–10 are the preparation of Back panel; operation 11, front-back assembly; operations 12 and 13 are the preparation of the right sleeve; operation 14 assemblies of body and right sleeve; operation 15 assemblies of body and left sleeve; operation 16 and 17 side seam and hemming (bottom and sleeve); operation 18, bar tacking; and operation 19, heat transfer.

There are numerous benefits of lean layout or production cell run on single-piece movement or with reduced bundle size of 3–4 pieces per bundle. Some of the benefits in lean context are:

- low WIP inline,
- shorter throughput time,
- shorter line changeover time,
- quick identification of any quality issues or abnormalities,
- QR to customer demand,
- facilitates one-piece-flow and higher velocities (reduced lead time),
- reduces extra handling (less labor),
- reduces transportation time and distance,
- requires less floor space (minimum of 25% space-saving),
- reduces in-process inventory (decreased overproduction and idle materials),
- reduces waiting time (less labor), and
- increases operator awareness of production needs and problems (reduced defects).

Typically, such production cell layouts are complete with sewing–checking–ironing–packing. Fig. 9.12 depicts a schematic diagram of a complete LEGO layout of a style. Please refer to Fig. 9.13 for a practically implemented LEGO layout for sewing in an apparel manufacturing floor.

Figure 9.13 LEGO Module sewing layout.
Source: Courtesy: Corrigo Consultancy Pvt. Ltd, India.

9.4.3.4 Agile manufacturing

In recent years, globalization has resulted in increased reachability of consumers to goods and services. This has opened up new horizons to the economies and the local competition of yesteryears has acquired the global stage with a cutthroat pricing war and an aim to be the first in reaching the consumer irrespective of any geographical boundaries. There is no single organization that remains unaffected by this competition brought on. Technological interventions especially digital technology responsible for connecting the world have further fueled up this competition. This has presented with newer options and solutions to the consumers, and there is a tendency in consumers to get the products and services now at the most competitive price. The never-ending customer demand pushes industries to adopt newer ways of working to fulfill customer requirements, and manufacturing is also not an exception to this phenomenon. A significant transition in terms of the evolution of newer manufacturing systems has been witnessed globally, and the same can be realized with craftsmanship-based manufacturing to the state-of-the-art technology power automated manufacturing (Gunasekaran, McGaughey, & Wolstencroft, 2001). In the dynamic business environment with fluctuating demands, it is imperative to be not only prepared to face the turmoil environments but also lead to a competitive edge. Agile manufacturing is also an effort to survive in competitive environments; in simple words, it may be referred to as the practices followed by the organizations to anticipate and respond to change.

As far as the functioning of agile manufacturing systems is concerned, these systems are significantly different from the traditional systems, where traditional systems are little static, stand-alone, and hierarchical and agile manufacturing setups are dynamic in responding to customer needs with faster learning and quick decision making. Such an agile system-based organization that is designed for both stability and dynamism operates as a network of teams against the decision trickle-down approach from the top as followed in traditional manufacturing environments (Ebrahim, Krishnakanthan, & Thaker, 2018). Such agile manufacturing organizations are capable to respond quickly toward achieving value-creation as well as value-protection. The agile manufacturing concepts have been discussed in detail in Chapter 14.

9.5 Success parameters of a manufacturing system

9.5.1 Flexibility

The ability to adjust to changes is a prime requirement. Meeting the customer expectations in terms of quantity and product specifications is a real challenge to the traditional way of manufacturing as it obsoletes the concept of huge WIP and standardized setup with some predefined process parameters. Product diversity with the ability to process smaller batches in the dynamic highly unpredictable environment is the key to success. Agile manufacturing—based systems are the most appropriate solutions in such situations.

9.5.2 Cost-effectiveness

Achieve profitability is one of the ultimate goals of any business. This is essential for survival as well as for growth. Waste (muda in lean terms) is a must to achieve cost-effectiveness in a manufacturing environment. Adopting flexible manufacturing systems may bring necessary cost-effectiveness to the process.

9.5.3 Quality

In simple words, conformance to the customer expectations in a given product or service is referred to as quality achieved. As far as customer satisfaction is concerned, there is no alternative to a quality product. A faulty product or service may severely damage the brand image and reputation of an organization.

9.5.4 Speed

The consumer of current times needs the product or service just at this moment. Technological interventions have bridged the gap between the manufacturer and the consumer. Online purchase solutions have ensured speedy delivery of the products without any much waiting time. This pattern of shortened lead times has pushed for the essential close and timely coordination in the supply chain. To compete on speed a company must have close contact with both suppliers and customers. The artificial intelligence and machine learning applications in real-time process monitoring are more accurate and it ensure the speedy supply of goods and services to the consumers.

9.6 Conclusion

Every business and industrial organization is concerned with the creation of goods and services in some form, utilizing workers, machines, and materials. It is very critical to adopt a suitable manufacturing system depending on the specific product requirements. In a lean manufacturing environment the key focus has to be on reducing the waste irrespective of the type of waste. Different manufacturing systems have their pros and cons, but it is of utmost importance that the system is managed scientifically by applying the engineering approach. The process design and standardization, a sequence of operations and procedures, technical specification of machine, equipment, and work aids, and work division and allocation should be well thought of and logical to ensure the optimum yield.

References

Beaumont, D., Thibert, J., & Tilley, J. (2017). *Same lean song, different transformation tempo*. McKinsey & Company. Retrieved from https://www.mckinsey.com/business-functions/operations/our-insights/same-lean-song-different-transformation-tempo.

Bischak, D. P. (1996). Performance of a manufacturing module with moving workers. *IISE Transactions, 28*, 723−734.

Chandler, A. (1994). *Implementing self managed work teams: Facilitator's guide*. Alan Chandler Associates.

Choudri, A. (2002). The agile enterprise. In J. B. Revelle (Ed.), *Manufacturing handbook of best practices: An innovation, productivity, and quality focus* (pp. 29−30). New York: St. Lucie Press/CRC Press.

Ebrahim, S., Krishnakanthan, K., & Thaker, S. (2018). *Agile compendium—Enterprise agility*. McKinsey & Company.

Gilbert, C. (1989). *Modular manufacturing: Sizzle or steak? Apparel research notes. 8*. Arlington, TX: American Apparel Manufacturers Association.

Gunasekaran, A., McGaughey, R., & Wolstencroft, V. (2001). Agile manufacturing: Concepts and framework. In A. Gunasekaran (Ed.), *Agile manufacturing: The 21st century competitive strategy*. Oxford: Elsevier Science Ltd.

Hague, J. (1995). *Work and technology. Restructuring manufacturing*. Manchester: Textile Institute.

Jana, P. (2004). *Production management goes back to the future*. Retrieved from https://www.just-style.com/: https://www.just-style.com/analysis/production-management-goes-back-to-the-future_id93130.aspx. (Accessed January 2014).

Jana, P. (2012). *Fat truths and untruths about lean*. Retrieved from https://in.apparelresources.com/management-news/production/great-knowledge-divide-fat-truths-lean/. (Accessed June 2015).

Jana, P. (2013). *The great knowledge divide: Unit production system*. Retrieved from https://apparelresources.com/technology-news/information-technology/the-great-knowledge-divide-iv-unit-production-system/. (Accessed July 2017).

Jana, P., & Tiwari, M. (2018). *Value engineering. Industrial engineering in apparel manufacturing—Practitioner's handbook* (pp. 232−233). New Delhi: Apparel Resources Pvt. Ltd.

Kulers, G. B., & DeWitt, J. W. (1990, May). Modular goes mainstream. *Apparel Industry Magazine*, pp. 44−52.

Merriam-Webster (2004). Manufacture. Retrieved from www.merriam-webster.com: https://www.merriam-webster.com/dictionary/manufacture. (Accessed May 2018).

Online Etymology Dictionary. (2011). *Manufacture (n.)*. Retrieved from www.etymonline.com: https://www.etymonline.com/word/manufacture. (Accessed May 2018).

Piore, M. J., & Sabel, C. F. (1986). *The second industrial divide: Possibilities for prosperity* (Vol. 4). New York: Basic Books.

Roser, C. (2017a). *All about lean*. Retrieved from https://www.allaboutlean.com/bucket-brigade-1/. (Accessed 24 April 2020).

Roser, C. (2017b). *All about lean*. Retrieved from https://www.allaboutlean.com/rabbit-chase/. (Accessed 17 May 2020).

Solinger, J. (1988). *Apparel manufacturing handbook: Analysis, principles and practice* (2nd ed.). Bobbin Blenheim Media Corp.

Somers, M. H. (2015). *Resource productive operations*. McKinsey & Company.

Spear, S., & Bowen, H. K. (1999). Decoding the DNA of the Toyota production system. *Harvard Business Review*, *77*(5), 96−106.

Stewart, T.A. (2007). *The wealth of knowledge: Intellectual capital and the twenty-first century*. United Kingdom: Crown.

Textile World. (1991). *Ergonomic related citations by OSHA in the textile and apparel industries*. Textile World.

The Economist. (1994, Dec 17). The celling out of America. *The Economist*, pp. 71−72.

Wang, J., Schroer, B., & Ziemke, M. (1991). *Understanding modular manufacturing in the apparel industry using simulation. Proceedings of the 23rd winter simulation congress* (pp. 441−447). Baltimore, AZ: IEEE.

Wickramasinghe, C. (2020). *Discussion on modular manufacturing* (P. Jana. Interviewer).

Lean quality management

Rajesh Bheda
Rajesh Bheda Consulting Pvt. Ltd., Gurgaon, India

> *Inspection does not improve the quality, nor guarantee quality. Inspection is too late. The quality, good or bad, is already in the product. Quality cannot be inspected into a product or service; it must be built into it. — W. Edwards Deming*

10.1 Introduction

Today's consumer wants to associate with brands and manufacturers who assure quality. Quality is an all-pervasive concept, and it cannot just be limited to product quality alone. It has to reflect in all aspects of a business. No enterprise can dream of adopting Lean management practices without being fully committed to quality.

This chapter is mainly devoted to Statistical Quality Control (SQC) for the apparel industry. It would focus on how to control the processes through Statistical techniques like Statistical Process Control (SPC) and how to make a judgment about the quality of the produced lots using Acceptance Sampling and Introduction to concepts of Six Sigma and Lean Sigma.

To start with let us look at the definition of quality. The most popular definition of Quality is "fitness for use" professed by Joseph Juran (Juran & Godfrey, 1998). This three-word definition is quite comprehensive. It covers both aspects of the quality of design and quality of conformance (Montgomery, 2013).

> Quality of design: There are various levels or grades of quality in all the produced goods and services. Quality of design is the technical term used for this. These differences are the result of intentional design differences among the types of products like automobiles. These differences include the material type, specifications of components, reliability as well as different accessories used in the manufacturing.
> Quality of conformance: Every product has some predefined specifications that need to be followed. Quality of conformance reflects on the extent of conformity to the specifications required by the design. It is influenced by multiple factors such as training of workers, the manufacturing process followed inspection, and worker motivation. Deviation from the specification or characteristic adversely affects quality. In other words, it can be said that quality is inversely proportional to variability.

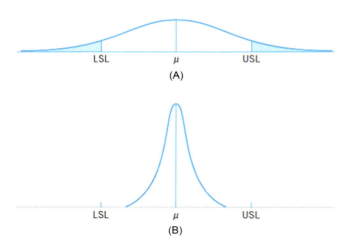

Figure 10.1 (A) Product variability with a cyclic pattern, (B) product variability with cyclic pattern eliminated.
Source: From Montgomery, D. C. (2013). *Introduction to statistical quality control* (p. 204). Hoboken, NJ: John Wiley & Sons, Inc.

The definitions imply that if variability in the important characteristics of a product decreases, the quality of the product increases (refer to Fig. 10.1). Jack Welch, the former chief executive officer of General Electric, has observed that "*your customer doesn't see the mean of your process, he only sees the variability around that target that you have not removed*" (Watson, 2001).

This variability can be easily identified by the impact on customers. Time, money as well as assigning manpower for repairs and warranty claims can be controlled largely by taking care of quality levels. It can be said that quality is inversely proportional to variability. Furthermore, it can be communicated very precisely in a language that everyone, particularly managers and executives, understand money.

This also takes us to the following definition of quality improvement.

Taguchi, Chowdhury, and Wu (2005) proposes a different view of quality, according to Taguchi, the meaning of quality should not be just limited to the "conformance to specifications," but it should be seen in a wider perspective. The loss due to poor quality is not just of the manufacturer at the time of production but the loss to the consumer and the loss to the society as a whole. Taguchi defines quality as "*the loss imparted by the product to the society from the time the product is shipped*." A quantitative evaluation of loss by functional variation of a product was proposed as a "Quality loss function" (Taguchi et al., 2005).

10.1.1 The meaning of quality and quality improvement

Quality improvement is the reduction of variability in processes and products.

Waste can be assessed as excess variability in the performance of several processes. As a result of high variability, money, time, as well as efforts spent on repairs get wasted. Thus quality improvement can be said to be directly proportional to waste reduction. Statistical methods play a vital role in quality improvement as it helps in analyzing the variability. Data are usually classified as attributes and variables in the field of quality engineering.

If the measured value resembles the desired value for that quality characteristic, it is known as nominal or target value for that characteristic. These target values are usually defined by a range of values. The largest allowed value for a quality characteristic is called the upper specification limit, whereas the smallest allowed value for a quality characteristic is called the lower specification limit.

Built-in-Quality, Right-First-Time (RFT), First-Time-Through (FTT), Quality Function Deployment (QFD), and Zero Defect (ZD) are some of the popular terms used in lean as well as in quality management systems (QMSs). Many times these terms are used interchangeably in the industry environment. All these terms have the same ultimate goal to ensure customer expectations by achieving the best quality in the end product.

Built-in-quality: Built-in-quality is a result of a proactive approach to continuous improvement based on Lean philosophy. This is the outcome of a healthy culture of attaining the best quality in the product all the time. It can also be referred to as a product essentially maintaining the quality by default throughout the development, as it is unique characteristics of the culture where quality is ensured (or assured) without much control (checks and inspections).

Right-first-time: RFT focuses on achieving the targeted quality or defect-free product as and when an activity, an operation, or a process occurs, and the same practice is followed throughout the value chain. By achieving the RFT the wastage of resources is eliminated in correcting or repairing the faulty products. RFT focuses on ensuring all the operations or activities undertaken in the right manner, for the first time, and at every time. This is an important aspect of lean philosophy, as by doing so, several Muda are either eliminated or reduced from the process.

First-time-through: FTT (also referred to as FTY First-time-yield) is a measure of production efficiency and quality. It is also used as a key performance indicator to measure the quality or the performance of a process in terms of achieving defect-free production. FTY also reflects the amount of rework in a process (the quantum of pieces that were not made correctly for the first time).

Quality function deployment: QFD is a Japanese concept developed in the 1960s as a form of cause-and-effect analysis. QFD aims to ensure the quality of the product with a key focus on Voice of Customer (VoC) that is needs and expectations of the customer. The requirements of VoC are carefully identified and organized in a structured manner, and accordingly, efforts are put in a planned manner to achieve the same (ASQ, 2010). The QFD has been developed and practiced in different industries in different ways, and hence, there is no single Industry-standard version or methodology for QFD (Cohen, 1988).

Zero defect: ZD is a powerful quality improvement philosophy. Achieving ZD is the ultimate objective of any process, and this is an ideal situation where everything is done perfectly right the first time and always. The concept of ZD was first introduced by the great Quality Guru Philip Crosby in 1979 (in his famous book titled "Quality is Free") through his 14 Step Quality Improvement Process. ZD was a management-led program that is referred to as a philosophy or a movement aiming to minimize the number of defects as much as possible. To achieve a stage of ZD, a data-driven rigorous system of total quality management is required to be followed. In real practical work environments, ZD does not mean any or nil defects but it is taken as the defects are unacceptable and activities or tasks must be performed right the first time. Quality management approaches like Six Sigma aims to achieve a ZD stage.

10.2 Lean philosophy and quality management

The demand-driven dynamic business environments have resulted in increased competition, where products are required to reach the consumer at the earliest, with the most competitive price, and without any defect. Further, due to the high competition and global connectivity with online mediums such as E-commerce, the customers are exposed to increased product options. This has made the customer more aware of the product and its quality. This has also forced the firms to adopt global sourcing to take business advantages (Su, Gargeya, & Richter, 2005). This has forced the manufacturers for process improvements to achieve the products RFT and reduction of resource losses (Moin, Doulah, Ali, & Sarwar, 2017). In such a scenario, applications of lean play a critical role in making the products with the best quality, best price, and at the earliest.

In recent years, there has been an increased focus on productivity enhancement, improving product quality, improving the working environment, etc. to meet internal and external consumer expectations through lean-based continuous improvements (Reeb & Leavengood, 2010). Initially, quality management was considered as part of the management and decision-making level, but later it became a key focus area of production and operations. In recent years, quality management is practiced with more comprehensive methodologies such as Total Quality Management (TQM), Six Sigma, and Lean, and these are collectively applied to improve productivity and quality. While Six Sigma uses Define, Measure, Analyze, Improve, and Control (DMAIC) extensively, Plan-Do-Check-Act, Root-cause analysis (RCA), Ishikawa Diagram, etc. that are applied to identify and correct the defects at its source are used in TQM, Six Sigma and Lean. Such techniques are an integral part of quality management as well as lean management.

Emphasizing on the importance of delivering the product with the right quality, Locher (2008) discusses the "Rule of tens," where the cost incurred by a defect (occurred at the initial stage of product development but got unnoticed and moved with the end product) increases multifolds at the later stage when the same defect is identified by the consumer (Locher, 2008). Uddin and Rahman (2014) while applying DMAIC methodology for defect minimization in the sewing floor state that defect minimization is the first step toward reducing operating costs and improving the quality. Reduced reworks result in reduced cycle time while eventually leads to productivity improvement (Uddin & Rahman, 2014).

Schonberger (2014) discusses the symbiotic relationship between quality management and lean (Schonberger, 2014). The combination of quality management

Lean quality management

and lean is not a new idea, and the same was also discussed around 38 years back in the book "*Japanese Manufacturing Techniques*" (Schonberger, 1982).

A combination or merger of Just-in-Time (JIT) lean-oriented approach with quality management is depicted in Fig. 10.2. It can be observed from the figure that it

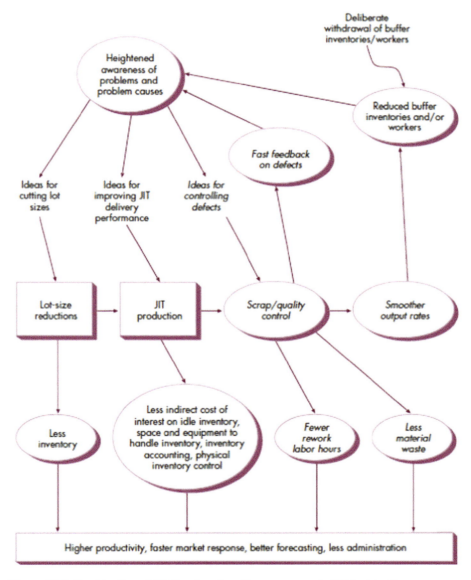

Figure 10.2 Combination of JIT (lean) and quality management. *JIT*, Just-in-Time. *Source*: From Schonberger, R. J. (2014). Quality management and lean: A symbiotic relationship. *Quality Management Journal*, 21(3), 6–10.

starts with the awareness of problems and its causes. It feeds both lean and quality equally. Also, the action of reducing the buffers and inventories (see at the top right of the Fig. 10.2) is aimed to bring visibility as reduced inventories can make issues and problems visible and make the system transparent (also refer the discussion in Chapter 8: Process Balancing). The terms or actions mentioned on the left side (related to lot size reduction and JIT production) are related to the lean approach, whereas the terms or actions mentioned in the right side (defect control and fast feedback on defects) are ingredients of the quality management. Eventually, both lean and quality management are resulting in enhanced productivity, increased responsiveness, and improved forecasting with a lesser requirement of administration to an organization (Schonberger, 2014).

Bacoup, Michel, Habchi, and Pralus (2018) confirm a closed relationship between QMSs (ISO 9001) and lean. It was stated that both the QMS and lean management are complementary and reinforcing. Despite having the same goal of improving the production process, in reality, both the systems are treated and applied in parallel which results in wastage of resources. To avoid wastage of the organizational resources, the synergistic combination of the QMS and lean management should be applied as a lean quality management system (LQMS) (Bacoup et al., 2018). Micklewright (2010) in his famous book "Lean ISO 9001" states that applying two systems (QMS and lean management) serving the same purpose can cause huge wastes. Micklewright emphasizes on the need for collective efforts by quality managers and lean managers to understand and eliminate the redundancies in these systems (Micklewright, 2010).

According to Clark, Silvester, and Knowles (2013), lean management systems are a means to create a culture of continuous quality improvement (CQI). Lean-based CQI systems are essential to meet the twin challenges of increasing quality and reducing costs. CQI systems combine systematic process improvement to improve the quality of service and a commitment to value and develop the skills to deliver services (Clark et al., 2013).

Several published works consider lean as synonymous to not only with JIT and Toyota Production System but with Six Sigma and QMS (Hall, 1983; Holweg, 2007; Monden, 1983; Schonberger, 2007). Though few researchers consider JIT, Six Sigma, and QMS not same as lean but an integral part of lean (Dahlgaard & Dahlgaard-Park, 2006; Forza, 1996; Hines, Matthias, & Rich, 2004; Sánchez & Pérez, 2001; Shah & Ward, 2003; Shah & Ward, 2007).

With this, it can be concluded that lean and quality management are complementary to each other with the same objective. The integration of lean and quality management is essential to make improvement impactful and sustainable. To an organization, both the lean and quality management philosophies play an instrumental role in achieving higher levels of excellence. Customer satisfaction in all aspects (product quality, price, timely delivery, etc.) is the prime concern for any business, as it determines the success of any business. This can only be achieved through continual improvement which is driven synergistically by a lean and QMS. We shall focus on the SQC for the apparel industry in this chapter.

10.3 A brief history of statistical methods in quality

There is a long history involved with the statistical methods used in the application of quality improvement activities. The statistical control chart concept was introduced by Walter A. Shewhart of the Bell Telephone Laboratories in 1924. This is usually deemed as the inception of SQC. Later on, by the end of the 1920s, Harry G. Romig and Harold F. Dodge of Bell Telephone Laboratories developed statistically based acceptance sampling as an alternative to 100% inspection. SQC methods were widely used by the middle of the 1930s at Western Electric which is the manufacturing arm of the Bell System. World War II saw an increased acceptance and use of SQC concepts in manufacturing industries.

The 1950s and 1960s saw the emergence of reliability engineering, the introduction of several important textbooks on SQC, and the viewpoint that quality is a way of managing the organization. In the 1950s, designed experiments for process and product improvement were first introduced in the United States. These practices were institutionalized by the Japanese, and it was only in the late 1970s or early 1980s when many Western companies discovered that their Japanese competitors had been systematically using designed experiments since the 1960s for process improvement, evaluation of new product designs, improvement of reliability, new process development, and field performance of products. This resulted in further interest in statistically designed experiments and the need to spread these methodologies across other industries. There has been a profound growth in the use of statistical methods for quality and overall business improvement across the world since 1980.

10.4 What is statistical process control?

SPC is focused on controlling various processes in the manufacturing cycle. Producing acceptable products with consistency and quality can be enhanced by using the SPC analysis. It involves the use of various charts that helps in the identification of processes if they are under control or are out of control.

Any variation from the target value is noted, and it is drawn in the form of a line graph. These graphs also have control limits drawn on them. If any observation beyond the control limit, it can be said that the process is getting out of control and will not be able to meet specified quality requirements. In such cases, a corrective action needs to be initiated. Corrective action is also warranted if the results of the inspection show that the process is under control but there is a trend indicating that it is likely to get out of control in the future (Bheda, 2004).

10.4.1 Types of charts

To determine if the recent set of readings represents convincing evidence that a process has changed or not from an established stable average, control charts act as a statistical test. It also checks the extent of variation.

Some control charts use a list of items for each measurement. The sample average values tend to be normally distributed resulting in construction and interpretation of the control charts. The centerline of a chart is the process average. The control limits are set at plus or minus three standard deviations from the average. The most commonly used control charts are the X-bar and the R control charts described below (Schenkelberg, 2015).

10.4.1.1   X-bar and R control chart

The X-bar and R chart or Shewhart charts use a subgroup of entries for each sample, and the mean of the sample and its range is plotted on two charts.

Plotted points, X-bars (refer to Fig. 10.3), are the average of the sample with n readings, where n is the subgroup size. The centerline, labeled as X-double bar, is the average of the sample averages. UCL and LCL are the upper control limit and lower control limit, respectively. The sample readings are collected, averaged, and plotted with dots. The connecting lines provide an easier way to interpret the data.

The range of the sample is denoted by R (refer to Fig. 10.4). The centerline, labeled as R-bar, is the average of the sample ranges. The y-axis should include values suitable to R values and provide a convenient way to plot and read the charts. These two charts, used correctly, can provide process stability and helps to provide the necessary feedback to allow for process improvement.

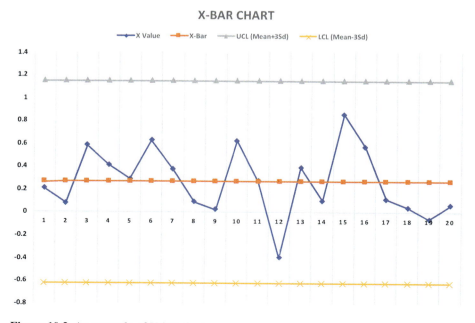

Figure 10.3 An example of X-bar chart.

Lean quality management 319

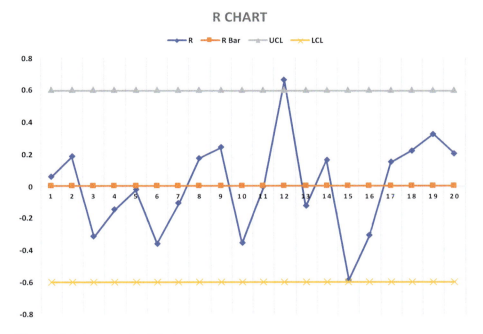

Figure 10.4 An example of R chart.

10.4.1.2 Other types of charts

This is a short description of the basic control charts, with details of when the particular chart is advantageous and a few assumptions to consider.

c chart: c chart works with attribute data when nonconformities or defects are counted per unit. The sample size is greater than one and is constant.
Assumptions:
The c chart relies on the Poisson distribution and can be used when:

1. The potential for nonconformities is quite large. For example, the paint on a car is a very large area and defects can be very small. Similarly, in garments or its components, there can be several defects at various locations.
2. The possibility of a single defect is small and constant.
3. The sample inspection process is consistent from sample to sample.

u chart: The u chart is used to monitor attribute count type data when the number or sample size of inspection units may vary.
Assumptions:
The u chart relies on the Poisson distribution and can be used when:

1. The potential for nonconformities is very large.
2. The possibility of a single defect is small and constant.
3. The sample inspection process is consistent from sample to sample.

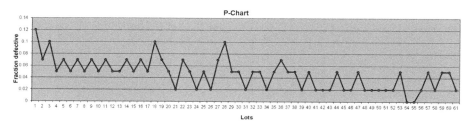

Figure 10.5 Sample p chart from a factory in Gurgaon, India.

p chart: The p chart caters to the proportion of nonconforming/defective units in a sample. The individual unit inspections result in the judgment of the pass/fail of the product. This approach needs a method to determine if each unit meets requirements and there should be more than one unit per sample.

False alarms need to be minimized, and for that, the sample size does not have to be constant yet large enough to expect at least one failed unit.

Assumptions:

The p chart (refer to Fig. 10.5) is dependent on the binomial distribution and has the following assumptions:

1. The chance of failure for each unit is the same, p.
2. Each unit is independent of other units in the process.
3. The sample inspection process is consistent from sample to sample.

Thus SPC enables the use of statistical principals and techniques at every stage of production. SPC aims to control quality characteristics on the methods, machines, products, and equipment, both for the company and operators with the Seven Quality tools.

Two kinds of variations occur in all manufacturing processes, which cause subsequent variations in the final product. The first is known as the common cause of variation and consists of the variation inherent in the process as it is designed. It may include a variety of fabric handling during the cutting or sewing operations or machine setting, properties of raw materials, etc. The second kind of variation is known as a special cause of variation and happens less frequently than the first. This is called a special cause or assignable cause.

10.4.2 Statistical process control implementation

The first reported large-scale implementation of SPC in the apparel industry was carried out by Liz Claiborne Inc. The implementation was started by Liz Claiborne in 1997 on a pilot basis with a Florida-based supplier, with factories in Columbia. The garment rejects rate at the factories plummeted from over 20% in the mid-1990s to 3% in 1997, below 2% in 1998 and to 0.4% in the first half of 1999 as a result of the SPC implementation. This factory was used as a global showcase for SPC implementation, and important vendors were invited from across the globe to witness SPC in action. Liz Claiborne's failed shipment index dropped by 33% between 1996 and 1999s because of the subsequent rollout of SPC implementation in Liz supplier factories across the globe (Bheda, 2004).

The production and quality staff in the apparel industry mainly rely on their technical expertise and experience to solve quality-related defects or variations and are relatively less exposed to the statistical concepts. Due to this, SPC implementation is likely to face resistance within the organization. Thus it is necessary to completely convince the top management of its implementation so that they support and ensure its implementation.

SPC implementation can be carried out using the following stages:

1. SPC briefing and training: The quality manager, supervisors, and operators on the critical operations should be briefed about the basic principles of SPC as well as the methodology of SPC implementation.
2. Identification of production line for pilot implementation and preparatory meeting: This shall involve a meeting between Production Manager, Pattern Master/Technical Supervisors, and Quality Manager/Supervisor. The objective of this meeting is to identify the critical operations of a garment style to be produced. These operations, if not controlled, are likely to contribute to a high rate of defects. The meeting also caters to the ways to minimize/eliminate defects by process modification and to take the desicion on the type of control chart that can be installed on these critical operations.
3. Installation of control charts at the critical operations: This stage revolves around training the operators/inspectors at the critical operation and the method to construct control charts and how to interpret them.
4. Interpretation and corrective action: Installation of the control chart is not enough at it only tells you the status of the process. It is important to have procedures in place for initiating corrective action should the process go out of control. Cause-and-effect diagrams can be drawn here as they are quite useful for analyzing the root causes of defects and initiating corrective action.
5. Continuous monitoring of progress and reviewing the improvement: The process is likely to get stabilized, and the defect rate is likely to drop with the implementation of control charts. It is important to monitor the progress and ensure that the implementation efforts do not hit roadblocks.
6. Deciding on implementing SPC in other areas of production: Organizations can now decide on the future of the SPC implementation and rollout the same throughout manufacturing processes using the experience of the pilot phase.

In the past decades, quality management practices in the apparel industry have undergone a significant change in factories across the globe. There is also no doubt that the quality of apparel produced has improved significantly in recent times. However, still, there is a large group of factories that struggle to produce quality right the first time. This is mainly because their processes are not under control and variation is high and unpredictable. A prominent example of the implementation of SPC in the Apparel industry is cited below.

SPC methods are widely implemented across different factories. In one of the factories, SPC was implemented in Trouser sewing lines for process performance improvement. The project included theoretical and on-job training schemes for different quality team members to understand the SPC concept and its implementation procedure. Significant improvements in the sewing section were achieved post implementation. The four months analysis before the implementation of the SPC tools showed that the average alternation percentage was 9.14%.

After implementing the SPC tools, the average alternation percentage reduced from 9.14% to 6.4%, a 30% reduction in the alteration rates (Abtew, Kropi, Hong, & Pu, 2018).

10.4.2.1 Critical success factors in statistical process control implementation

Certain critical success factors form an essential prerequisite to the successful implementation of any SPC initiative, which are

1. getting complete commitment from senior management,
2. comprehensive training of the middle managers and supervisors,
3. spreading awareness of the potential benefits of SPC,
4. putting robust processes into place to correctly measure the right "variance factors" with the correct technical know-how,
5. putting adequate measurement systems in place,
6. having a clear understanding and follow-through action in terms of which processes to prioritize, and
7. ability to read and interpret control charts accurately and take follow-through corrective action (Antony & Masaon, 1999).

10.5 Acceptable quality level

One subset of SQC that is widely used in the apparel manufacturing industry is the concept of Acceptance Sampling. Acceptance sampling is an important field of SQC that was popularized by Dodge and Rooming and originally applied by the US military for the testing of bullets during World War II.

Acceptance sampling plans enable us to distinguish between the acceptable and the unacceptable lots if the proportionate sample is randomly drawn from a lot. This would ensure that the sample would represent the quality level of the lot, and based on this, the acceptance decision can be made. And from hereon comes the concept of Acceptable Quality Level (AQL) as it is popularly known in the industry.

Having known that 100% inspection may not be the best thing to do, the next question is if not 100%, how much to inspect? There are two options available. The first option is to decide the sample size based on a fixed proportion of the lots that will be inspected (for examaple sample size of 5% or 10% of the lot size) to arrive at the acceptance decision of the whole lot, or the second option is to use the Acceptance Sampling procedure to arrive at a sampling plan for given AQL and make an acceptance decision. The first option is arbitrary, and it does not have any scientific basis. It is therefore important to have a reliable and scientific method of arriving at such a decision, and one should be aware of the extent of risk involved

Lean quality management   323

in such decisions. Acceptance sampling is a scientific technique, and it also tells us the probability of making a wrong judgment while using it (Bheda, 2014).

What is AQL? As Pradip V. Mehta describes, "*The AQL is the maximum percent defective that for sampling inspection can be considered satisfactory as a process average.*" In layman's language, this means, when a buyer specifies a particular AQL for sampling inspection (refer to Table 10.1), it is an indication that as long as the percentage of defective garments in the shipments (lots) supplied by a manufacturer is lower than the AQL, most of the shipments will be accepted. Process average means the average percentage of defective products (percent defective) in the lots submitted for the first inspections. Assume a true percent defective level of six lots of garments is 2.3, 2.7, 2.4, 2.6, 2.8, and 2.2, respectively, the process average will be 2.5% defective. The method of arriving at the process average in apparel factories is explained in the latter part of this chapter

Based on the extensive work by the US military during and past-World War II, the US Government issued the standard for sampling procedures and tables for inspection called MIL-STD-105 D in 1963. This was further modified in 1989 as MIL-STD-105 E and redesignated as ANSI/ASQC Z 1.4 in February 1995. For all the practical purposes, MIL-STD 105 D and ANSI/ASQC Z 1.4 are almost similar. For acceptance sampling inspection in the garment industry, most buyers refer to the tables (refer to Table 10.1) from either of these standards. Though the garment industry generally uses a normal level on inspection, the standards also provide from reduced and tightened inspections based on the past performance of the supplier.

Most international apparel retailers use AQL 1.5 and 2.5 levels for making acceptance decisions of finished lots of garments. It must be noted that there are AQL charts that provide sampling plans at AQL 0.65 and even lower. These are used for high-value products where the cost of a wrong acceptance decision can be very high.

10.5.1 *Working of the acceptance sampling plans*

The apparel industry mainly uses single-sampling plans for acceptance decisions. However, a few buyers also use a double-sampling procedure. In a single sample based on the AQL table, you randomly draw a sample consisting of a specified number of garments from a lot. For example, as shown in Table 10.1, the sample

Table 10.1 Acceptable quality level (AQL) table.

AQL chart				
Lot quantity	**Audit quantity**	**AQL 1.5**	**AQL 2.5**	**AQL 4.0**
1−32	32	0	0	0
33−1200	32	1	2	2
1201−3200	50	1	3	3
3201−10,000	80	2	5	5
10,001−35,000	125	3	7	8
35,001−1,50,000	200	5	10	10

size or audit quantity for a lot size of 10,000 pieces would be 125. Similarly, the sample size for a lot of 3000 pieces would be 50 units. The sample plan also provides the number of maximum allowed defective pieces. If the defective pieces are equal or less than the allowed number, the lot is accepted, and if the number of defective pieces is greater than the allowed limit, the lot is rejected. One may say that as the acceptance sampling is scientific, ideally speaking, it must lead to 100% reliable results. In other words, it must always lead to acceptance of lots containing a lower defective level than AQL and must reject all the lots that contain more defective products than AQL. But this is not possible, as the acceptance decision is made only based on small sample drawn from the lot based on probability and it carries a risk of making a wrong judgment.

The acceptance decisions based on AQL inspections contain two kinds of risks as detailed below:

1. **Producer's risk**: The chance of rejecting a good lot that contains equal or less percent defective than AQL.
2. **The Customer's risk**: The chance of accepting a bad lot that contains more defective than the largest proportion of defects that a consumer is willing to accept a very small percentage of the time. It is also known as Lot Tolerance Percent Defective (LTPD) or represented as Rejecting Quality level.

The kind of risk customer and producers faces in terms of making a wrong decision while using an acceptance sample plan can be understood by the Operating Characteristic Curve (OC Curve) of a sample plan. OC curve of a sample plan indicates the chance of acceptance or rejection of lots with varying degrees of defective level. Fig. 10.6 shows the OC curve of sample plan $n = 200$ and $c = 10$ for a lot size of 10,000 pieces at AQL 2.5. The y-axis on the graph indicates the probability of acceptance of the lot, whereas the x-axis indicates the percent defective level of the lots. As can be seen, the lots containing 2.5% defective merchandise are likely

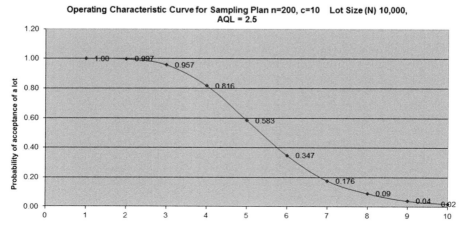

Figure 10.6 Operating characteristic (OC) curve for sampling plan.

to be accepted 95% of the time, and there is a possibility of it getting rejected 5% of the time (producer risk). The readers will be surprised to know that a lot containing 5% defective (twice as compared to the AQL) also has the chance of acceptance of 58.3 times out of 100 inspections. The customer's risk with this sample plan is about 10% where a lot containing 8% defective may get accepted. This point beyond which the customer would not like the lots to be accepted by sampling inspections is called LTPD. With a simple procedure, an OC Curve (refer to Fig. 10.6) for every sampling plan can be drawn to understand how that plan discriminates between good and bad lots. The people who want to go further deep in the subject can specify their AQL and LTPD and find out an appropriate sample plan for their needs.

10.5.2 Ensuring success of acceptance sampling plans

The answer to these question is very simple but it would needs specific efforts. You need to ensure that your average percent defective level is below the AQL prescribed by your buyer. What does this mean? It means the true percent defective level of the lots submitted for AQL-based inspection must be less than the AQL. For this purpose, an organization has to measure its current average percent defective level (process average). This can be achieved by conducting sampling inspections of the lots before the inspection by the customer. In such a case, all the pieces in a sample drawn from the lot are inspected to arrive at the percent defective level of respective lots. If an organization does this for about 300 consecutive lots and calculates the average of the percent defective of all lots inspected, it would give a good idea of the "process average." Assuming your process average is lower than the AQL level, then there can a very minimal chance (generally less than 5%−10%) of your shipment getting rejected. If your process average is greater than AQL level, you need to work toward, if not eliminating, reducing the generation of defect level at the source so that the process average becomes lower than the AQL level. In case, if process average remains higher than the AQL level, the chances of your shipments failing to pass AQL-based inspection are higher depending on the process average.

The best way to reduce the process average could be to analyze the kind of defects noticed in the inspection and their occurrence (frequency). A Pareto analysis as shown in Fig. 10.7 can be very useful. Once you know which the most frequently occurring defects are, it is possible to go to the source of these defects through RCA or cause and effect analysis. After the root causes of the defects are identified, countermeasures need to be developed for each of these causes. The implantation of the countermeasures should help reduction of variability and elimination of special causes resulting in defects. The effect of the application of the countermeasures needs to be monitored to see their effectiveness. This will gradually reduce the defect generation and bring the process average at the desired level.

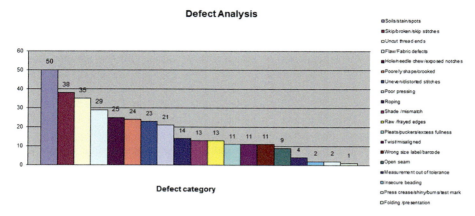

Figure 10.7 Pareto analysis of defects.

10.6 Six Sigma

Six Sigma (6σ) can be defined as a set of techniques and tools for process improvement. It was introduced by Bill Smith, an American engineer while working at Motorola in 1980. However, Six Sigma gained fame when Jack Welch made it central to his business strategy during his stint as CEO at General Electric in 1995. By its nature, Six Sigma views all work as processes that can be defined, measured, analyzed, improved, and controlled, abbreviated as the DMAIC approach. Six Sigma uses qualitative and quantitative techniques or tools to drive process improvement. Such tools include SPC, control charts, failure mode and effects analysis (FMEA), and process mapping aimed at eliminating defects. Six Sigma quality performance can be considered as 3.4 defects per million (DPM) opportunities (accounting for a 1.5-sigma shift in the mean).

Six Sigma follows a specific five-step problem-solving approach, DMAIC. The DMAIC framework applies control charts, measurement systems capability studies, designed experiments, process capability analysis, and many other statistical tools.

1. Define phase: This phase defines the strategic direction of the organization, develop the problem statement, define the resources, evaluate the key organizational support, identify the team, and develop the project plan and milestones along with a well-defined process map.
2. Measure phase: Measures are set for the strategic objectives of the organization. This phase also involves defining defect and opportunity, developing a data collection plan, a detailed process map of appropriate areas, validating the measurement system, collection of data, determining process capability, and setting a baseline.
3. Analyze phase: This phase aims to define performance objectives, identify value/nonvalue-added process steps, and identify sources of variation.
4. Improve phase: Some of the main objectives of this phase is to identify the opportunities for improvement and convert them to Six Sigma projects for improvement, develop

Lean quality management 327

potential solutions, perform the design of Experiments, define operating tolerances of the potential system, and validate potential improvement by pilot studies.

5. Control phase: This phase aims to set up a management control action of continuous reviews on the improvements made on the project. The goal is to define and validate the monitoring and control system, implement SPC, determine process capability, develop standards and procedures, and develop the transfer.

Business organizations have been very quick to understand the potential benefits of Six Sigma and to adopt the principles and methods. Between 1987 and 1993, Motorola reduced the defective rate on its products to 3.4 DPMO. This success led to many organizations adopting the Six Sigma approach.

Six Sigma has spread well beyond its manufacturing origins into areas including health care, many types of service business, and government public service. The reason for the success of Six Sigma in organizations outside the traditional manufacturing sphere is that variability is everywhere, and where there is variability, there is an opportunity to improve business results.

The Six Sigma philosophy thus aptly converges with Lean philosophy aiming at the relentless elimination of waste through the involvement of all employees in process improvement and standardization of processes.

10.7 Conclusion

In the past few sections, we have discussed the meaning of quality, its relationship to variability, and how statistical methods play a key role in all quality improvement initiatives. This was followed by a glimpse into SQC and the various control charts, followed by Six Sigma and its evolution from defect elimination and basic variability reduction to organizationally tie these initiatives to projects and activities that improved business performance through cost reduction. We have also discussed "Acceptance sampling" methodology, that validates or invalidates the quality of merchandise based on inspections conducted on proportionate quantity, randomly picked from the bulk lot. This concept widely used in the apparel manufacturing industry.

It is imperative to understand here that building quality into the design and managing it at the various process levels has to be seen as an integral of one of the several strategic initiatives taken by the Top Management. While quality management division has the responsibility for evaluating and using quality-cost information for identifying improvement opportunities in the system, and for making these opportunities known to higher management. It is important to note, however, that the quality function is not solely responsible for quality. After all, the quality organization does design, manufacture, distribute, or service the product. Thus the responsibility for quality is distributed throughout the entire organization.

The philosophy of Deming, Juran, and Feigenbaum implies that responsibility for quality spans the entire organization. However, there is a danger that if we adopt the philosophy that "quality is everybody's job, then quality will become nobody's job" (Montgomery, 2013). Because quality improvement activities are so broad, successful efforts require deep and intrinsic top management commitment. It can be concluded that the lean-based optimization and continuous improvement approach which is based on a firm foundation of quality management is an essential requirement. After all, each of the philosophies and efforts is aiming toward customer satisfaction, and hence, it is imperative to apply lean management and quality management collectively as a LQMS.

References

Abtew, M. A., Kropi, S., Hong, Y., & Pu, L. (2018). Implementation of statistical process control (SPC) in the sewing section of garment industry for quality improvement. *AUTEX Research Journal, 18*(2), 160−172.

Antony, J., & Masaon, B. (1999). *Key success factors for the implementation of SPC.* Quality America. Retrieved from https://qualityamerica.com/LSS-Knowledge-Center/statisticalprocesscontrol/key_success_factors_for_the_implementation_of_spc.php. (Accessed December 2019).

ASQ. (2010). *What is quality function deployment (QFD)?* American Society for Quality. Retrieved from https://asq.org/quality-resources/qfd-quality-function-deployment. (Accessed January 2020).

Bacoup, P., Michel, C., Habchi, G., & Pralus, M. (2018). From a quality management system (QMS) to a lean quality management system (LQMS). *The TQM Journal, 30*(1), 20−42.

Bheda, R. (2004). *Statistical process control in apparel manufacturing.* Stitch World.

Bheda, R. (2014). *What is acceptable quality level.* Rajesh Bheda Consulting. Retrieved from https://www.rajeshbheda.com/pdf/rbc-article9.pdf. (Accessed October 2019).

Clark, D. M., Silvester, K., & Knowles, S. (2013). Lean management systems: Creating a culture of continuous quality improvement. *Journal of Clinical Pathology, 66*(8), 638−643.

Cohen, L. (1988). Quality function deployment: An application perspective from digital equipment corporation. *National Productivity Review, 7*(3), 197−208.

Dahlgaard, J. J., & Dahlgaard-Park, S. M. (2006, May). Lean production, six sigma quality, TQM and company culture. *The TQM Magazine, 18*(3), 263−281.

Forza, C. (1996). Work organization in lean production and traditional plants: What are the differences? *International Journal of Operations & Production Management, 16*(2), 42−62.

Hall, R. (1983). *Zero inventories.* McGraw-Hill.

Hines, P., Matthias, H., & Rich, N. (2004). Learning to evolve: A review of contemporary lean thinking. *International Journal of Operations & Production Management, 24*(10), 994−1011.

Holweg, M. (2007). The genealogy of lean production. *Journal of Operations Management, 25*(2), 420−437.

Juran, J. M., & Godfrey, A. B. (1998). The quality control process. In J. M. Juran, & A. B. Godfrey (Eds.), *Juran's quality handbook* (p. 4.20). New York, NY: McGraw-Hill.

Locher, D. A. (2008). *Value stream mapping for lean development: A how-to guide for streamlining time to market.* New York, NY: Productivity Press.

Micklewright, M. (2010). *Lean ISO 9001—Adding spark to your ISO 9001 QMS and sustainability to your lean efforts*. Milwaukee, WI: American Society for Quality.

Moin, C. J., Doulah, A. B., Ali, M., & Sarwar, F. (2017). Implementation of an operating procedure for quality control at production level in a RMG industry and assessment of quality improvement. *The Journal of the Textile Institute, 109*(4), 524–535.

Monden, Y. (1983). *The Toyota production system: Practical approach to production management*. Norcross, GA: Industrial Engineering and Management Press.

Montgomery, D. C. (2013). *Introduction to statistical quality control*. Hoboken, NJ: John Wiley & Sons, Inc.

Reeb, J. E., & Leavengood, S. (2010, October). *Introduction to lean manufacturing. EC 1636.*

Sánchez, A. M., & Pérez, M. P. (2001). Lean indicators and manufacturing strategies. *International Journal of Operations & Production Management, 21*(11), 1433–1452.

Schenkelberg, F. (2015). *Introduction to Control Charts (FMS Reliability)*. Accendo Reliability. Retrieved from https://accendoreliability.com/introduction-to-control-charts/. (Accessed June 2019).

Schonberger, R. J. (1982). *Japanese manufacturing techniques: Nine hidden lessons in simplicity*. London, UK: The Free Press.

Schonberger, R. J. (2007). Japanese production management: An evolution—With mixed success. *Journal of Operations Management, 25*(2), 403–419.

Schonberger, R. J. (2014). Quality management and lean: A symbiotic relationship. *Quality Management Journal, 21*(3), 6–10.

Shah, R., & Ward, P. T. (2003). Lean manufacturing: Context, practice bundles, and performance. *Journal of Operations Management, 21*(2), 129–149.

Shah, R., & Ward, P. T. (2007). Defining and developing measures of lean production. *Journal of Operations Management, 25*(4), 785–805.

Su, J., Gargeya, V. B., & Richter, S. J. (2005). Global sourcing shifts in the U.S. textile and apparel industry: A cluster analysis. *The Journal of the Textile Institute, 96*(4), 261–276.

Taguchi, G., Chowdhury, S., & Wu, Y. (2005). Introduction to the quality loss function. In G. Taguchi, S. Chowdhury, & Y. Wu (Eds.), *Taguchi's quality engineering handbook* (pp. 169–179). Wiley-Interscience.

Uddin, S. M., & Rahman, C. M. (2014). Minimization of defects in the sewing section of a garment factory through DMAIC methodology of six sigma. *Research Journal of Engineering Sciences, 3*(9), 21–26.

Watson, G.H. (2001, November). Cycles of learning: Observations of Jack Welch. *Six Sigma Forum Magazine, 1*(1), 13–17.

Lean human resources

Chandrark Karekatti
Ananta Garments Ltd., Dhaka, Bangladesh

> We get brilliant results from average people managing brilliant processes. We observe that our competitors get average (or worse) results from brilliant people managing broken processes. — Fujio Cho

11.1 Introduction

Most of the time in manufacturing environments, we witness that the work and its practices change but people do not. And many times, we come back to Square One, as we do not achieve the expected improvements. Why...? maybe because an important and key element, "Involving people" is missing. We generally do not give due attention to the "social and people" aspect of the systems, and put most of the efforts, energies, and time in the "Technical nitty-gritty" only. Murli (2018) rightly states that lean transformations are not just engineering transformations but the cultural transformation. It is important to involve human resource (HR) in the lean transformations, though many times it is not given due attention and HR plays a passive role. HR should focus actively, and early engagement of HR in the lean transformation may help in the timely implementation of the plans (Murli, 2018).

Taira (1996) discusses an important difference in the approach of utilizing HRs between mass production and lean production. To the mass producers, lean production is like a borrowed concept. The author states the transition from mass production to lean product as controversial and traumatic in the context of the human resource management (HRM), industrial relations, and work processes. With mass production, the thrust of HR is well supported by autocratic Taylorism that believes in the employment at will. Here the key focus is on the maximum division of labor and increasing the speed through reduction of cycle times. Taira (1996) cites that the key to mass production is "the complete and consistent interchangeability of parts and the simplicity of attaching them." It is true for human changeability as well, where the workers are to be reduced to "brainless and voiceless automatons." The lean approach is just the opposite of this idea of mass production. The true lean approach transfers the maximum tasks and responsibilities to the workers, wholly adding value to the end product. This also expects a proactive role of workers in not only identifying the defects and problems in the process but a vital role in working out the solutions as well. This requires a significant and drastic reorganization of the management process and a revolutionary redistribution of power and dignity (Taira, 1996).

Lean philosophy emphasizes on people involved in every stage of the transformation journey. In various literature, we see a great focus on HR practices. Some of the notable practices include cross-functional teams, training and skill development, job rotations, teamwork with multiskilling, people-driven problem-solving, quality circles, etc. as the core components of lean the transformation program (Farris, Aken, Doolen, & Worley, 2009).

Lean implementation projects in a garment factory, as experienced by the author, are typically related to waste elimination in core production processes. However, conceptually lean manufacturing should include all organizational activities, including manufacturing, marketing, accounts, administration, HR, utilities, etc. Though few garment factories are adopting lean principles beyond the core manufacturing processes like product development, warehouse, accounts, and supply chain management (to some extent by the virtue of Entreprise Resource Planning (ERP) systems), HR function mostly does not play any role in lean implementation and is untouched and unchanged by their company's commitment to lean.

Traditionally, the HR function in a garment factory is preliminary concerned with recruitment, compensation, employee records, and performance review. However, practically, their task was mostly limited to record-keeping for most of the aforementioned functions. Global competition made factories adopt lean systems for their core manufacturing processes. The traditional HR setup was not technically qualified or trained to understand the cross-functional requirements of the lean system like self-inspection, work cells, flow manufacturing, and Kaizen event. Neither did it had the technical know-how to ascertain its (HR's) role and responsibilities in the new lean culture. Moreover, the role of the HR function in the lean environment was rarely addressed by consultants while implementing lean projects in an apparel factory. Thus with limited industry know-how (here the context is apparel manufacturing function) in core HR functions in a lean environment and with a lack of knowledge in the public domain, HR remains more or less untouched in lean implementation projects. In this chapter, the author describes the strategic role that HR needs to play in developing a sustainable lean culture in the organization.

11.2 Lean human resource management

It is a proven fact that adopting lean practices is vital to survive and thrive for an organization. Irrespective of any field or industry, the lean concepts are equally applied in the different parts of the globe. In the last 70 years, there have been several success stories where the lean transformations proved instrumental in bringing new waves of success and operations excellence. In today's competitive business environment, it is essential to adopt a lean approach to remain competitive. However, it is important to note that this success is just not due to the technical aspects of lean, but the drive and boost from the HR. On the other side, there is a huge unused HR potential, and lean recognizes it as one of the Muda (waste) in the system. The zeal, courage, and commitment from everybody involved playing a decisive role in any improvement mission. After all lean is a culture and possesses a much wider spectrum than just tools and techniques. If not utilized in the right manner, the same unused (or misused) workforce potential may lead to inefficiencies and even damage the organization. HR can play an important, creative, productive role in channelizing this unutilized or underutilized workforce potential. Several studies confirm the importance of the HR role in achieving the desired success of lean implementations. Ramarapu, Mehra, and Frolick (1995), based on the results of 105 studies (between 1980 and 1993) on lean implementations, identified "management commitment and employee participation" as one of the five critical categories for lean implementation. Cross-training and education, team decision-making, management participation and commitment, and employee suggestions were identified as four elements associated with this category of management commitment and employee participation (Ramarapu et al., 1995). Marodin and Saurin (2013) discuss the factors affecting the lean implementations based on their study of 102 published research works in the years between 1996 and 2012. It was revealed that management support and/or commitment was mentioned as an important factor in the 55% of the studies analyzed (Marodin & Saurin, 2013). Olivella, Cuatrecasas, and Gavilan (2008) discuss the importance of work organization in lean production, though work organization is not given due attention in lean production. It is recommended that the work organization must be customized according to local labor laws, culture, and collective agreements. The term "lean work team" is referred to the teams that follow self-quality control, task rotation, and standard work, whereas the term "high-performance work organizations" is referred to the organizations that have self-directed work teams, total quality, quality circles, and task rotation. Herein each of the traits mentioned the central theme is participative culture. The published work by leading authorities in the area of lean manufacturing confirms to following central concepts (Olivella et al., 2008) with human participation:

- Standardization, discipline, and control,
- Continuing training and learning,
- Team-based organization,
- Participation and empowerment,
- Multiskilling and adaptability,
- Common values, and
- Compensation and rewards.

Human resource (HR) is one important area of assessment while developing a lean assessment tool. It is important to have a HR strategy for the adequate functioning of lean systems. Liao and Han (2019) discuss the importance of HRs in bringing the success of lean implementations and the interaction effect of HRM and lean production and emphasized on the need for the ability- and motivation-focused HRM to achieve better results of lean initiatives. Apart from this, several pieces of research establish the vital role of HR for a successful lean transformation (Bamber & Dale, 2000; Longoni, Pagell, Johnston, & Veltri, 2013; Rothstein, 2004; Wood, 2005).

According to Birdi et al. (2008), based on extensive research on 308 firms for 22 years, empowerment and employee development (maybe through training, and teamwork) directly result into benefits, whereas the operational lean processes own their own do not have such benefits (Birdi et al., 2008). Delbridge (2005) lean manufacturing is an approach that incorporates work organization, operations, logistics, HRM, and supply chain relations (Delbridge, 2005).

MacDuffie (1995) explores the role of HR in the organizational logic of a production system more deeply and states that in a lean system, high-commitment HR practices and low inventory consistently outperformed the mass production plants. MacDuffie attempted to capture the nature of integrated systems and categorized the measures in two categories: (1) high-involvement work systems or work practices, and (2) high-commitment HR practices as shown in Table 11.1 (MacDuffie, 1995).

Spear and Bowen (1999) in *Decoding the DNA of the Toyota Production System*, based on their 4-year-long study covering 40 plants in the United States, Europe,

Table 11.1 MacDuffie's measures of work systems and HRM policies.

"High-involvement" work systems	A high percentage of the workforce involved informal work teams.
	A high percentage of the workforce involved in employee involvement groups.
	A large number of production-related suggestions received per employee.
	A high percentage of production-related suggestions implemented.
	Frequent job rotation within and across teams and departments.
	Production workers responsible for quality inspection and data gathering.
"High-commitment" HRM policies	Hiring criteria that emphasize openness to learning and interpersonal skills.
	Pay systems contingent upon performance.
	Single status workplace (common uniform, common parking, common cafeteria, no ties).
	High levels of initial training for recruits (workers, supervisors, and engineers).
	High levels of ongoing training for experienced employees.

Adopted from MacDuffie, J. P. (1995). Human resource bundles and manufacturing performance: Organizational logic and flexible production systems in the world auto industry. *Industrial and Labor Relations Review*, 48(2), 197–221.

and Japan, discuss the unspoken rules of the Toyota Production System that are responsible for providing Toyota a competitive edge. The authors capture four key rules, and one of the rules (rule no. 1) is "How people work." This rule talks about performing work in a specified manner in terms of its content, sequence, timing, and outcome. Here the role of specialized training becomes vital. At Toyota, people are considered as the most significant corporate assets. It gives the utmost importance to impart knowledge and skills to build necessary competitiveness. An important aspect of HR development at Toyota is that each of the managers is trained in such a manner that he is capable of performing the task of everyone he/she supervises and also able to teach about solving the problems. This model is applied at all the levels of the organization, and it makes contributions from everybody in developing HRs at Toyota (Spear & Bowen, 1999).

Tracey and Flinchbaugh (2009), while discussing the role of HR in lean transformations, state that lean is beyond 5S and U-shaped layouts, it is all about people, culture, and leadership. However, many times the role of HR is passive in lean transformations. Adopting lean practices beyond core manufacturing (areas like supply chain management, product development, and even accounting) has transformed several organizations, but HR remains untouched as far as its contribution is concerned. Based on their research, the authors suggest five key variables to engage HR in lean transformations as indicated in Table 11.2 (Tracey & Flinchbaugh, 2009).

Jekiel (2011) highlights the wastage of human talent in the organizations, and it leads to the failure of their efforts. The abilities of people are generally hidden and unseen, hence it remains fully unutilized. This provides a tremendous opportunity for an organization to achieve greater heights through the identification and utilization of talent in a structured manner. Various organizations in the United States and Europe could not witness the same success of lean philosophy as witnessed by Toyota. Despite investing significant resources, efforts, and time, it was difficult to achieve sustainable continuous improvement. The author identifies key reasons for failed lean efforts as:

1. Applying lean as a set of tools
2. Changes require new ways to work
3. The balance of power creates resistance
4. Lack of HR involvement

There is no doubt that HR plays a vital role in the success of lean transformations. Jekiel argues for the need for cultural change in terms of redesigning HR processes for continuous improvement where the HR department works as a business partner to help and stimulate other departments toward achieving continuous improvements. HR can play an active role in many aspects including cultural implementations, designing newly required processes, and handling resistance to leadership changes (Jekiel, 2011).

Forrester (1995) discusses the implications of lean manufacturing for HR strategy and states that in a lean manufacturing environment and its processes require interrelated policies that virtually cover every aspect of personnel policy and

Table 11.2 Key variables to engage human resource in lean transformations.

Variable	Key focus *(in the context of the lean environment)*
Development of teams as a supporting structure	• Teams should have a common language, common principles, and common tools. • Common drive (as per vision, metrics, and goals). • Designing the workaround visually to improve transparency, visibility, and common agreement. • Teams should be equipped with the capability and skill to manage themselves.
Calculation and communication of metrics	• The process owners should own the scoreboard and metrics to make it easy to maintain and update. • Metrics should be predictive as much as possible to support day-to-day decision-making. • Management support to the metrics, in terms of review (with clear who will review, when will review, what will be reviewed, and how to respond as next course of action). • Metrics must point a steady and consistent direction towards the ideal state.
Communication across boundaries	• Improving communication (related to material and information flow) across boundaries, such as departments and functions. • Communication must be vertical, horizontal, and two-way.
Communication to employees regarding their role	• There should be clarity about each employee's roles and responsibilities. • Training of employees on communication and discussion techniques (most of the time training happens only on the technical part of lean focusing on "Go and implement Lean"). • Changing of roles of employees as the organization progresses in the lean journey. • Maintaining role clarity and integration throughout the lean journey.
Acknowledgment and celebrations of successes	• There should be a sense of recognizing success and accomplishment. • As the lean is an endless journey, clear milestones should be set and defined, accordingly, the progress should be communicated and celebrated. • If possible, the employees should be rewarded but at the same, time it should be ensured that it is done in such a balanced manner that rewards should not hamper or retard the future growth of the lean journey.

practice. The formation and utilization of teams (which is considered as the central word in lean philosophy) is a visible change. The cultural effects are visible in other areas as well including a waste reduction (at macro- and microlevel), improved maintenance practices (from reactive maintenance practices to proactive maintenance practices), multifunctional new product development, etc. The author emphasizes understanding the need for integration to create more fundamental synergy between business and people issues (Forrester, 1995).

Bonavia and Marin-Garcia (2011) discuss the integration of HRM to lean production. Based on a research conducted on 79 establishments in Spain, it was confirmed that most of the companies who use lean production practices also take care to train their workers using lean practices as well as improving their employment security, though it was not true for the pay for performance system. The combination of HR and lean resulted in boosting productivity and reduced inventory levels but did not affect the other performance variables (Bonavia & Marin-Garcia, 2011). Thirkell and Ashman (2014) emphasize playing the central role by HRM in lean thinking implementation that is in line with the organizations' goals. Here employee attitude and behaviors act as a mediating variable through which HRM influences performance outcomes. The authors cite some shortcomings of HRM that can undermine the lean thinking as, poor selection of change agents and improvement teams, lack of engagement and buy-in from an individual and teams, silo thinking, lack of necessary skills and expertise, poor communication, and inadequate performance management systems (Thirkell & Ashman, 2014).

Heinzen and Höflinger (2017) discuss the role of HRs product development (LPD). The analysis of the impact of HR practices was done on two aspects: (1) skill and motivation of employees, and (2) performance in lean product development. It was found that the training, internal development, and performance appraisal contribute significantly to the employee performance in LPD, whereas the performance compensations did not show any impact (Heinzen & Höflinger, 2017). Dibia and Onuh (2010) consider HRs as the "real quality" and the heart of lean philosophy. For a sustainable, effective, and efficient lean environment, motivated, self-directed, intelligent, and skilled HR is required. The researchers found that in recent years, the impact of a lean transformation is slowing down and dwindling due to the absence of intellectually unique HRs. Hence, there is a need to make this foundation pillar of lean (HRM) stronger. The researchers also recommend for good leadership and standard cooperate culture along with professional discipline for a sustainable lean culture in an organization (Dibia & Onuh, 2010).

The people aspect of lean is called cultural enablers by Shingo and guided by humility and respect. The traditional white coat leadership is replaced by lean management, where leaders exhibit humility (because they do not know it all); they are curious about how things work (because they do not know); they act as facilitators (not telling what to do); and most importantly, they need to be a student first (before they can teach only when they know), the control is replaced by perseverance (Patel, 2016). The cultural HR transformation also prescribes long-term planning through the constancy of purpose, abolishing the merit system and incentive pay in favor of performance, and encouraging management by facts over management by objectives (Patel, 2016).

From this discussion, it is evident that HR is the heart of the lean culture. The tools and techniques of lean (or the technical or engineering aspects) can only deliver the stainable improvements with the HR that is empowered, self-directed, skilled, and intellectual. Hence, the soft-side of lean must also be given due attention to make continuous improvements sustainable.

11.3 Lean and downsizing

Many times, adopting lean practices is referred to as downsizing the manpower. The common drive behind this thought is "lean means more output from lesser resources," and it is generally taken as a reduction in workforce and a fear of losing jobs. This is generally witnessed with the organization that fails to understand the real meaning of being lean. Kinnie, Hutchinson, and Purcell (1998) discussed the relationship between downsizing and the lean organization. The authors argue that downsizing may always do not mean lean. Lean aims for waste (Muda) elimination, including the Muda of unused human potential or talent. It is observed that downsizing may result in some improvements toward achieving immediate financial targets (or cost reduction objectives), but such improvements are not sustained in other areas improved customer service and increased competitive advantage. In such a situation rather than going for downsizing, the focus could be on redeploying people into a lean promotion function (Kinnie et al., 1998). Here, as suggested by Doherty and Horsted (1995), the role of HR becomes very critical in managing the change in successfully addressing the issues with individual and organizational perspectives. It is worth mentioning the statement by Doherty and Horsted (1995), "This requires the development of HR strategies which dovetail the use of internal and external interventions, to give equal support and assistance to individual's transition by offering them opportunities to develop themselves for the benefit of the organization, whatever that may be in the future".

Womack (1996) addresses this issue and links it with work psychology. There is no doubt that lean practices are more efficient, hence it requires lesser people to deliver more. Here, according to Womack, the management has following two choices:

1. Lay-off the workers;
2. Find new works and new markets by speeding up the product development.

The second choice is more important as the first option is resulting in termination of the livelihood of the employees. Naturally, by design lean is all about active cooperation and teamwork. Many firms witness success initially and then go for downsizing that results in inures in their lean mission. Firms should ensure not only the job guarantee while lean transitions but also the career progression. In a lean environment, the creation of horizontal work teams threatens traditional career paths (Womack, 1996). The idea of "alternating career" is proposed by Womack and Jones (1994), which focusses on assuring employees about the requirement of their primary skills by the organization as an important success element. Here, the organization takes the responsibility of continually upgrading the skills through rotation from team assignments to the functional assignments. This results in developing production work to process experts (Womack & Jones, 1994). Here is a strong statement from Womack and Jones (1994) "Companies must pursue every option for preserving jobs as they create lean enterprises."

11.4 The role of HR in lean implementations in the apparel industry

The role of HR in a lean implementation project can be divided into following two categories:

- Organization-wide lean implementation initiatives.
- Engagement in employee participation initiative.

11.4.1 Organization-wide lean implementation initiatives

To start with, the organization should take initiatives to train HR in understanding lean philosophy and its functionality in their factory. In current scenario, HR has little participation in lean project training. The top management needs to ascertain HR's role in developing a sustainable lean environment across the organization. HR should also be empowered to develop an ecosystem that is conducive for sustainable lean initiatives. The following are few important functions, where HR participation is required.

11.4.1.1 Develop a team that enables lean environment and work as supporting structure of lean

One of the major HR functions is recruitment. It is not only important that the HR scout for candidates with technical expertise and experience in lean projects, but they should also investigate potential candidates having personality traits that align with the lean culture. A few of the basic traits are the ability to communicate cross-functional teams, work in teams, design improvements project, create and follow standards, work across organizational boundaries, and maintain positive attitude. HR should check if these traits are present in some form in the potential employees. While selecting exiting employees for lean projects, HR should carefully evaluate team members for these traits. Specific training programs can be arranged to enhance these traits for existing/new employees. However, it should be remembered that these traits can be enhanced to a certain extent by training, but if they are not present, they cannot be taught. It is observed that often experience candidates do not have traits and aptitude required in a lean environment. Therefore HR should develop recruitment tests/procedures that can identify these traits. Another intervention required from HR is to develop cross-functional expertise within the workforce. HR should map these requirements in discussion with production/process heads. HR should further facilitate cross-functional competency development by competency mapping and on-job training. The main purpose of developing cross-function expertise is to have a pool of flexible workforce. The cross-function expertise and flexible workforce have two aspects. First, the workers must receive training on the new methods of production/machines/operations or multiskilling on similar operations/machines, for example, multi-skilled operators on specific overlock operations/lock stitch operations as per identified requirement. It can also involve training of a few qualified marker man/helpers on automated machines, which do not require specific machine handling expertise. The second aspect is to develop calibrated cross-functional expertise among employees, like the one required in the lean practice of "operator self-inspection," where production workers are responsible for checking the quality of their output. If operators are given this responsibility, of "self-inspection," they need training on quality procedures like Statitstical Process Control (SPC), buyer's quality requirement, and defect identification. In factories, where Total Productive Maintenance is practiced, certain routine machine maintenance (operator's machine) check is required to be done by the machine operator. Here, it is necessary to train the concerned operators in the machine routine check procedure. Both of these skills reduce the need for support personnel on the floor (quality inspectors and maintenance personnel) and reduce machine downtime while workers wait for the support personnel to provide the required services. In the latter case, it helps in early (failure) detection and reduces premature breakdowns.

11.4.1.2 Identifying Key Result Areas, setting and tracking departmental and individual Key Performance Indicator

Naturally, the owners, top management have specific business objectives for implementing the lean project. HR needs to qualitatively and quantitatively understand the organization's vision and work with business and operation heads to identify key performance areas that will help to achieve this vision. Based on the Key Result Areas (KRA), HR along with concerned process heads should determine the Key Performance Indicators (KPI) for each department and key designations. The KPI selected should help to achieve the identified KRA and hence the lean vision of the organization. KPI targets should be mutually agreed by the management and the department heads. Target selection should be based on SMART (Specific, Measurable, Attainable, Relevant, Time-based) methodology and accompanied by baseline mapping exercise. The KRA and KPI should be owned by process owners/department heads, for example, heads for production cell team, merchandising team, and store or supply chain team. KPI should be daily tracked by department heads and top management should conduct a quarterly review meeting to ascertain the direction of lean initiates. Example of KRA

Lean human resources

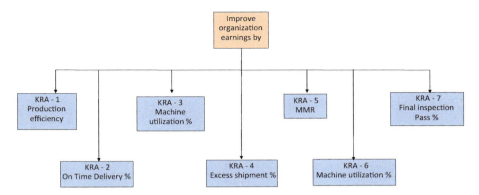

Figure 11.1 Company vision and Key Result Areas (KRA) aligned with vision.

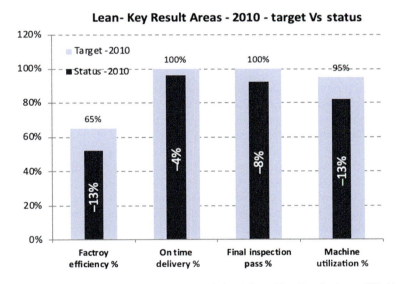

Figure 11.2 Key Performance Indicators (KPI) derived from Key Result Areas (KRA).

aligned with the policy objective of the factory is shown in Fig. 11.1. Fig. 11.2 shows KPI derived from the KRAs. Fig. 11.3 shows the KPI tracking report for one manufacturing parameter.

Employee yearly performance appraisal should be based objectively on their performance concerning the KPI target. It is recommended that the HR to rethink the traditional wage grid/incentive system in the lean paradigm. This is because, certain cross-boundary work responsibilities and cross-functional expertise that were never considered/required in traditional manufacturing systems are necessary for a lean environment. Few examples are given as follows:

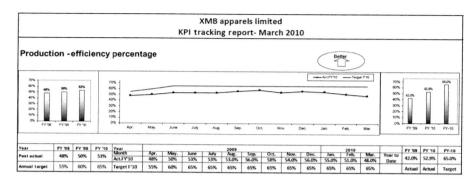

Figure 11.3 Key Performance Indicators (KPI) derived from Key Result Areas (KRA).

1. Cross-functional expertise for self-inspection of own produce quality.
2. Cross-functional expertise for basic routine machine maintenance checking.
3. Preference for multi-skilled operators.
4. Capability in complying with right first time quality system.
5. Ability to contribute to continuous improvement initiates like Kaizen projects.
6. Expertise to participate in performance improvement initiates, like Single Minute Exchange of Die (SMED), Andon, and flow manufacturing.

The transaction to lean manufacturing was aimed to reduce waste and to make the organization leaner and more productive. The leaner organization helps in reducing production manpower and hence production cost. Ideally, lower costs help to improve the competitive position of the factory and thereby improve sales. In the face of global competition, this transaction was necessary to stay competitive, avoid layoffs, and possible closures. However, given sustainable organizational development, HR should take a balanced view of the existing wage grid and incentive system in their factory. This is because, in long terms, without reasonable monetary benefits (compared to traditional production system) and in the absence of job security, getting workers/employees to participate in productivity improvement initiates would be difficult, as they will passively resist the improvement, which can potentially push (few of) them out of a job.

11.4.1.3 Communication among organization members, particularly across organizational barriers

For an effective lean implementation, factories need to improve communication across boundaries, that is, within and between departments. The lean system works on quick response principles and factories not investing in seamless information flow cannot leverage the advantages of various lean initiates. It is important that in a lean environment, communication across boundaries should be both top-bottom and bottom-top. Apart from clear and objective communication of lean vision to employees, top management should demonstrate that they "walk the talk." After lean implementation, workers and employees should see the change in thinking and work practices of top management. This will motivate more workers/employees to do the same. Bottom-top communication plays an important role as it provides valuable, timely information about the changes that are taking place, and about new barriers that arise as progress is made. Horizontal communication must occur directly from the source of the information to the decision-maker/checkpoint, which can initiate action in response to the information flow. When starting a lean project, cross-functional collaboration and cross-boundary communication are essential. When different work cells, groups do not effectively align to collaborate, the objective of system-wide waste elimination fails. HR should train top and middle management in this aspect. Another aspect is top-down communication regarding employee's role in lean implementation. This communication should be clear, objective, and defined.One major reason for poor communication is lack of defined channels of formal communication and poor communication skills of shop floor employees. It is necessary to train employees in communication techniques. Most of shop floor employees do not understand how to ask questions and how to get feedback.

Many organizations begin their lean journey by training many employees in lean. Employees are then directed to implement lean without any role clarity. This is a very bad approach. HR should ensure that the individual's role is defined at the end of lean training. It is also important to explain to employees that their roles will change as an organization moves forward on lean maturity. It is strongly recommended to have a dedicated lean cell, which should include an HR-lean coordinator, apart from other lean team members like Industrial Engineer (IE) and other technical persons. The HR-lean coordinator will serve as a nodal point for collecting/communicating information on lean projects. This HR representative will track lean initiates and maintain team records, performance scorecard, KPI records, etc. The HR-lean coordinator will communicate closely with shop floor lean experts/lean project teams and carry out HR-related responsibility for lean projects.

11.4.1.4 Acknowledging, felicitating, and celebrating success in the lean journey

HR benchmark important milestone must acknowledge achievements and specific contributions by employees/teams. It is important to felicitate the achievers by organizing annual events like "Annual Lean Day" and felicitate achievers with a memento or "Lean Trophies." This acknowledgment and celebration should be done in addition to annual performance appraisal as explained earlier. This celebration is important because, lean being a continuous improvement project, employees need to be made aware of milestones and major accomplishments. Also, these annual events provide an opportunity for raising awareness among employees regarding the performance gap of the future course of action.

Lean works best in an environment that fosters open, honest communication, and mutual respect. Mendomi is a Japanese expression where the workers are treated as a family. While having discussions and decisions, consider others as a family. The company takes care of its workers and does not fire them. In turn, the company expects a lot of loyalty and commitment. In an apparel manufacturing environment, the supervisors and line managers often use personal rapport with workers to convince or motivate to achieve target rather than relying on data and analysis.

11.4.2 Engagement in employee participation initiatives

A quality circle is a participative management technique, where a team of employees, mostly volunteers, drawn from the shop floor are mentored and guided by a team leader to drive a program for overall organizational development. The team is trained to identify, analyze, and implement corrective/preventive action.

11.4.2.1 Starting a quality circle project

While developing a quality circle team, it is better to maintain a small team size, around 5−10 employees; this helps to manage the team and fix accountability. It is recommended to have a manageable number of quality circle teams concurrently working on different problems in the factory. Small continuous improvements carried by several quality circle teams can develop a sustainable quality culture in the organization. Since this is an organization-wide initiate, this initiative needs to be nurtured and assisted by top management, wherever required. It has been observed that a close HR involvement in the quality circle project can help to improve employee participation. The quality circle team, after deciding the corrective action plan, should present the action plan to the management and colleagues, for possible improvement suggestions; this helps in getting their buy-in and in smooth implementation of the action plan. As the quality circle team matures and gains expertise in identifying the problem, analyzing it, and problem-solving methodology, it becomes self-sustaining, requiring very little assistance from top management.

When starting a new quality circle project in the factory, employees and team members need to be trained in various problem-solving tools. Also, since several teams will be working simultaneously on quality circle projects, high employee participation can improve the effectiveness of the project. HR can contribute here, first by organizing effective training for team members on relevant problem-solving tools and by improving employee participation through awareness sessions and employee engagement programs. HR should formulate a system to consider specific achievements of quality circle teams in the annual performance appraisal process.

Quality circle projects should be selected from shop floor day-to-day problems. It is not advisable to select projects that require sophisticated technical know-how or design development expertise. The problems can be taken up from the already identified problems or problem-bank based on management goals. Problems can be also selected from the current priority problems of the team members. The problems that need extensive technical know-how or design development or any skill set which is not within the capacity and capability of the team is not a suitable problem for the quality circle project. Selecting such problems will make the team extensively dependent on external know-how/help and a team member will soon lose focus and get disinterested in the project. Typically, quality circle

Lean human resources   345

projects relate to performance/quality improvement through methods improvement, line layout, balancing, work aids, skill development, occupational safety, etc. The team leader must direct the team's efforts in identifying the project that can help in organizational development. Team leaders must constantly direct teams' efforts in identifying the wastes, like over production, defects and rework, unnecessary transportation, unnecessary motion, excess inventory, waiting time and over processing. After selecting the problem, the leader should get it approved by the management. Each project should be registered with HR with a valid registration number.

11.4.2.2 Selecting team members and project

Since the quality circle project is an organizational development initiative, it is an initiative to ensure maximum possible employee participation. HR should carefully evaluate the team members for personality traits as explained earlier in this chapter. Following guidelines can help to formulate the quality circle team:

1. Restrict each team size to 5−10 members.
2. Team members selected should include nonexecutives like operators, supervisors, checkers quality controllers.
3. Typically, they should be from problem areas, having firsthand knowledge and understanding of the problem.
4. It is necessary to ensure that team members have adequate knowledge in problem-solving or have educational/ mental aptitude to be trained on various problem-solving tools.
5. Members should have synergy for teamwork.
6. It is advisable to select team members, who can by themselves identify, select critical problems concerning departmental/organizational goals.
7. Team members should be open to taking the help of their immediate line chiefs/production managers, quality managers as per the problem requirement.
8. It is also advisable to have a cross-functional team, by including members from other departments/work areas as per the requirement of the project.

HR should help to formulate a team and assigning team leaders. The team leader should ideally be a work area head/technical expert and one who carries respect among team members. HR along with team leader should prepare a project charter, describing the project title, team members, baseline, and project action plan. The project file should be submitted to top management for the quarterly quality circle review meeting.

11.4.2.3 Quality circle project review meeting

The quality circle team should periodically meet for progress review. It is advisable to fix a weekly/monthly schedule for the review meeting. This is important in the garment industry due to the prevalent firefighting culture on the shop floor. The team leader should attend each meeting and note down the meeting progress. The meeting should focus on problem identification, root cause analysis, and progress on the problem-solving action plan. The team leader should facilitate to remove inter-departmental bottlenecks by lesioning with senior management.

Weekly progress reports should be noted by the leader in the project file. HR coordinator should attend this meeting and check with team members for any training needs, etc.

11.4.2.4 Quick win problem-solving

Since the team members have firsthand information and understanding of the problem, there is no need for extensive measurement and analysis of data using sophisticate statistical tools. On the other hand, efforts should be directed to find quick, inexpensive solutions that are within the capacity of the quality circle team. The quick win problem-solving approach is an effective tool under shop floor conditions prevalent in the apparel industry.

A quick win is a tool that helps the user in selecting an appropriate action plan from among a set of action plans. It can be effectively used by quality circle teams for quick and effective solution selection. Here, only that solution is selected which can satisfy certain laid out conditions, such as:

1. It is cheap to implement.
2. It is easy to implement.
3. It is quick to implement.
4. It is reversible.
5. It is within the team's authority.

Here, the solution must select is cheap to implement, reversible, and within the capacity of the team. Otherwise, with several quality circle teams working simultaneously, a pile-up of proposals waiting for top management approval will slow down the progress of quality circle initiates. Table 11.3 shows a shop floor example of quality circle initiates with quick win solution mapping. As the project progresses, HR should ensure that the project outcome is properly documented by each team. Each project file should be submitted to top management during the quarterly quality circle review meeting. A successful project should be used in developing/revising SOPs/work instructions, wherever necessary.

11.5 Harada method

Apart from the seven wastes that were discussed in Chapter 3: Fundamentals of Lean Journey, "Underutilization of People's talents" is coined by many as the eighth waste. While the first seven wastes can be eliminated or reduced by the tools discussed in earlier chapters, the eighth waste can be eliminated or reduced by better utilizing and handling employees. The Harada method helps to eliminate the waste of skills and talent. Developed by a Japanese high school teacher, Takashi Harada, this method deals with the human side of lean and enables employees to be successful by identifying and embracing a goal or task that helps them move forward. The Harada method's focus is self-reliance—individuals should improve themselves so much that success follows on its own (Marathe, 2013). This can be achieved by improving an

Table 11.3 Quick win solution mapping for quality circle initiate.

Quality circle problem defined	Quick win-problem solution	Is it cheap to implement	Is it easy to implement	Is it quick to implement	It is reversible	Is it within team's authority	Image
Vertically integrated factory—fabric not received on time— high time lost in fabric pre-cutting activities— (fabric inspection/testing), especially for speed orders. Fabric inspection results at fabric unit and that at garment unit do not match.	Joint inspection of fabric by team of checkers taken from fabric and garment unit. They will declare joint inspection report. There will be no repeat inspection at garment unit	Yes	Yes	Yes	Yes	Yes	
Marker pen used at waist belt attach operation/chalk mark at back pocket marking.	Use pattern at sewing and eliminate chalk marking on panels.	Yes	Yes	Yes	Yes	Yes	

individual's spirit, skills, physical conditions, and daily life. The method includes 33 questions for self-reliance, a 64 field charts, and other diagrams.

The 33 questions (attributes) that gauge a person's self-reliance are listed in Table 11.4. Every individual rate themselves on a scale of 1 to 10 and becomes more self-aware by assessing themselves according to these descriptors and can better understand where they need to focus their time and energies.

After assessing their skills with the 33 questions, individuals know what areas in which to improve. The next step is to analyze their goals and purposes with the long-term goal form. The step to follow are:

- Set a long-term goal.
- Analyze the purpose of the goal.
- Decide measures for the goal.
- Check tangible and intangible perspectives on the goal.
- Analyze success and failure.
- Analyze problems and solutions.
- List activities to reach the goal and set intermediate deadlines.

Fig. 11.4 is an example of the long-term goal form for a sewing operator.

The next is an open window 64 charts. The open window 64 chart takes the goal sheet to the microlevel. It is a specific framework that helps individuals to develop 64 mini-tasks and routines to support the goal defined in the long-term goal sheet. Refer to Fig. 11.5, the chart consists of 81 boxes. At the center, the ultimate goal is mentioned; it is surrounded by eight smaller tasks that the individual believes are required to support the central goal. Next, for each of those eight smaller tasks, identify eight even smaller goals. Therefore there will be 64 steps that need to be followed and finished to achieve the ultimate goal. Those 64 steps form the name of this template. Fig. 11.5 shows the open window 64 charts for the same sewing operator explained in long-term goal form.

Table 11.4 Thirty-three descriptors of self-reliance in the Harada method.

1. Accountable	9. Determined	17. Innovative	25. Prepared
2. Adaptable	10. Ethical	18. Inspired—love to work	26. Realistic
3. Authentic	11. Flexible	19. Inquisitive	27. Responsible
4. Brave	12. Highly skilled	20. Empowered	28. Self-managed
5. Capable	13. Honest	21. Knowledgeable	29. Strategic
6. Caring	14. Imaginative	22. Motivated	30. Strong-will drive
7. Confident	15. Independent	23. Organized	31. Supportive
8. Creative	16. Initiative/ Proactive	24. Personable	32. Trustworthy
			33. Visionary

Lean human resources   349

Goal: Be the best pocket attaching operator in the left front of a men's shirt					
Name of goal-setter: Urmila					
Planned date of completion >	July 31, 2019	Date target set >	April 1, 2019	Date target achieved >	
Service activity					
Final targets	Ultimate target		To attach 500 pockets a day with fewer than 2% defects		
	Intermediate target		To attach 425 pockets a day with fewer than 2% defects		
	Definitely achievable target		To attach 350 pockets a day with fewer than 2% defects		
	Current target		To attach 300 pockets a day with fewer than 2% defects		
Interim targets	To sew more than 500 pieces per 8-h shift with fewer than 2% defects				

Four perspectives on goals and targets	1. Higher productivity for the organization.	• I get an incentive payment/bonus • My skill rating is upgraded in annual appraisal
	Tangible	
Society & others		Myself
	1. My family gets more time to spend with me.	• I will receive appreciation • I will require lesser time for rework • I will get more time for family
	Intangible	

	Analysis of success	Analysis of failure
Mental		
Skill		
Health		
Lifestyle		
	Possible problems	Success
Mental	I might wilt under pressure	I will maintain my focus
Skill	I am slightly weak at precision stop	I will work on my hand–eye–leg coordination
Health	I have to be careful about backache	I will exercise daily to ensure avoid backache
Lifestyle	I will cut down my online social networking	I will ensure eight-hour sleep every day
Routine activities (List in order of importance)		Key deadlines (list in order of occurrence)
Activities	**Date activity to be performed**	
Will practice one sewing skill exercise daily		
Will practice one sewing skill exercise daily		
Will maintain a daily record of time taken for each skill exercise		
People helping me to achieve my targets		1. My supervisor 2. My family 3. Industrial engineer
Help that people are giving me to achieve my targets		1. Overall guidance 2. Tutoring ban y industrial engineer 3. Encouragement

Figure 11.4 Long-term goal form for a sewing operator.

The continuous improvement practice of looking back and thinking about how a process or personal shortcoming can be improved is "self-reflection." In the Toyota Production System, Hansei or reflection meetings typically are held at key milestones and at the end of a project to identify problems, develop countermeasures, and communicate the improvements to the rest of the organization so mistakes are not repeated. Thus Hansei is a critical part of organizational learning along with kaizen and standardized work. It sometimes is compared to "check" in the plan—do—check—act improvement cycle.

Watching a movie together	Engage in stimulating conversations	Watch old family albums together	Precision stop and pivot clockwise	Short burst sewing with precision stop	Edge topstitch using compensating presser foot	Ensure SPI and back-tack programming are set correctly	Ensure no rework by correct thread tension & bobbin thread refilling.	Dispose of and next pick up in rhythmic motion
Spent quality time with children	**More time with family**	Going out together with family	Precision stop and pivot anti-clockwise	**Practice sewing skill exercises in detail**	Ply separation and pick up from fabric stack	Pivot the piece in a single action without fumbling	**Attach 500 pockets 08-h shift with less than 2% defect**	Pick up and place shirt-front to the machine at a fixed position
Do gardening	Help in household chores	Play with pets	A short burst of straight followed by curve sewing	Curve sewing followed by straight sewing	Plaid or stripe matching dynamically	Sew every length in a single burst	Slide and position below presser foot motion	Pick up pocket and place on front mark
Sleep for 8 h	Practice pranayama (breathing exercise)	Practice yoga	More time with family	Practice skill exercises in detail	Attach 500 pockets in 08-h shift with less than 2% defect	The smooth functioning of the machine with vibrations within the limit	Adequate natural light in the work-area and appropriate lighting at needlepoint	Controlled and suitable working conditions
Read book	**Mental fitness**	Early to bed and early rise	Mental satisfaction	**Be the best shirt-front patch pocket setter in the factory**	Ensure Ergonomically correct workplace	Following the Principles of Motion Economy	**Ensure ergonomically correct workplace**	Appropriate arrangement of breaks between work
Listen to music	Take up a hobby	Exercise 30 min daily	Physically fit	The better growth prospect in the factory	Recognition in the factory	Ensuring occupational safety	Adhering to the standardized work method	Ergonomically designed workstation and ensuring correct posture
Neck exercise	Lower back exercise	Upper back exercise	Punctual	Respectful	Ask for self-evaluation	Help others in need	Talk with humility	Volunteer your service if required
Eye exercise	**Physical fitness**	Ankle exercise	Know what's your worth	**The better growth prospect in the factory**	Accept challenge	Keep your supervisor in the loop	**Recognition in the factory**	Under-promise and over-deliver
Pinch grip exercise	Wrist exercise	Lower back exercise	Learn from mistakes	Work on communication skills	increase awareness about job & industry	Work as a team	Be firm to oppose injustice	Be humble in accepting compliments

Figure 11.5 Open window 64 charts for an example of a sewing operator. Copyright Prabir Jana & Manoj Tiwari 2020.

11.6 Conclusion

HR is an important pillar of lean culture and should play a proactive role in contributing to the individual and the organization's continuous improvement. It is true that the lean systems cannot sustain or survive without an adequate HR, as the lean transitions are more of cultural changes (than the technical changes), and to make these changes as value-added improvements, the role of HR becomes vital. The development of the right attitude and professionalism is an important part of people's empowerment in a lean culture. In the mature lean environments, employees should feel that their voices and suggestions are being heard, and it impacts on the process. This results in collaborative ownership and enhances the individual commitment toward the overall improvement of the organization.

While this chapter discusses different capacity building methods for the people that can effortlessly execute lean philosophy in the organizations, evaluation of managers according to lean includes five aspects, all of them hard to measure but designed to ingrain the Toyota DNA into managers (Roser, 2020). These are:

- Kadai Souzouryoku for ability and creativity to identify problems.
- Kaidai Suikou for the ability to carry out tasks.
- Soshiki Management for organizational management abilities.
- Jinzai Katsuyou for ability to utilize and develop people.
- Jinbou for the trust of the subordinates.

The people know the best about their work and their work environment, which makes them the most useful resource for developing effective and efficient systems. And here is the concluding statement of this chapter, as rightly stated by Womack and Jones :

> *"It has become conventional wisdom that higher levels of management should learn to listen to the primary work team since they know the most about how to get the job done."*

References

Bamber, L., & Dale, B. G. (2000). Lean production: A study of application in a traditional manufacturing environment. *Production Planning & Control: The Management of Operations, 11*(2), 291–298.

Birdi, K., Clegg, C., Patterson, M., Robinson, A., Stride, C., Wall, T. D., & Wood, S. (2008). The impact of human resource and operational management practices on company productivity: A longitudinal study, September Personnel Psychology 61(3):467–501 *Personnel Psychology, 61*(3), 467–501.

Bonavia, T., & Marin-Garcia, J. A. (2011). Integrating human resource management into lean production and their impact on organizational performance. *International Journal of Manpower, 32*(8), 923–938.

Delbridge, R. (2005). Workers under lean manufacturing. In D. Holman, T. D. Wall, C. W. Clegg, P. Sparrow, & A. Howard (Eds.), *The essentials of the new workplace: A guide to the human impact of modern working practices* (pp. 15−32). John Wiley & Sons, Ltd.

Dibia, I.K., & Onuh, S. (2010). Sustaining the human resource 'the Real Quality' in lean production system. In *International Conference on Education and Management Technology (ICEMT 2010)* (pp. 297−300). IEEE.

Doherty, N., & Horsted, J. (1995). Helping survivors to stay on board, January 12 *People Management*, *1*(12), 26−31.

Farris, J. A., Aken, E. M., Doolen, T. L., & Worley, J. (2009). Critical success factors for human resource outcomes in Kaizen events: An empirical study. *Internation Journal of Production Economics* (117), 42−65.

Forrester, R. (1995). Implications of lean manufacturing for human resource strategy. *Work Study*, *44*(3), 20−24.

Heinzen, M., & Höflinger, N. (2017). People in lean product development: The impact of human resource practices on development performance. *International Journal of Product Development*, *22*(1), 38−63.

Jekiel, C. M. (2011). *Lean human resources: Redesigning HR processes for a culture of continuous improvement*. New York: CRC Press.

Kinnie, N., Hutchinson, S., & Purcell, J. (1998). Downsizing: Is it always lean and mean? *Personnel Review*, *27*(4), 296−311.

Liao, W.-C., & Han, T.-S. (2019). Lean production and organisational performance: Moderating roles of ability- and motivation-focused human resource management. *International Journal of Human Resources Development and Management*, *19*(4), 335−354.

Longoni, A., Pagell, M., Johnston, D., & Veltri, A. (2013). When does lean hurt? − An exploration of lean practices and worker health and safety outcomes. *International Journal of Production Research*, *51*(11), 3300−3320.

MacDuffie, J. P. (1995). Human resource bundles and manufacturing performance: Organizational logic and flexible production systems in the world auto industry. *Industrial and Labor Relations Review*, *48*(2), 197−221.

Marathe, S. (2013). *The Harada method: Templates to measure long-term goal achievement − Part 2 Of 2*. Retrieved July 13, 2020, from ISIXSIGMA: <https://www.isixsigma.com/methodology/lean-methodology/the-harada-method-reduce-the-eighth-waste-part-1-of-2/>.

Marodin, G. A., & Saurin, T. A. (2013). Implementing lean production systems: Research areas and opportunities for future studies. *International Journal of Production Research*, *51*(22), 6663−6680.

Murli, J. (2018). *Human resources and lean; It really is about people*. (Lean Enterprise Institute) Retrieved January 2020, from lean.org: <lean.org/leanpost/Posting.cfm?LeanPostId = 860>.

Olivella, J., Cuatrecasas, L., & Gavilan, N. (2008). Work organisation practices for lean production. *Journal of Manufacturing Technology Management*, *19*(7), 798−811.

Patel, S. (2016). *Lean transformation: Cultural enablers and enterprise alignment* (1st Ed.). Florida: CRC Press.

Ramarapu, N. K., Mehra, S., & Frolick, M. N. (1995). A comparative analysis and review of JIT "implementation" research. *International Journal of Operations & Production Management*, *15*(1), 38−49.

Roser, C. (2020). *AllAboutLean.com*. Retrieved July 13, 2020, from <https://www.allaboutlean.com/dfma-1/>.

Rothstein, J. S. (2004). Creating lean industrial relations: General motors in Silao, Mexico. *Competition & Change, 8*(3), 203−221.

Spear, S., & Bowen, H. K. (1999). Decoding the DNA of the Toyota Manufacturing System. *Harvard Business Review*, 97−106.

Taira, K. (1996). Compatibility of human resource management, industrial relations, and engineering under mass production and lean production: An exploration. *Applied Psychology: An International Review, 45*(2), 97−117.

Thirkell, E., & Ashman, I. (2014). Lean towards learning: Connecting Lean Thinking and human resource management in UK higher education. *The International Journal of Human Resource Management, 25*(21), 2957−2977.

Tracey, M. W., & Flinchbaugh, J. W. (2009). How human resource departments can help lean transformation. In A. f. Excellence (Ed.), *Sustaining lean* (pp. 5−10). New York: CRC Press.

Womack, J. P. (1996). The psychology of lean production. *Applied Psychology: An International Review, 45*(2), 119−122.

Womack, J. P., & Jones, D. T. (1994). From lean production to the lean enterprise. *Harvard Business Review* (March-April), 93−103.

Wood. (2005). Organisational performance and manufacturing practices. In D. Holman, T. D. Wall, C. W. Clegg, P. Sparrow, & A. Howard (Eds.), *The new workplace: A guide to the human impact of modern working practices* (pp. 197−218). New York: John Wiley & Sons, Inc.

Total productive maintenance

12

Rashmi Thakur[1] and Deepak Panghal[2]
[1]Department of Fashion Teachnology, National Institute of Fashion Teachnology, Mumbai, India, [2]Department of Fashion Technology, National Institute of Fashion Technology, New Delhi, India

12.1 Introduction

While one talks about quality and Total Quality Control in Lean Philosophy, it is often said to be dependent on the process. Nakajima more appropriately stated it to be dependent on equipment. Production has mainly turned out to be unmanned, owing to robotics and automation in industry. Equipment in production has become complex owing to automation. This further affirms that production output, as well as quality, depends on the equipment. However, the rate at which production is getting automated and thus unmanned, maintenance is not. This further requires maintenance organization and philosophy to be upgraded for the facilitation of modification in maintenance methodology and enhancement of maintenance skills as well, thus leading to Total Productive Maintenance (TPM). TPM is an equipment maintenance system across companies. This chapter briefly reviews the basic concepts of TPM and majorly focuses on the evolution of TPM over a period specifically in the form of smart maintenance, which is the need of the hour across all the industries to cope up with Industry 4.0 (Nakajima, 1989).

12.1.1 History and evolution

The term TPM has been derived through stages of evolution in maintenance philosophy right from the age of World War II when Japanese production started gaining attention because of the level of quality they offered, thus challenging the western countries. The West had started the concept of maintenance with Preventive Maintenance, which was later adopted by the East and called to be Productive Maintenance, Maintenance Prevention, and Reliability Engineering. TPM is a modified concept of maintenance to engulf the Japanese industrial movement and suit their way of production. As the industrial revolution gradually percolated across the globe in other western and Asian countries, TPM got widely adopted (Nakajima, 1989).

The evolution of TPM can be given in brief through the schematic diagram below in Fig. 12.1. Various types and terms of maintenance coined over some time have been elaborated in a later section of this chapter. The present scenario looks

---

Lean Tools in Apparel Manufacturing. DOI: https://doi.org/10.1016/B978-0-12-819426-3.00005-9
© 2021 Elsevier Ltd. All rights reserved.

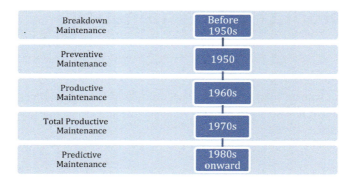

Figure 12.1 Schematic diagram of TPM. *TPM*, Total productive maintenance.

up to TPM with the inclusion of Predictive Maintenance or condition-based maintenance concept. Predictive maintenance can be achieved through the analysis of data in the form of machine parameters which are collected from machines during its running condition.

TPM features preventive maintenance, maintainability improvement, economic efficiency, and autonomous maintenance (AM) carried out by operators (termed as small group activities), whereas productive maintenance does not include AM. Preventive maintenance further narrows down to only the inclusion of economic efficiency as its feature (Nakajima, 1989).

12.1.2 Definition

Nakajima (1989) briefs TPM to be inclusive of five elements listed below:

- Aiming at maximizing equipment effectiveness
- Establishing a system of preventive maintenance for the entire life span of an equipment
- Implementation of TPM in various departments including engineering, operations, and maintenance
- Involving all employees of the organization from workers to management at the top level
- Based on the promotion of preventive maintenance through small group activities

12.2 Basic concept of TPM

Every production business needs maintenance of production equipment continuously. Maintenance is, therefore not a new term for production stakeholders, but the way the concept of maintenance has evolved from the simple equipment breakdown maintenance to the integrated production maintenance is fetching interest by the organizations and researchers both. This is a comprehensive, integrated, and systematic approach of continuous improvement and elimination of equipment, and production efficiency losses by involving the team-based active participation of employees are Total Productive Manufacturing (Pomorski, 2004). This approach focuses on improving the overall production as well as the work culture of organizations. TPM works with the goals of obtaining "overall plant/performance effectiveness" by adopting strategies like running the machines even at lunchtime, having zero customer complaints, reducing the manufacturing costs, and maintaining an accident-free and employee-friendly environment at the workplace (Saha, Akash, & Biswas, 2017). The approach aims toward reducing the emergency and unscheduled maintenance (Masud, Jannat, Khan, & Islam, 2007).

TPM works on the principle of Preventive maintenance and goes much beyond merely an equipment maintenance process; instead, it competitively aims toward getting production equipment with improved and efficient design and characteristic model (Kumar, Kumar, & Rawat, 2014; Saha et al., 2017). This approach involves the operators and human resources in the operational decision making so that they develop a sense of ownership toward the production process and work cohesively toward system improvement. This leads to the most optimum utilization of resources and enhanced system efficiency.

12.2.1 Pillars of TPM

TPM has eight sections or stages, which are also known as pillars of TPM (Ahuja & Khamba, 2008; Díaz-Reza, García-Alcaraz, & Martínez-Loya, 2019). All the eight pillars are closely interrelated and overlapping but are differentiated from each other based on their objectives to be accomplished. A brief discussion of the pillars follows:

1. AM (Jishu Hozen): This stage of TPM focuses on developing a sense of ownership in operators for the production process. Operators are trained for the basic maintenance functions like "cleaning, lubrication, adjustments, visual inspections, and readjustments production equipment" (Díaz-Reza et al., 2019; Singh, Gohil, Shah, & Desai, 2013).
2. Focused improvement (Kobetsu Kaizen): In this stage, with the help of 5S techniques, continuous improvement in the production process is emphasized. Kaizen approach focuses on waste elimination, defect reduction, increased efficiency, and enhanced flexibility in the production process.
3. Planned maintenance: This stage has the underlying combination of activities like preventing the machine failure through time-based maintenance (preventive maintenance), need-based equipment correction (predictive maintenance), training for AM, and cost reductions and corrective maintenance (Díaz-Reza et al., 2019; Jasiulewicz-Kaczmarek, 2016).
4. Quality maintenance (Hinshitsu Hozen): This stage focuses on providing an assured supply of quality products to customers, thereby making a shift from QC to quality assurance (Díaz-Reza et al., 2019).
5. Education and training: This is one of the fundamental pillars of TPM because it deals with making the manpower involved in the production process, aware of the need for efficient equipment maintenance. Operators are trained for the fundamentals of maintenance and the procedures for the same.

6. Office TPM (OTPM): Aim of the OTPM is to effectively organize the workplace and work procedures, and to have efficient offices to reduce functional losses (Ahuja, 2009; Díaz-Reza et al., 2019).
7. Safety, Hygiene, and Environment: This stage highlights the importance of a safe and clean place for working. Every effort is made to provide a healthy work environment to employees, and if there is any deviation at production place, that is immediately attended and corrected.
8. Developed management: This section of TPM deals with the management of the knowledge and experience acquired in the maintenance of existing machinery and equipment, which further helps in designing new and improved products.

12.2.2   Types of maintenance

The concept of TPM has evolved and passed through several phases, which explicit the need for this concept in production. The different phases gave origin to different types of maintenance; a few of them are reiterated below (refer Table 12.1):

Table 12.1 Types of maintenance.

S. No.	Type of maintenance	Salient features
1.	Breakdown maintenance	Follow the reactive approach, where machines or equipment are repaired upon failure or performance breakdowns and were prominent before the 1950s (Ahuja and Khamba, 2008)
2.	Preventive maintenance	Post-1951 emphasis was given on preventing equipment failures to enhance equipment efficiency and life. Time-based equipment maintenance activities such as lubrication, cleaning, corrections, and replacements are done (Ahuja and Khamba, 2008; Telang, 1998)
3.	Corrective maintenance	The concept is introduced in 1957; wherein emphasis is given on equipment improvement in terms of its design, structure, safety features, reliability measures to come out with maintenance and failure-free equipment. Corrective actions are usually taken upon equipment and machinery as and when they show signs of deterioration (Ahuja and Khamba, 2008)
4.	Predictive maintenance	It works on the principle of need-based correction or maintenance. Certain standards are set for the efficient functioning of equipment. As soon as the equipment starts showing signals of deterioration, measures of maintenance are initiated, and only when the equipment or machine starts showing complete damage signals it is removed

12.2.3 Six major losses in the absence of TPM

TPM works with the objective of elimination or minimization of losses in the production system to have improved production efficiency. In the absence of TPM, six major losses are identified, which can significantly lower the efficiency of the production system (Ahuja & Khamba, 2008; Gupta, Sonwalkar, & Chitale, 2001). Nakajima (1988) was the first one who categorized the six major losses in a production system. These losses further go ahead to calculate the overall equipment effectiveness (OEE) of a system:

1. Equipment failure: Breakdown losses relate to equipment failure and need consistent monitoring and vigilance to be checked upon.
2. Setup and adjustment losses: Due to poor maintenance and equipment failures, production plants need to bear machinery setup and adjustment losses.
3. Idling and minor stoppage losses: Due to poor management of operators and the absence of a sense of belongingness among employees, the production process faces several periods of idle machinery, which otherwise can be run efficiently. Also, due to poor continuous vigilance, machinery and equipment get faults or damages, which lead to minor stoppage losses.
4. Reduced speed: Due to poor maintenance practices, unskilled operators, and other factors, the actual speed of the machine is lesser than the design speed.
5. Defect in the process: Inefficient production systems lead to defective products, which eventually require rework and enforce a huge monetary burden on the company.
6. Reduced yield: Many times reduced yield occurs during the initial stages of production that is till the time stabilization occurs.

The first two hampers the availability of the system; the third and fourth describe the performance rate of the system, and the last two reduces the quality rate of the system. Later, Agustiady and Cudney (2016) endorse the same. However, Smith and Hawkins (2004) have included the more losses to it and classified into four categories, namely, (1) planned losses, (2) inactivity time loss, (3) efficient development loss, and (4) quality loss. Effective implementation of TPM, therefore leads to eliminating all such losses, which hampers production efficiency.

12.2.4 Maintenance performance indicator

Maintenance performance indicators are used to evaluate the impact of maintenance work on process performance. The performance evaluation of equipment is usually carried out by estimating the failures, occurred in the past. Failure information may be sought from sources like F-tags, Machine history log, and Minor stops.

F-tags: These faults are identified by the AM or TPM team in the very beginning phase of cleaning and then during their subsequent routine inspections and cleaning visits. The most common record sheets adopted for the purpose are cleaning map, defect map and defect chart, category spreadsheet, log sheet, task certification sheet, and failure analysis sheet.

The faults identified in the machines over a while are recorded every time. TPM records the faults in the form of F-tags in record sheets. Subsequently, the repair information is also recorded to have an accurate analysis. Usually, the information is recorded in terms of F-tag number, the date and time of failure and date and time of repair, first signs of failure, and repair completion times. The basic purpose of fault and repair information recording is, so that whenever required, the basic required information about the equipment may be retrieved appropriately. The most basic expected information includes:

- The number of failures
- The uptime information for the tool
- The mean time between failures (MTBF) and mean time to repair (MTTR)

The equipment reliability and quality of maintenance tasks can be measured through MTBF. Similarly, the failure complexities, spare parts availability, and skill of the support staff influence the MTTR.

The minor stops: There are some equipment and machinery in the production plants, which record minor faults in routine (almost daily). Such minor problems or faults are very easily noticed by the technicians but are usually left unattended, and thereby contributing to minor losses daily, which accrue to big losses later. TPM focuses on recording and eliminating such minor or unrecorded faults as well to eliminate losses. Usually, the information for minor stops is recorded either in log sheets or in the form of drawing or picture, and so on.

12.2.5 Overall Equipment Effectiveness (OEE) and its calculation

Overall Equipment Effectiveness (OEE) is a metric to measure the success of TPM and through a structured framework; it helps in identifying the areas that need improvement. OEE associates the success measurement of TPM with the extent of the elimination of six major losses. OEE is a widely accepted quantitative tool to measure how efficiently a system is working. The term OEE was first coined by a Japanese citizen Seiichi Nakajima in his book, namely, "TPM tenkai" (Nakajima, 1982). He was the pioneering founder of the TPM system.

OEE can be used to understand the impact of improvements on the performance of a system and compare them with the initial scenario. The OEE identifies the losses categorized under (1) availability (2) performance rate, and (3) quality. The OEE for a machine/system can be calculated as:

$$OEE = Availability \times performance\ rate \times quality.$$

Over a period of time, various researchers have given different approaches for the calculation of OEE. In the present section, the significance and application scenario of the approaches given by four leading researchers, namely, Nakajima (1988 and 1989), Ames, Gililland, Konopka, Schnabl, and Barber (1995), De Ron and Rooda (2005), and Wauters and Mathot (2007) are discussed, based on a study conducted by Gamberini Rita, Luca, Francesco, and Bianca (2017). They studied a real-time case scenario to understand the impact of all the four approaches given by these researchers. In conclusion, it is remarked that Ames et al. (1995) approach is

Total productive maintenance

most significant when the manufacturing system is of types, namely, Flexible Manufacturing System (FMS), Flexible Assembly System (FAS), Cellular system, or Automated System, following Industry 4.0 principles. On the other hand, approaches given by Nakajima (1988, 1989), De Ron and Rooda (2005), and Wauters and Mathot (2007) are having significance when the manufacturing system is of types, namely, job shop, manual assembly, and FAS having integration in between manual and automated system.

The Nakajima (1988, 1989) approach is the most significant in the case of the apparel manufacturing system as to date it is a manual assembly system. The OEE calculation for an apparel manufacturing unit is illustrated below via an example:

A shirt manufacturing unit has a shift time of 10 hours which includes lunch break of 1 hour. There are 6 batches in each shift with a cycle time of 1.5 minutes per garment. Machine setting time for thread changing and bobbin change together takes 40 sec, which is carried after every 05 pieces of garment. Total number of 250 garments are produced in a shift, out of which 16 garments are rejected as defective pieces. The OEE calculation for this particular case scenario for the shift is given below:-

$$OEE = Availability \times Performance \times Quality$$

$$Availability = \frac{Available\ Production\ Time\ (APT)}{Scheduled\ Production\ Time\ (SPT)}$$

$$Number\ of\ times\ machine\ setting\ to\ be\ done = \frac{250}{5} = 50\ times$$

$$Total\ Machine\ Setting\ time\ (min) = \frac{(50 \times 40)}{60} = 33.3\ min$$

$$APT = SPT - Total\ Machine\ Setting\ Time = 540 - 33.3 = 506.7\ min$$

$$Availability = \frac{506.7}{540} = 0.938 \quad (i)$$

$$Performance\ rate = \frac{Actual\ number\ of\ pieces\ produced}{Number\ of\ pieces\ that\ can\ be\ produced\ in\ APT}$$

$$No.\ of\ pieces\ that\ can\ be\ produced\ in\ APT = \frac{APT}{Cycle\ time\ per\ piece} = \frac{506}{1.5} = 337.3$$

$$Performance\ rate = 250/337.3 = 0.741 \quad (ii)$$

$$Quality\ rate = \frac{Number\ of\ accepted\ pieces}{Actual\ number\ of\ pieces\ produced}$$

$$Quality\ rate = \frac{(250-16)}{250} = 0.936 \quad (iii)$$

$$OEE = Eq.\ (i) \times Eq.\ (ii) \times Eq.\ (iii)$$

$$OEE = 0.938 \times 0.741 \times 0.936 = 0.651\ i.e.\ 65.1\%$$

12.3 Benefits and challenges of TPM

12.3.1 Benefits of TPM

TPM provides several tangible and intangible benefits to the contemporary production scenario. The tangible benefits get reflected in various manufacturing activities like (Ahuja & Khamba, 2008):

1. Enhanced productivity due to a reduction in machine breakdowns and equipment failures. Productivity improvement is also due to the availability of options for additional capacity and product design customization.
2. Improved quality due to much stable and efficient production, improved product design, and the possibility of quick changeovers.
3. Cost improvisation due to the adoption of life cycle costing, efficient and improved production systems, reduced wastages, and efficient maintenance.
4. More efficient, reliable, and fast delivery and improved line availability of skilled workers.
5. Creates a safe working environment with zero accidents at the workplace and the elimination of toxic and harmful substances from the working zone.
6. Better Employee engagement and involvement, and worker's knowledge and skill upgradation.

Apart from the tangible benefits, several intangible benefits are also observed due to the adoption of TPM which includes improved overall image of the organizations which leads to increase in sales, organized and clean shop-floor areas

Total productive maintenance 363

because of trained and motivated workers, development of the sense of ownership among employees thereby making themselves responsible for losses and wastages.

12.3.2 Challenges in implementation of TPM

Manufacturing units find TPM implementation quite challenging. There are several cases where companies fail to implement TPM in their production processes. Reasons for the same are manifold, a few of them are discussed as follows (Ahuja & Khamba, 2008):

1. Lack of communication in the organization due to which employees lack clarity about the goal of TPM.
2. Poor or partial implementation of TPM.
3. Organizational resistance to change due to cultural or social factors or ignorance.
4. Lack of awareness and clear understanding among the management about TPM.
5. Incorrect methodology for TPM execution.
6. Improper coordination between a human resource, technical, and management policies.
7. Inadequate reward or recognition mechanism.

Companies therefore need to understand that systems like TPM require a long-term commitment and a clear focus for implementation. Top management should first have the clarity of objectives and methodology for implementation of TPM and the same must be communicated to all the employees.

12.4 TPM in the textile and apparel manufacturing industry

Lean philosophy carries a basic idea of waste elimination. Any activity or material, which does not add value to the end product, has to be treated as waste. As known textile and apparel manufacturing industry is a labor-intensive sector, thus making it a challenging task to improve productivity with enhanced quality and efficiency. Further to this, post-quota removal, the global market is open for trade, thus increasing the competitiveness which has continuously been growing. The challenge lies in reducing the cost of production without compromising the quality along with fair trade practices, to provide competitive pricing for the products. Under such scenarios, the Textile and Apparel Manufacturing industry sector has realized the importance of lean manufacturing, thus adapting the same. Lean manufacturing has aided as an essential tool in this sector, and several industries in this sector have attained excellence through its adaption. Numerous work has been reported to showcase the benefit of adapting Lean tools in the area of TPM (Wickramasinghe & Perera, 2016; Alauddin, 2018; Anwar & Deep, 2017; Karim & Rahman, 2012; Maralcan & Ilhan, 2017; Masud et al., 2007; Saha et al., 2017; Senthilkumar & Thavaraj, 2014; Shakil & Parvez, 2018; Solanki, Yadav, Yadav, & Yadav, 2017).

OEE is one of the important parameters to evaluate the status of TPM in a manufacturing unit, it being the function of availability, quality rate, and performance rate. Karim and Rahman (2012), in their work, have measured the OEE to evaluate the performance of the sewing section of a readymade garment sector. It was observed that the performance rate and quality rate for every sewing line studied were much better than the availability. The availability of every sewing line was found to be reasonably low. The study suggested the unit to measure OEE of every sewing line on a daily, weekly, and monthly basis to minimize equipment breakdown, downtime due to setup, defect, and minor stoppages. Shakil and Parvez (2018) carried a similar study for the improvement of OEE in sewing by analyzing the layout, process flow, and batch size. The layout was changed to reorganize the process flow eliminating backflows and reducing transportation time. Furthermore, it was observed that batch size had a significant effect on waiting time and transportation time. So the batch size was optimized with a summation function of transportation time and waiting time. The changes brought in layout and batch size led to a reduction in transportation time by 30.95% and an increment in OEE by 3.75%. Saha et al. (2017) also worked in a similar line with an attempt to improve OEE by taking up certain changes through TPM for a garment manufacturing unit. Maralcan and Ilhan (2017), in their study, have utilized tools on productivity measurement such as Takt Time, Key Performance Indicators (KPI) Tree, and OEE. KPI Tree study was carried out for all the processes, OEE evaluation was carried out for the stenter frame machine, which is usually a bottleneck in such units and Takt time was exemplified for QC Department. Masud et al. (2007) attempted to increase the productivity of the RMG factory by applying TPM. All the departments of the factory were studied and OEE improved to 65% from 59%, through the elimination of losses by using Why Why Because of Logical Analysis (WWBLA) tool. OEE has been a quantitative measure to evaluate productivity, even in the textile machine manufacturing industry. One such study had connected TPM with safety and had come with suggestive solutions through Root Cause Analysis (Solanki et al., 2017). Another research conducted for the sewing section of an apparel manufacturing unit reported an increase of OEE to 86.85% from 85.36%, attained through line balancing and monthly training, thus increasing the productivity of the unit (Alauddin, 2018).

Wickramasinghe and Perera (2016) attempted to evaluate the effectiveness of TPM by carrying out a survey based on multiple parameters governing TPM, selected through a literature survey. Correlation and regression analysis was carried out to establish the relation of TPM with various factors taken under consideration. TPM and cost-effectiveness were found to be positively related and have statistically significant relationships. TPM and product quality were also found to be positively related and have statistically significant relationships. Volume flexibility and on-time delivery were also found to have similar relations with TPM.

An app-based production monitoring system (PMS) was developed for apparel SME in India. This work aimed to move from registers, challans, and manual datasheet−based monitoring to an app-based platform, thus making the process error-free and more efficient. Devising this method led to a cost-effective solution of PMS by reducing the usage of paper, reducing time spent on data handling, and record keeping. It enabled real-time monitoring of sewing status, cutting, and finishing at any point in time. The data collected this way also helped with ready plots and graphical analysis to analyze daily production. The graphical analysis further confirmed the impact of TPM in apparel production planning (Anwar & Deep, 2017).

Performance efficiency levels of these sewing machines were studied before and after the TPM implementation. There is an improvement in lead time along with a 15%–30% improvement in the performance is found. The emotional balance of the employees may also influence the effective implementation of the TPM. A study came up with the development of a methodology for a TPM procedure in the apparel manufacturing factory. The study was conducted in a T-Shirt production layout where five different sewing machines were chosen. The methodology helped in improving the efficiency of sewing machines, thus minimizing machine downtime and improving product quality. The sewing machine showed performance improvement from 15% to 30% (Senthilkumar & Thavaraj, 2014).

According to Karekatti (2014), TPM has not given due attention in apparel manufacturing if the same is compared with other industries such as automobiles. The apparel manufacturing industry is still following planned and preventive maintenance practices and yet to achieve the advantages of TPM at its full potential. In the apparel industry, maintenance is still perceived as a cost center and efforts are toward minimizing the maintenance expenses rather than seeing the holistic view of enhancing the overall organization performance. The author emphasizes adopting AM practices to control accelerated deterioration. This practice of AM is also referred to as Clean–Lubricate–Inspect–Tighten that involves basic maintenance procedures quickly performed daily by operators aiming to restore the machine condition (Karekatti, 2014).

The scope of implementing AM in an apparel manufacturing setup was gauged by collecting end line quality inspection data for sewn products. Oil stains, soiling, and loose threads were some of the major issues observed. Such defects can be addressed by AM. A checklist (refer to Fig. 12.2) containing checkpoints (focusing the observed issues) was prepared to minimize such defects in the sewing process. At the same time, to monitor the minor maintenance issues, white tags were posted. It was observed that many times the machine mechanic had to attend a machine to address minor issues such as adjusting thread tension, seam puckering issue, needle change, skip stitch issue, however, such issues can be handled by the operators themselves with some basic training and orientation. In this way, a capacity building of sewing machine operators can be done enabling relieving the machine mechanics to perform specialized maintenance functions (Karekatti, 2014).

It is recommended that there should be a separate AM checklist according to machine types (refer Fig. 12.3) as well as shop-floor capabilities addressing the specific maintenance requirements. Machine operators should perform the activities as indicated in the checklist. Using the white tags, the concerned operator should flag the defects or issues encountered till the time these are not resolved. Application of tags (white tag for minor maintenance issues and red tag for major maintenance issues, refer Fig. 12.4) is an important practice as it helps in tracking and recording

Sr. No.	Activity	Description
1	Initial clean-up	Thorough clean-up of dust and dirt on equipment, and implementation of lubrication, and machine parts adjustment; discovery and repair of malfunctions and abnormalities in equipment.
2	Eliminate sources of contamination	Prevent causes of dust and dirt and scattering; improve places which are difficult to clean and lubricate, and reduce the time required for clean-up and lubrication.
3	Formulate clean-up and lubrication standards	Formulate visual standards so that it is possible to steadily sustain clean-up, lubrication and machine parts adjustment in a short period.
4	Improve overall inspection	Training in check-up skills through manuals; exposure and restoration of minor equipment defects through overall check-ups.
5	Autonomous check of machines	Thorough study, formulation and implementation of autonomous check-up sheets.
6	Standardize workplace rules and procedures	Standardize all types of the job management items and devise systematization of upkeep management • Standards for clean-up, check-ups, and lubrication • Standards for physical distribution in the workplace • Standardization of data records • Standardization of guides, pressure foots, trimmers, and measuring tapes.

Figure 12.2 Generic checklist of activities for autonomous maintenance in apparel manufacturing.

the issues, as well as creating awareness on the shop-floor. The line supervisor should closely monitor the filled checklists and report the abnormalities to the maintenance in charge.

At the same time while the capacity building of machine operators for AM, it is important to sensitize the operators on quality management systems, sharp tools policy (including broken needle handling procedures). Furthermore, the sustainability of such initiatives can be ensured by auditing their conformance periodically (Karekatti, 2014).

12.5 Smart maintenance

Lean management, in general, and TPM in specific, as discussed in the previous section of this chapter, is a well-established concept applied in the manufacturing sector as well as in the service industry. However, with the intervention of Information and Communication Technology in industry, Smart Factory has lately evolved and organizations adopting it, popularly use the term Industry 4.0 coined at the 2011 Hannover Fair, which refers to the fourth industrial revolution to come

Total productive maintenance

Autonomous maintenance checklist — Sewing section

Sr. No.	What to check	Shift start	Shift breaks	Shift end	1	2	3	4	5	6	7	8	9
	Tools and machine CLIT												
	Cleanliness												
1	Machine head clean	Yes	Yes	Yes									
2	Table top clean												
3	Side table clean												
4	Cleanliness of thread guides												
5	Cleanliness of scissors												
6	Cleanliness of tube light covers												
	Lubrication and inspection												
7	Loose oil seals												
8	Loose screws												
9	Loose wires												
10	Loose pressure foot lifter												
11	Loose thread stand												
12	Loose pressure foot												
13	Loose pedal												
14	Loose parts - others												
15	Abnormal sound - Hook set												
16	Abnormal smell												
17	Insufficient light												
18	Oil level	Yes											
19	Missing parts	Yes											
20	Broken parts	Yes											
21	Thread trimmer working	Yes											
22	Fabric trimmer working	Yes											
23	Low air pressure (for m/c with pneumatic suction for edge cutter)	Yes											
24	Functional scissor	Yes											
25	No skip stitches	Yes											
26	UBT functioning correctly												
	Quality												
27	SPI	Yes	Yes										
28	Needle number	Yes											
29	Needle quality	Yes											
30	Cloth under needle bar	Yes	Yes	Yes									
31	WIP covered in poly bags	Yes	Yes	Yes									
32	Operations clarity	Yes	Yes										
	Checked by in-charge												

Figure 12.3 Autonomous maintenance checklist for the sewing section.

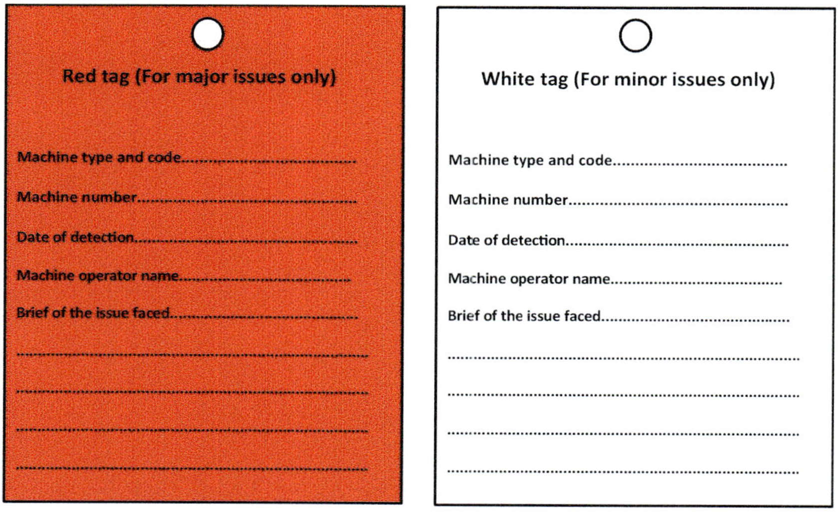

Figure 12.4 Red tag and white tag.

(Rüttimann & Stöckli, 2016). To utilize the synergistic effect of Industry 4.0 on Lean Manufacturing, the establishment of symbiosis between the two is inevitable. Several pieces of literature have been reported on the same and authors from various fields unanimously agree with the term coined Lean 4.0 to indicate the amalgamation of conventional lean manufacturing with the Industry 4.0 approach (Adam, Chola, & Jens, 2016; Mayr et al., 2018; Prinz, Kreimeier, & Kuhlenkötter, 2017; Rüttimann & Stöckli, 2016; Wang & Wang, 2018; Xing & Marwala, 2018). The technologies of Industry 4.0, such as Cloud Computing, Digitalization, Big Data Analytics, Machine Learning (ML), Industrial Internet of Things (IIoT), Additive Manufacturing (AM), Augmented Reality (AR), and Virtual Reality (VR), to name a few, are going to reinforce the Lean principles and aid the growth of any business (Golchev, 2019).

Several researchers and experts in the area have thrown light on specific Industry 4.0 tools being merged with TPM as a technological intervention. Moving toward smart maintenance, that is, utilizing smart manufacturing tools for TPM, is going to enhance the goal achievement for TPM in any industry, as defined by Nakajima (1988). "E-maintenance" is defined as "maintenance support which includes the resources, services, and management necessary to enable proactive decision process execution" (Macchi, Martínez, Márquez, Granados, & Fumagalli, 2012). As recommended by Mayr et al. (2018) the lean philosophy of TPM could utilize smart manufacturing tools like Cloud computing and Big Data Analytics. Predictive maintenance, in particular, among all other pillars of TPM has got a scope of tools like IIoT, ML, and Big Data to be utilized (Golchev, 2019). It is well understood that along with the monitoring of the status and availability of machines in a manufacturing unit, TPM also involves training and education of operators.

Thus human–machine interface (HMI) needs to be focused upon as well to enable strong interaction between humans and machines. The efficient implementation of maintenance methodologies can also be ensured through a combination of AR and VR along with head-mounted displays. Visual communication has a longer and more efficient impact on operators than verbal communication. Furthermore, the data analytics of planned maintenance, relating the condition parameters of machines and the probability of defaults, can enable prediction and thus facilitating predictive maintenance, thus making lifetime expectancy prognosis more accurate. Also, digitalization can eliminate the chance of data missing and mishandling, thus enabling seamless communication among different stages of production and departments (Mayr et al., 2018). Macchi et al. (2012) briefly explain various technological tools which can assist in E-maintenance and its growth and list them as smart sensor, RFID, power handheld computing devices, SCADA, computerized maintenance management system, reliability, and maintenance management systems, to name a few. Fig. 12.5 shows a framework of the process flow for an Intelligent Predictive Maintenance.

Though the conceptual understanding of the scope of Smart maintenance is clear and well explored, implementation of the same in real-time is scarcely reported. The implementation done in this area mainly consists of automotive, electronics goods, mechanical tools, and parts manufacturing industry. The challenge lies in understanding the nature in which they both, Lean Manufacturing and Industry 4.0, will interact and further, formulation of methodologies to implement the smart manufacturing tools for TPM and its improvisation. Though one can foresee the

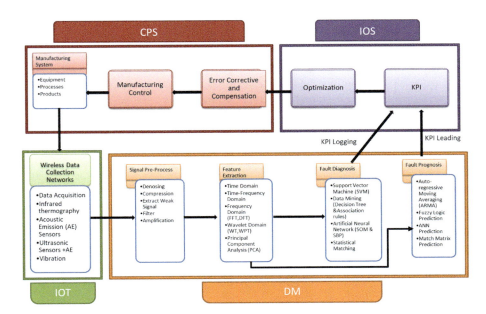

Figure 12.5 Framework of Intelligent Predictive Maintenance (IPdM) systems in Industry 4.0.

scope of utilizing AR, VR, and AM for effective and enhanced functioning of TPM, the utilization of the same has yet not been exploited largely (Bengtsson & Lundström, 2018).

12.5.1 Technological intervention

A few scenarios to showcase the adoption of smart maintenance in various manufacturing industries (not necessarily the apparel manufacturing sector) have been reported in the following sections.

12.5.1.1 Artificial neural network and genetic algorithm

Altay, Ozkan, and Kayakutlu (2014) report a study conducted for aviation industry. The industry had to survive through the global economic crisis during the merger period. The industry was supposed to provide a fleet with low pricing but with utmost safety, which in itself carries the utmost cost. So, elimination of excessive maintenance costs without compromising the safety of the fleet had become the main goal for the aviation industry. This laid to the need of this study to provide a model for correctly predicting the need for predictive/preemptive maintenance.

The study proposed two models for the prediction of maintenance by using two different approaches. The models were devised to predict the failure of aircraft by their type and age. One model utilized the Genetic Algorithm while the other model used Artificial Neural Network for prediction of failure time. The models were developed by utilizing the data of 60 aircraft, which survived 532 failures. Both the proposed models had a good correlation between the actual and the predicted failure schedules of the aircraft. However, the ANN model: Back Propagation algorithm was reported to yield slightly more accurate results and a better degree of correlation. The authors suggested adopting the ANN method using other metaheuristics, transformations, and hidden Markov chains for further improvements. This study has been reported to be a guide for predictive maintenance for both commercial and public airlines (Altay et al., 2014).

12.5.1.2 Cloud computing: big data storage and analytics

To abreast with the increase in large volume data being collected through the automated procedure in the majority of the manufacturing industry at present, Big Data Analytics is going to play a significant role and enable the maintenance personnel with an ability to study the trends and forecast the maintenance problem as well as propose on-time solutions. Karim, Westerberg, Galar, and Kumar (2016) proposed a concept of Maintenance Analytics to utilize Big Data Analytics specifically for maintenance management. They suggested four different time-dependent perspectives of it. They are Maintenance Descriptive Analytics, Maintenance Diagnostic Analytics, Maintenance Predictive Analytics, and Maintenance Prescriptive Analytics.

Uhlmanna, Laghmouchia, Geiserta, and Hohwielera (2017) illustrate a concept for decentralized data analysis through the usage of electronic components,

Raspberry Pi, and MEMS vibration sensors. The vibration data captured by vibration sensors are analyzed by Raspberry Pi. These components enable the implementation of wireless networks for distributed data analysis. The network is connected to a cloud, which stores the results of data analysis and manages the data management system, vis-a-vis PostgreSQL. Cloud services can be utilized for condition monitoring, maintenance planning, and report generation of the current condition of the system, through smart mobiles. The concept discussed in this work enables analysis of data on the sensor nodes through ML algorithms, which offers opportunities to implement future monitoring solutions for different machine setup.

The final results can be transmitted to the cloud server, which can later be provided for different stakeholders involved in the production, such as maintenance planners (Uhlmanna et al., 2017).

Roy, Stark, Tracht, Takata, and Mori (2016), in their comprehensive review, have elaborated on the prospects of continuous maintenance in light of optimizing through life cost. They affirm the fact that the Internet of Things (IoT) and cloud computing are two major popular tools to abreast shortly with Industry 4.0. While IoT enables capturing online data continuously from various equipment and processes, Cloud computing enables storing this large dataset, sharing the data, managing the data, and better human and machine interaction. The six foundation skill sets required for continuous maintenance research has been listed as component level in-service degradation science, product level in-service degradation science, modeling based on material, modeling based on design, features, manufacturing process parameters for different environmental and use conditions.

A research group has attempted to utilize the Machine Learning model for predicting the maintenance of machines in an apparel-manufacturing unit. Vibration analysis was identified as the machine working parameters to relate to their performance and working conditions. Most of the breakdown in these electromechanical machines can be anticipated through vibration analysis. Vibration analysis enables us to anticipate the existence of misbalance, misalignment, loose mounting, and so on. For the collection of the vibration readings, equipment consisting of the accelerometer (three-axis), gyroscope (three-axis), and a temperature sensor was used. The data from this device were then collected externally through IoT gateway from all the critical machines, identified earlier, for 2 weeks throughout the shift of 8 hours. The machines used for the collection of data were Single Needle Lock Stitch machine (SNLS), Double Needle Lock Stitch (DNLS) Machine, and Button Attachment machine. These collected data were then classified into three classes as OK, About to Fail, Failed. They were then divided into training and test data. The class of Failed was then classified into types of failures. Different algorithms used for supervised ML were attempted to check the accuracy of prediction. KNN algorithm was found to give the best accuracy and thus was utilized to come up with the prediction model for maintenance (Verma & Verma, 2019).

Through feature selections analysis of ML, this research came up with an additional conclusive remark that gyroscope reading is a better indicator for SNLS machine failure than the reading of accelerometer (which otherwise is extensively used for the same). This study, however, had a challenge that the data collected were less for the class of Failed due to less of machine failure. Therefore the scope of implementing this model across all types of machines in an apparel manufacturing unit was limited. A dataset, if attempted to be collected for a longer duration, can lead to a more robust predictive model through big data analytics and enable reduction of machine breakdown through predictive maintenance for all variations of machine type and machine failure.

12.5.1.3 Data acquisition and mathematical modeling

Okogbaa, Huang, and Shell (1992) presented a framework of database design for a predictive preventive maintenance system (PPMS). It is particularly important to have an efficient database management system to handle the fast pace of information flow with a growing dataset and to support decision evaluation models. Otherwise, PPMS cannot function successfully. Designing a database for maintenance management system is simpler than the complex nature of developing and implementing it. For making such a database, elements like operating history of the equipment, maintenance specifications, maintenance resources, cost, and schedule are required. This literature acts as a guideline for the design and development of the framework by elaborating on different stages and elements of database design and development for smart manufacturing units.

Mathematical modeling is found to be helpful in coming up with a tangible output, thus making the noticeable difference postimplementation of TPM tools. This makes the employees aware of their involvement in TPM and the benefits out of it. Lawrence (1999) has explained the same through four mathematical models in the maintenance field. These models provide objective assessment and quantification. One of the four examples involved a linear programming model (LPM) to schedule maintenance activities in an aluminum smelting plant. This LPM was carried out to increase the completion of maintenance work orders on time, at the lowest cost. The model was suggested to be evidence for showcasing the benefits of increasing cross-training of maintenance and production employees, on the improvement of maintenance department performance. It also enabled allotment of overtime on a high priority basis. The LPM model also enabled elimination or reduction of overtime for maintenance personnel. Thus implying on-time performance on work order.

Analytical Hierarchical Process was applied in the automotive industry to develop a model for maintenance decisions. The attempt was made to have clarity and robustness in the maintenance system to avoid failing machines/equipment. Furthermore, to have a dynamic and adaptable maintenance system that would utilize collected data and analyze them to support decisions accordingly. The developed system had three stages which could manage multiple criteria decision analysis, conflicting objectives as well as subjective judgments. The developed model was validated on two criteria: functionality and usefulness. It was concluded that the unidentified faults reduced from 30% to 5%. Also, maintenance tradesmen were able to do more skilled jobs as the tool operators could be trained for the

Total productive maintenance 373

less-skilled job related to maintenance. Thus justifying the usefulness of the developed model (Labib, O'Connor, & Williams, 1998).

12.5.1.4 Fuzzy logic

Sonmez, Testik, and Testik (2018) have attempted to utilize fuzzy logic for OEE calculation. This attempt was made considering the limitations of the traditional method of OEE calculation with accuracy in scenarios involving uncertainty. Two types of uncertainty were taken into consideration here, production speed and stoppage duration measurements. The study also reported the interval arithmetic methodology to be preferred over point estimate for low accuracy cases, to best guess interval estimates of operators. The work came up with software development for practitioners to have accurate OEE calculations. Proposed methods would allow the practitioners to quantify uncertainty with limits.

Similar manner, Foulloy, Clivillé, and Berrah (2019) attempted to improve the OEE by formulating a decision-making process to enable information for the correct decision. A temporal performance expression model was built up. The fuzzy approach was adopted to express both symbolic and numerical information associated with the processes. The developed model was built upon the basis of two pillars. The first pillar is related to three temporal dimensions of performance expressions—instantaneous, trend, and predictive. Second pillar concerning three semantics of performance expressions—physical measurement, performance measurement, and performance evaluation. The model was validated by applying it in the Fournier company to check the reduction in production cost through OEE monitoring on a daily, weekly, and monthly basis. The main aim of the study carried out was to provide the right information in the correct format at the right time.

12.5.1.5 Human–computer interaction

Lee, Kao, and Yang (2014) have highlighted the need for reinforcing the HMI for better interaction to enable efficient collection and utilization of data toward smart manufacturing and smart maintenance. Their article briefs about the usage of cyber-physical system (CPS)-based manufacturing and its advantages along with one case study carried out in a heavy-duty equipment vehicle factory. Authors affirm the possibility of conventional machines to turn into self-aware and self-learning machines if seamless interaction is established with surrounding systems in a smart factory. This will help in overall performance and maintenance management. It has been reported that self-learning machines are yet to be implemented in manufacturing industries like textile and apparel. Authors have attempted to explain the self-aware machine systems based on industrial big data analysis and further discussed the feasibility of it being installed/implemented successfully, through case study explanation. Advanced analytics, along with cloud computing and CPS, will enable in achieving wider information and convert machines to be self-aware, thus preventing potential performance issues.

Yazdi, Azizi, and Hashemipour (2018), in their study, have designed and implemented an intelligent material handling system for pick and place mechanism for a manufacturing system in an Small and Medium Enterprise (SME). This control system has utilized an agent-based algorithm as control architecture. The system includes several sensors that are installed at different points of the conveyor belt and other moving parts for actuation and execution of the process. Through HMI, the time study of the process flow has been carried out continuously online. The validation of the developed system has been done through the calculation of availability, performance, and OEE of each device in the work station. Thus calculated parameters were then attempted to be optimized to maximize OEE, to improve productivity. This system was, however, developed on a prototype model by using educational components for sensing and actuation. But this can be taken up forward in a commercial scenario as well.

Sly (2018), in his study, developed and deployed a web-based system for inventory management, which is conventionally managed through the Kanban tool of lean management. This leads to a reduction in line-side inventory, reduction in material handling labor, and also reduction in line-side part shortages. The term coined for this methodology was named eKanban. It was found to be cost-effective as well as efficient in the reduction of inventory levels. It also enhanced accountability and inventory visibility. This toll would be more helpful for companies being labor-intensive and dealing with multiple suppliers and vendors.

12.5.1.6 Optimization approach

Vlasov et al. (2018) have reported a model for the optimization of predictive maintenance of electric motors through wireless sensor networks. The sensor networks allow real-time analysis of the state of equipment. This approach also enables the stability of functioning. If the electric motor, running for a long time, is not remedied on time, it may lead to severe failure. The rolling or friction bearing in these motors have a certain life span. The framework developed to communicate through the sensor network enables collecting information about the vibrations through a vibration sensor, thus detecting the defect on time. It was observed that after the installation of this system 3 out of 155 cases of defects were only reported, rest were communicated before the occurrence, thus implying 98% efficiency of the developed system in this regard. Several such sensors and communication mechanisms can be achieved through IIoT and data collection. This can later enable process optimization by considering the critical checkpoints. This will thus enable indicative maintenance implementation with higher efficiency. However, it may also be noted that any failure in the functioning of sensors may lead to malfunctioning of the communication network, thus leading to failure in the predictive maintenance framework (Lee et al., 2014).

The design of experiments was carried out to come up with a regression model for the calculation of OEE. Data were collected from 50 ancillary automobile companies. These companies' maintenance team was unaware of the meantime to failure (MTTF) and mean time to repair. Furthermore, the response surface plots were used to optimize the factor taken into consideration; they are availability, performance rate, and quality rate to maximize the OEE value. This study indicates that OEE will be significantly improved if the focus is given on performance rate improvement (Relkar & Nandurkar, 2012).

12.5.1.7 Simulation

A maintenance management model was developed using stochastic modeling, integrating quantitative methods, and qualitative concepts. The model assisted in constructing relationships among the parameters like maintenance cycle, personnel allocation, human recovery, and system's tolerance time. Furthermore, a simulated experiment was carried out to explore other parameters supplementing the model, latent human error, critical human error, system's tolerance time, and recovery rate. The simulation experiment was carried out through a feed-water system. This maintenance management model delivered a ready reference for maintenance managers in the form of a reference table, which provided a combination of the number of available maintenance personnel and maintenance cycle time (Wang & Hwang, 2004).

12.5.2 Challenges of smart maintenance

The existence of smart maintenance in a smart manufacturing system is no doubt to bring a synergistic effect and enable efficient production. The preceding sections have holistically captured the essence of smart maintenance in manufacturing. However, there are cyber threats involved with smart maintenance, which is the major challenge to overcome, for avoiding malfunctioning of maintenance systems in a manufacturing unit. The integration of the Internet and network with conventional manufacturing platforms has made the manufacturing system complex, thus complicating the task of making the system threat free. Zarreha, Wana, Leea, Saygina, and Janahia (2019), in their review, have summarized specific cyber threats on the various aspects of TPM. The review also highlights the effect of different types of cyber-physical threats on OEE. Cyber threats potentially disrupt the production processes and even bring a hazardous situation to the manufacturing system. A few specific examples cited in this literature serve as eye-openers. A supplier of Apple Co was affected by a virus attack, which made it delay the shipment of the company's products. There had been a situation where the quality of part was compromised in AM by alteration of printing direction. This defect could pass QC without being detected at all. However, certain cyber threats that are more dangerous need not have a direct impact on productivity and thus the effect may not be seen instantly. Still, they target the integrity of the system, which will indirectly affect productivity. Therefore formulating a robust defense policy against all cyber threats has become very important. The defense policy should ensure low recovery or repair time from the attack. So any industry adopting smart maintenance should ensure to have a plan for recovery from such an attack and do the repairs in minimum time to enable the availability of the system.

The manufacturing setup utilizing E-maintenance needs to be sensitive toward the usage of remote diagnostic technologies for the successful functioning of the former. Monitoring of manufacturing process, which is condition-based, includes certain apparent organizational problems while installing sensors that observe such apparent situations and maintaining the history of collected data to predict imminent failures. At the same time, there is a need for restructuring the organizational setup and to train the human resources for maintaining the pace at which information is flowing. The literature has linked three categories of E-maintenance challenges. First related to maintenance type, second to tools, and lastly to activities (Aboelmaged, 2015).

Maintenance type includes four types of challenges:

- remote maintenance (e.g., security, risk management, and human resources issues)
- collaborative maintenance (e.g., networking and integration issues)
- distributed maintenance (e.g., optimization and synchronization issues)
- predictive maintenance (e.g., diagnostic and prognostic methods)

Maintenance tools include challenges associated with documentation issues like lack of bar code reader, handhelds, laptops, scanners, and databases. Maintenance activities include challenges related to inspection, monitoring, and system improvement (Aboelmaged, 2015).

12.6 Conclusion

Several pieces of literature have been reported on the basics of TPM; thus this chapter has kept it brief to majorly focus on the concept of smart maintenance/ E-maintenance. Various case studies and research reported in this chapter indicate that TPM implementation has significant improvement in manufacturing performance. The chapter majorly briefs the fact that there exists and is going to exist, a paradigm shift in the methodology of TPM practices in any manufacturing unit. The two major factors, that is, increasing global competitiveness and the Industry 4.0 revolution, are tending to take maintenance in the direction of smart maintenance/ E-maintenance, in conjunction with the conventional maintenance methods as well. With the advent of Smart Maintenance, predictive maintenance, which had evolved much later, is gaining importance to optimize maintenance cost without compromising on productivity and quality. Though there are barriers related to human psychology and culture for successful implementation of TPM in any factory, visible proven results across all sectors are influencing human resources at the workplace to adopt and practice it. The chapter summarizes the usage of various analytical tools to come up with the rationale behind the maintenance schedule, majorly for preventive and predictive maintenance. Tools like Data analytics, Artificial Neural Network, Genetic Algorithm, Fuzzy Logic, Mathematical Modeling, and Simulation have been summarized through certain existing cases that have been reported in the literature. An attempt has been made to throw insight on possible technological intervention in any manufacturing unit to keep the pace of TPM at par with the pace of manufacturing, which is enhancing day by day owing to global competitiveness and automation.

References

Aboelmaged, M. (2015). E-maintenance research: A multifaceted perspective. *Journal of Manufacturing Technology Management*, 26(5), 606–631.

Adam, S., Chola, E., & Jens, W. (2016). Industry 4.0 implies lean manufacturing: Research activities in industry 4.0 function as enablers for lean manufacturing. *Journal of Industrial Engineering and Management*, 9(3), 811–833.

Agustiady, T. K., & Cudney, E. (2016). *Total productive maintenance strategies and implementation guide*. CRC Press Taylor and Francis Group.

Ahuja, I. P. S. (2009). Total productive maintenance. In M. Ben-Daya, S. O. Duffuaa, A. Raouf, J. Knezevic, & D. Ait-Kadi (Eds.), *Handbook of maintenance management and engineering* (pp. 417–459). London: Springer.

Ahuja, I. P. S., & Khamba, J. S. (2008). Total productive maintenance: Literature review and directions. *International Journal of Quality & Reliability Management*, 25(7), 709–756.

Alauddin, M.D. (2018). *Process improvement in sewing section of a garments factory—A case study*. (M.S. thesis for Engineering in Industrial and Production Engineering).

Altay, A., Ozkan, O., & Kayakutlu, G. (2014). Prediction of aircraft failure times using artificial neural networks and genetic algorithms. *Journal of Aircraft*, 51(1), 47–53.

Ames, V.A., Gililland, J., Konopka, J., Schnabl, R., & Barber, K. (1995). Semiconductor manufacturing productivity—Overall equipment effectiveness (OEE) guidebook—Revision 1.0, Technology Transfer #95032745A-GEN, SEMATECH.

Anwar, S., & Deep, P. (2017). App based production monitoring system for apparel small and medium sized enterprises. *Amity Journal of Operations Management*, 2(2), 28–43.

Bengtsson, M., & Lundström, G. (2018). On the importance of combining "the new" with "the old"—One important prerequisite for maintenance in Industry 4.0. *Procedia Manufacturing*, 25, 118–125.

De Ron, A. J., & Rooda, J. E. (2005). Equipment effectiveness: OEE revisited. *IEEE Transactions on Semiconductor Manufacturing*, 18, 190–196.

Díaz-Reza, J. R., García-Alcaraz, J. L., & Martínez-Loya, V. (2019). *Impact analysis of total productive maintenance—Critical success factors and benefits*. Springer.

Foulloy, L., Clivillé, V., & Berrah, L. (2019). A fuzzy temporal approach to the overall equipment effectiveness measurement. *Computers & Industrial Engineering*, 127, 103–115.

Gamberini Rita, G., Luca, G., Francesco, L., & Bianca, R. (2017). On the analysis of effectiveness in a manufacturing cell: A critical implementation of existing approaches. *Procedia Manufacturing*, 11, 1882–1891.

Golchev, R. (2019). *Inter-dependencies between lean manufacturing and industry 4.0—A systematic state of the art literature review*. Politecnico Milano.

Gupta, R. C., Sonwalkar, J., & Chitale, A. K. (2001). Overall equipment effectiveness through total productive maintenance. *Prestige Journal of Management and Research*, 5(1), 61–72.

Jasiulewicz-Kaczmarek M. (2016). *SWOT analysis for planned maintenance strategy—a case study*. Retrieved from https://doi.org/10.1016/j.ifacol.2016.07.788.

Karekatti, C. (2014). *Arresting accelerated equipment deterioration in apparel factory* (pp. 18–22). Stitch World.

Karim, R., & Rahman, C.M.L. (2012). A performance analysis of OEE and improvement potential at a selected apparel industry, *Proceedings of the 6th international mechanical engineering conference and 14th annual paper meet, 28–29 September 2012*, Dhaka, Bangladesh.

Karim, R., Westerberg, J., Galar, D., & Kumar, U. (2016). Maintenance analytics—The new know in maintenance. *IFAC Papers Online*, *49*(28), 214−219.

Kumar, D., Kumar, D., & Rawat, R. (2014). Methodology used for improving overall equipment effectiveness by implementing total productive maintenance in plastic pipe manufacturing industries. *International Journal of Modern Engineering Research*, *4*(9).

Labib, A. W., O'Connor, R. F., & Williams, G. B. (1998). An effective maintenance system using the analytic hierarchy process. *Integrated Manufacturing Systems*, *9*(2), 87−98.

Lawrence, J. J. (1999). Use mathematical modelling to give your TPM implementation effort an extra boost. *Journal of Quality in Maintenance Engineering*, *5*(1), 62−69.

Lee, J., Kao, H., & Yang, S. (2014). Service innovation and smart analytics for Industry 4.0 and big data environment. *Procedia CIRP*, *16*, 3−8.

Macchi, M., Martínez, L.B., Márquez A.C., Granados, M.H., & Fumagalli, L. (2012). Value assessment of an E-maintenance platform. *2nd IFAC Workshop on Advanced Maintenance Engineering, Services and Technology, Universidad de Sevilla* (pp. 145−150). Sevila, Spain.

Maralcan, A., & Ilhan, I. (2017). Operations management tools to be applied for textile. *17th World Textile Conference AUTEX 2017 - Textiles—Shaping the Future, 254* (pp. 1−6).

Masud, A. K. M., Jannat, A. S., Khan, A. K. M. S. A., & Islam, K. J. (2007). Total productive maintenance in RMG sector a case. *Journal of Mechanical Engineering*.

Mayr, A., Weigeit, M., Kuhl, A., Grimm, S., Erll, A., Potzel, M., & Franke, J. (2018). Lean 4.0—A conceptual conjunction of lean management and Industry 4.0. 10.1016/j.procir.2018.03.292.

Nakajima, S. (1982). *TPM tenkai*. Tokyo: JIPM.

Nakajima, S. (1988). *Introduction to total productive maintenance*. Cambridge, MA, USA: Productivity Press.

Nakajima, S. (1989). *TPM development program*. Cambridge, MA, USA: Productivity Press.

Okogbaa, G., Huang, J., & Shell, R. L. (1992). Design database for predictive preventive maintenance system of automated manufacturing system of automated manufacturing system. *Computers and Industrial Engineering*, *23*, 7−10.

Pomorski, T. R. (2004). *Total productive maintenance concepts and literature review*. Brooks Automation Inc.

Prinz, C., Kreimeier, D., & Kuhlenkötter, B. (2017). Implementation of a learning environment for an Industrie 4.0 assistance system to improve the overall equipment effectiveness. *Procedia Manufacturing*, *9*, 159−166.

Relkar, A. S., & Nandurkar, K. N. (2012). Optimizing and analyzing overall equipment effectiveness through design of experiments. *Procedia Engineering*, *38*, 2973−2980.

Roy, R., Stark, R., Tracht, K., Takata, S., & Mori, M. (2016). Continuous maintenance and the future—Foundations and technological challenges. *CIRP Annals—Manufacturing Technology*, *65*(2), 667−688.

Rüttimann, B. G., & Stöckli, M. T. (2016). Lean and Industry 4.0—Twins, partners, or contenders? A due clarification regarding the supposed clash of two production systems. *Journal of Service Science and Management*, *9*, 485−500.

Saha, S., Akash S.M., & Biswas, T.K. (2017). Improving OEE of a garment factory by implementing TPM approach. *International Conference on Mechanical, Industrial and Materials Engineering 2017 (ICMIME2017) 28−30 December*, 1−6, Rajshahi, Bangladesh.

Senthilkumar, B., & Thavaraj, H. S. (2014). An evaluation of TPM implementation in clothing industry in India—A lean philosophy based approach. *International Journal of Industrial Engineering & Technology, 4*(6), 11−18.

Shakil, S. I., & Parvez, M. (2018). Application of lean manufacturing in a sewing line for improving overall equipment effectiveness (OEE). *American Journal of Industrial and Business Management, 8*, 1951−1971.

Singh, R., Gohil, A. M., Shah, D. B., & Desai, S. (2013). Total productive maintenance (TPM) implementation in a machine shop: A case study. *Procedia Engineering. 51*, 592−599.

Sly, D. (2018). Internet based eKanban/eKitting involving suppliers. *Procedia Manufacturing, 17*, 484−490.

Smith, R., & Hawkins, B. (2004). *Lean maintenance: Reduce costs, improve quality, and increase market share.* Elsevier Science.

Solanki, K., Yadav, R., Yadav, V., & Yadav, S. (2017). Study of implementation of total productive maintenance in textile machine manufacturing industries. *International Journal of Scientific Research in Engineering, 1*(3), 179−183.

Sonmez, V., Testik, M. C., & Testik, O. M. (2018). Overall equipment effectiveness when production speeds and stoppage durations are uncertain. *The International Journal of Advanced Manufacturing Technology, 95*, 121−130.

Telang, A.D. (1998). Preventive maintenance. *Proceedings of the National Conference on Maintenance and Condition Monitoring, February 14, Government Engineering College* (pp. 160−173). Thissur, India: Institution of Engineers, Cochin Local Centre.

Uhlmanna, E., Laghmouchia, A., Geiserta, C., & Hohwielera, E. (2017). Decentralized data analytics for maintenance in Industrie 4.0. *Procedia Manufacturing, 11*, 1120−1126.

Verma, R., & Verma, U. (2019). Development of Machine Learning Based Predictive Maintenance System for Apparel Manufacturing. *National Institute of Fashion Technology Delhi, Graduation Report.*

Vlasov, A. I., Grigoriev, P. V., Krivoshein, A. I., Shakhnov, V. A., Filin, S. S., & Migalin, V. S. (2018). Smart management of technologies: Predictive maintenance of industrial equipment using wireless sensor networks. *The International Journal of Entrepreneurship and Sustainability, 6*(2).

Wang, C. H., & Hwang, S. L. (2004). A stochastic maintenance management model with recovery factor. *Journal of Quality in Maintenance Engineering, 10*(2), 154−164.

Wang, K., & Wang Y. (2018). How AI affects the future predictive maintenance: A primer of deep learning, *IWAMA 2017: Advanced Manufacturing and Automation VII* (pp. 1−9).

Wauters, F., & Mathot, J. (2007). *OEE overall equipment effectiveness.* ABB Inc.

Wickramasinghe, G., & Perera, A. (2016). Effect of total productive maintenance practices on manufacturing performance: Investigation of textile and apparel manufacturing firms. *Journal of Manufacturing Technology Management, 27*(5), 1−26.

Xing, B., & Marwala, T. (2018). Introduction to smart maintenance, smart maintenance for human−robot interaction. *Decision and Control, 129*, 21−31.

Yazdi, P. G., Azizi, A., & Hashemipour, M. (2018). An empirical investigation of the relationship between overall equipment efficiency (OEE) and manufacturing sustainability in Industry 4.0 with time study approach. *Sustainability, 10*, 3031.

Zarreha, A., Wana, H., Leea, Y., Saygina, C., & Janahia, R.A. (2019). Cybersecurity concerns for total productive maintenance in smart manufacturing systems. *29th International Conference on Flexible Automation and Intelligent Manufacturing (FAIM2019).* Limerick, Ireland: Procedia Manufacturing.

Lean supply chain management

Prabir Jana
Department of Fashion Technology, National Institute of Fashion Technology, New Delhi, India

13.1 Introduction

A supply chain is the global network used to deliver products and services from raw materials (i.e., upstream) to consumers (i.e., downstream) through an engineered flow of information, physical distribution of material, and cash (Alber & Walker, 1998). Christopher (2016) defined supply chain as a network of organizations that are involved, through upstream and downstream linkages, in the different processes and activities that produce value in the form of products and services delivered to the ultimate customers. Fig. 13.1 represents simple apparel and textile supply chain.

The material flows downstream (refer to Fig. 13.1) with value added in each process and cash flows upstream, while the information flows in either direction. The decoupling point is a standard term given to the position in the material pipeline where the product flow changes from push to pull (Mason-Jones & Towill, 1999). The position of material decoupling point has a significant impact on inventory levels in the supply chain and responsiveness to demand. Similarly, the information decoupling point may be defined as the point in the information pipeline where the forecast is driven and market-driven information flows meet each other (Mason-Jones & Towill, 1999). The total time taken by the material from upstream to downstream is called supply lead time. Supply chain management could be seen as a management philosophy, implementation of management philosophy, or a set of management processes (Mentzer et al., 2001).

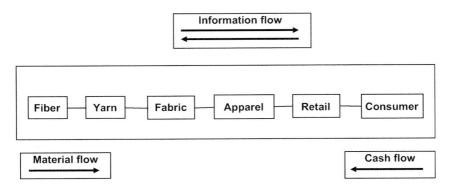

Figure 13.1 Apparel and textile supply chain (Jana, 2010).

Lean Tools in Apparel Manufacturing. DOI: https://doi.org/10.1016/B978-0-12-819426-3.00015-1
© 2021 Elsevier Ltd. All rights reserved.

The nonvalue-added activities in the supply chain started to be more than 90% of the supply lead time. Inventory reduction and improved logistics (Christopher, 2016), collaborative product development (CPD) (Womack & Jones, 1996), and partnership sourcing (Macbeth & Ferguson, 1994) are some of the methods to reduce lead time and improve supply chain efficiency. Fig. 13.2 represents a textile and apparel supply chain where horizontal lines represent the average process time while vertical lines (drawn to the same scale) represent the waiting time in the queues for each process chain (Scott & Westbrook, 1991). The representation shows two useful measures, the sum of horizontal lines is the process lead time and indicates responsiveness to a demand increase within the same stock constraints. Pipeline volume is determined by adding vertical and horizontal lines together and indicates the time taken to respond to decreases in demand given the same rate of manufacturing throughput.

Postponement of material conversion moves the decoupling point downstream enabling smooth upstream planning and pull effect in the supply chain.

13.2 Evolution of lean supply chain

The evolution of lean supply chain structure evolved toward the partnership approach in a departure from the adversarial relationship, reduction of lead time, concurrent product development against traditional one, reduction of nonvalue-added activities, need for postponement of conversion, and reduced inventory.

13.2.1 Demand amplification and bullwhip effect

The subject was pioneered by Jay Forrester at MIT while summarizing many of the phenomena associated with real-life supply chains. For example, what often appears

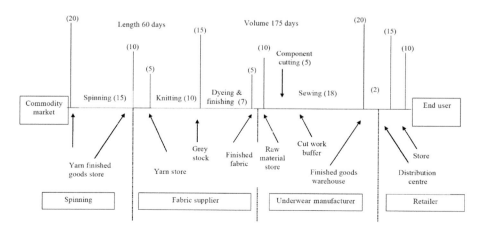

Figure 13.2 Value-added and nonvalue-added activities in textile and apparel supply chain (Scott & Westbrook, 1991).

as a small random ripple variation in sales at the marketplace is amplified dramatically at each level in the chain so that upstream companies are experiencing the classical "boom-bust" scenario with huge swings in manufacturing capacity required plus typically antiphase variations in stock levels (Forrester, 1961). As a rule of thumb, based on several supply chain studies, the demand amplification experienced is about 2:1 across each business interface. Hence, in a typical apparel and textile supply chain involving retailer—distributor—apparel manufacturer—fabric manufacturer/processor—and yarn-fiber supplier, the latter commonly is bombarded with swings 16:1 bigger than the marketplace. In consultancy practice, this highly undesirable phenomenon frequently is explained by reference to the "Forrester flywheel effect" or "bullwhip effect" in which every "player" is overordering against uncertainties in both the marketplace and in the supply chain. The term "bullwhip effect" was first coined by the logistics executives of Procter and Gamble (Lee, Padmanabhan, & Whang, 1997) when they experienced extensive demand amplifications for their diaper product "Pampers."

Several researchers cited different reasons behind the bullwhip effect; those are information and material delays (Towill, 1991) (Forrester, 1961), and irrational decision making by participants (Sterman, 1989). Although Lee et al. (1997) suggested four factors: demand forecast updating, order batching, price fluctuation, and rationing and shortage gaming behind the bullwhip effect, Paik & Bagchi (2007) concluded that the six factors that are statistically significant behind bullwhip effect are demand forecast updating, order batching, material delays, information delays, purchasing delays and level of echelons.

Although some research (Dejonckheere, Disney, Lambrecht, & Towill, 2003) showed that the bullwhip effect is not avoidable under order-up-to replenishment policy whatever forecasting method is used, some researchers (Disney & Towill, 2003) showed that, by implementing the vendor managed inventory (VMI) supply chain, both rationing and shortage gaming effect and order batching effect might be eliminated and the removal of one or more intermediaries led to the significant reduction of the bullwhip effect (Ackere, Larsen, & Morecroft, 1993).

Clothing retailers such as H&M, Zara, have broken the traditional production model by decreasing manufacturing lead time and producing clothing according to "fast fashion" trends, regardless of seasonality. The "fast fashion" can induce Muda (waste), Muri (overburden), and the bullwhip effect. As the factories are made to produce a greater variety of merchandise, which can lead to overworking and increased manufacturing errors. This desire for increased flexibility can lead to a "bullwhip" effect (Glatzel, Helmcke, & Wine, 2009). When suppliers are unable to maintain flexibility for their clients, orders are often duplicated and include "safety-stock" built into the order that results in excess waste.

In the early weeks of the COVID 19 pandemic when nonessential retail was quite suddenly shut, the retail demand suddenly dried up across the globe. The apparel buyers were faced with tough decisions: whether to honor orders in production and postpone upcoming orders or refuse incoming merchandise and cancel future orders. As the apparel manufacturers and upstream suppliers have already

384 Lean Tools in Apparel Manufacturing

curtailed down their production capacity, it is expected that the global apparel supply chain will be amid severe bullwhip effect (Cosgrove, 2020).

13.2.2 Product development

The importance of product development in an apparel supply chain is evident from the definition which says "Supply chain is the network of facilities and activities that performs the functions of product development, procurement of materials from vendors, the movement of material between facilities, manufacturing of products, distribution of finished goods to end consumers, and aftermarket support for containment" (Mabert & Venkataramanan, 1998). While in the car industry, new product development used to take 4−5 years (Womack, Jones, & Daniel, 1990), in the apparel industry, new product introduction every 4−6 weeks has become common in the so-called "fast fashion" (Mintel, 2007).

Any new product development (PD) aims to reduce cost and time overruns, limiting iterations, limiting redundant (nonvalue-added) activities, and maximize conversion (to production). According to Kurt Salmon Associates (KSA), a prominent global management consulting firm, almost 95% of product development cycle times are made up of nonvalue-added processes and more than 70% of the nonvalue-added activities can be eliminated (Parnell, 1999). Also, 67% of companies polled by KSA said that improvement of product development was their number one priority (Parnell, 1999).

The most comprehensive description of the apparel product development process was found in Plumlee's six-stage no-interval coherently phased product development model (Plumlee & Little, 1998). In Birnbaum's 101-step manufacturing cycle, the first 19 steps designed an in-house PD activity; then the factory collected information (receives tech-pack). PD and preproduction activity continued until the 86th step, where the order was ready to cut. A study (JBA, 1998) among European and US organizations revealed that out of 167 days of average supply lead time, product development consumed 104 days leaving approximately 25% of the lead time for manufacturing.

The most popular business model applied in the apparel industry is to keep design and color selection at company headquarters or home country and to outsource labor-intensive production offshore (Li, 2007). Once the color silhouette and fabric selection process are completed at headquarters, the development/costing request comes to the apparel manufacturers in India. In the contract manufacturing environment, the PD process is a tripartite activity involving buyers, apparel manufacturers, and raw material suppliers. Furthermore, "developing" often means "sourcing." There is a great deal of difference in terms of expertise required for "development" and "sourcing." The main developments done are developing fabrics (texture, design, and color), developing the pattern (silhouette), and developing and/ or sourcing accessories, developing the wash effect if required, and developing embroidery/value addition/handwork.

Fig. 13.3 shows the process flow of a product development flowchart in a contract manufacturing scenario with a dependency relationship. There are three clear

Lean supply chain management

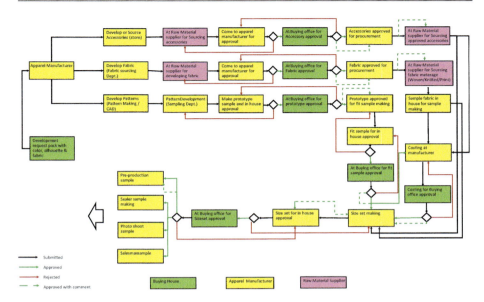

Figure 13.3 Product development process of a typical contract apparel manufacturer (Jana, 2010).

parallel process flows: fabric development, accessories development, and pattern development. All three process flow converges to one point at size set making, which requires all three inputs. When compared with Plumlee's six-stage product development process, for a basic product, a contract manufacturer would probably get involved in only two steps, that is, prototype and preproduction sample development (Jana, 2010). Other activities like line planning and research, concept development, marketing to the retail channel, and line optimization are generally done by retail organizations/brands themselves. In the case of fashion products, a small manufacturer will be involved in four stages except marketing to the retail channel and line optimization. In the case of a medium-size company making fashion products, having offices and design studios abroad may involve all six stages. Faster lead time, removal of nonvalue-added activities made product development an ideal place for exercising lean principles. Some of the tools applied to reduce the product development lead time in apparel manufacturing include the use of a critical chain approach to compress the lead time (Jana, Gupta, Joshi, & Knox, 2005), use of 3D virtual prototyping (Sareen, 2008) and CPD.

13.2.2.1 Critical chain

The critical chain uses the time compression (remove time within a process) technique. By adopting a critical chain approach it was possible to reduce the product development time by 45% (Jana, 2010), and by avoiding multitasking it was possible to reduce start-up loss in pattern making and sample making activity by 43%

(Jana et al., 2005). It was also found that elimination of iteration (nonvalue-added activity) and implementation of the critical chain had inherent potential to reduce manufacturing lead time by one third (Jana, 2010).

13.2.2.2 Virtual prototyping

In the contract manufacturing scenario every value-added activity (done by the upstream player) was followed by an approval activity (necessary nonvalue-added activity) by the downstream player. Due to geographical distance and other logistic constraints, approval activity took two to three times the original value-added activity (Jana, 2010). Thus virtual prototyping solutions offer not only faster time to market but reducing raw material cost as well (Papahristou & Bilalis, 2016). It reduces the waste of overproduction by reducing the number of iteration. Some companies can develop only one real sample before production skipping the proto, first, and color sample stages—that is a 50% time saving and 70% cost saving in some cases (Mageean, 2018). The electronic fitting can reduce product development time to 10−15 days or by 50% (Sareen 2008).

13.2.2.3 Collaborative product development

Some of the industry practices of collaborative product development (CPD) are given here. Gap Inc. shares its seasonal forecast color with Coats worldwide so that the correct embroidery thread can be developed. All Gap vendors are then directed to source their embroidery thread needs from local Coats suppliers (Coats, 1999). Levis shares its specifications with the accessory manufacturer and supplier Universal Fasteners to develop zipper and rivets, stud buttons collaboratively (AAMA, 1995). Freshtex, the specialty garment finisher shares forecasts and develops washes with leading jeans brands during the concept stage (Freshtex, 2020). Liz Claiborne and M&S collaborate with fabric manufacturers to develop designs based on forecasts.

13.2.3 Inventory management

Process mapping exercises in the textile and apparel supply chain in the US were conducted by KSA during 1986. For a ladies' night suit it was observed that a total of 66 weeks were required from fiber stage to consumer. However, out of the 66 weeks, only 11 weeks were spent on actual production, that is, spinning—weaving—wet processing—cutting—sewing—packaging—distribution; the rest of the time was inventory delay (Bruce, Moore, & Birtwistle, 2004).

The 55 weeks of inventory involves (refer to Table 13.1) activities that add no value, for example, goods in storage, but which existed for traditional reasons. During the 1990s, there had been a paradigm shift from cost-based to time- and value-based competition in global sourcing of merchandise, resulting in an ever-increasing and irreversible global trend toward "speed and replenishment sourcing of merchandise" (Abernathy, Dunlop, Hammond, & Weil, 1999). Increasing

Lean supply chain management

Table 13.1 Clothing pipeline inventories and works in progress (Lawson, King, & Hunter, 1999).

		Inventory	WIP (in weeks)
Fiber	Raw material	1.6	
	WIP		0.9
	Finished fiber @ fiber	4.6	
	Fiber @ textile	1.0	
	Total	7.2	0.9
Fabric	WIP—greige		3.9
	Greige goods@ greige	1.2	
	Greige goods@ finish	1.4	
	Finishing		1.2
	Finished fabric @ textile	7.4	
	Fabric @ apparel	6.8	
	Total	16.8	5.1
Apparel	WIP		5.0
	Finished apparel @ apparel	12.0	
	Ship to retail	2.7	
	Apparel @ retail distribution center	6.3	
	Apparel @ store	10.0	
	Total	31.0	5.0
	Total	55.0	11.0

importance was given to design innovation capabilities of the manufacturer and shifting of product development functions from retailer to manufacturer for an effective supply chain.

13.2.3.1 Postponement

Postponement is delaying supply chain activities until customer orders are received to customize products as opposed to performing these activities in anticipation of future orders (Hoek, 2001). The position of the material decoupling point has a significant impact on inventory levels in the supply chain and responsiveness to demand, and postponement of material conversion moves the decoupling point downstream (Mason-Jones & Towill, 1999), thereby enabling pull effect in the supply chain. Researchers also suggest that postponement has the potential to improve responsiveness while reducing inventory, transportation, storage (three kinds of wastage), and obsolescence costs (Yang, Burns, & Backhouse, 2004).

The concept of differentiation is important for the textile and apparel industry as it offers a variety of end products to customers. In the case of garment manufacturing, fabric dyeing and sewing represent two main points of differentiation. In dyeing, an irreversible change takes place about color in sewing, irreversible change takes place about style. In the apparel industry, Benetton used postponement techniques to improve its responsiveness to customer demands. By postponing the dyeing of its garments, Benetton was better positioned to respond to demands for popular colored clothing and reduce the excess inventory of less popular colors (Dapiran, 1992). Postponing the shipment of appliances to Sears until a customer order is received allowed Whirlpool to realize a significant reduction in inventory and transportation costs (Waller, Dabholker, & Gentry, 2000). During the year 1991–92, high street retailers in the UK like French Connection, Miss Selfridge, Wallis, Topshop ran a successful program of garment dyeing program (called distress rayon) using postponement technique, where greige viscose rayon fabric was cut and sewn by apparel manufacturers in India. The color-wise dyeing quantities are decided at the last minute enabling unprecedented flexibility to the retailers to adjust based on the latest sales trend. Design for the postponement is one of the variants of a general *design for excellence*, or more generally "Design for X," (Roser, 2020) the lean approach for applying lean in design.

13.2.3.2 Vendor managed inventory

VMI is essentially a distribution channel operating system whereby the inventory at the retailer is monitored and managed by the manufacturer. It includes several tactical activities, such as determining appropriate order quantities, managing proper product mixes, and configuring appropriate safety-stock levels (Chopra & Meindl, 2001). The rationale is that by pushing the decision making responsibility further up the supply chain, the manufacturer or vendor will be in a better position to support the objectives of the entire integrated supply chain resulting in a sustainable competitive advantage. Refer to Fig. 13.4, in the conceptual evolution of inventory management, VMI stands in the middle.

In VMI, the retailer still owns the inventory and the manufacturer simply manages it. Consignment selling (CS) can be considered the next step (after VMI), where the manufacturer owns the inventory and the retailer charges a percentage for providing shelf space and customers. In practice, both VMI and CS are being practiced in the industry often under the more popular acronym VMI.

TAL Apparel is a Hong Kong-based apparel manufacturer having factories in HK, Thailand, Malaysia, Taiwan, China, Indonesia, Vietnam, Mexico, and the USA. TAL collects point-of-sale data of J. C. Penny's shirts directly from its stores in the USA, then run the numbers through proprietary software to determine the number of different styles, colors, and sizes to make. One of its Asian factories manufactures the shirts and sends it directly to each J. C. Penny's store (bypassing the warehouses and corporate decision makers. TAL's New York office analyses the market, trends, sales data to design, and fine-tune the proprietary replenishment software. TAL provides similar services to Brooks Brothers and Lands' End.

Figure 13.4 Conceptual evolution of inventory management.

Wal-Mart believed to be the pioneer to start VMI with Procter and Gamble has supposedly moved to CS (Cid, Gordon, Kearns, Lennick, & Sattleberger, 2000). The enabling technology behind successful VMI is electronic data interchange that provides the manufacturer with essentially the same point of sales and inventory information.

13.2.3.3 Just-in-time

The idea of making a product for the customer at the time the customer orders it is fascinating because that is what the creators of lean always dreamed about. It is the ultimate just-in-time (JIT) (Onetto, 2014). JIT is a method where material arrives just on (in) time when it is needed. This is valid both for purchased or delivered material and material processed on-site. On-site, the moment a worker needs a part, it should arrive right where he needs it. What VMI does for retailers, JIT does almost the same for the manufacturers. In JIT scenario vendor is supplying inventory to manufacturers JIT, whereas in VMI vendor is maintaining (means already supplied before time) inventory at the retail shop. In VMI the manufacturer requires to predict (forecast!) the consumer demand, whereas in the manufacturing scenario the consumable vendor simply has to calculate the requirement for the manufacturer; this is the key difference. The following procurement of consumables like sewing thread, sewing needle, zippers, spare parts at apparel manufacturing follows some kind of JIT or VMI approach.

Table 13.2 summaries the benefits and challenges of following JIT/VMI with the above e-identified consumables. While the benefits are substantial and challenges are negligible for sewing thread (as it is possible to calculate the type and quantity of merchandise with certainty), the benefits for sewing needle may be negligible and challenges are significant (although the type of sewing needle is easy to decide based on fabric merchandise the quantity of replenishment is not easy to calculate). It is a common practice by sewing thread and zipper suppliers to supply and

Table 13.2 Benefits and challenges of JIT/VMI in apparel manufacturing (Jana, Banerjee, & Knox, 2003).

	Benefits			Challenges to predict and calculate	
	Cost saving	Space saving	Procurement time	Type	Quantity
Sewing thread	Substantial	Substantial	Substantial	Certain	Certain
Zippers	Substantial	Substantial	Substantial	Certain	Certain
Sewing Needle	Negligible	Negligible	Negligible	Certain	Uncertain
Spare parts	Negligible	Negligible	Substantial	Uncertain	Uncertain

Note: JIT, Just-in-time; VMI, vendor managed inventory.

maintain inventory at apparel manufacturers warehouse and raise invoices once consumed.

13.2.4 Supply chain structure

Although in the automotive supply network a tier system of suppliers exists, apparel and textile networks take the shape of matrix and not tiers (Massey, 2000). Fig. 13.5 explains completely disconnected (vertically integrated), partially covered, applied, and completely overlapped (horizontally spread) networks for apparel.

Although disconnected network works very similarly to vertically integrated organizations, it can work with lean inventories and preferred for commodity products. The completely overlapping networks in contrast result of "globalization" or "core competency" concept are very common for fashion merchandise with a highly flexible and responsive supply chain. These are agile but may not be very lean (in terms of inventory holding).

The supply chain network can also be classified based on the predominant flow pattern and can be referred to as I-V-A-T analysis where the form of the letter suggests the pattern of material flow. The I-shape is the unidirectional vertical enterprise. Historically, companies "managed" linked supply chain operations by ownership, that is, vertical integration (Harrigan, 1983), but rare to find now. Carmaker Henry Ford had a sheep farm that grew wool for car seat covers, General Motors made car paint, newspaper magnates owned forests and paper mills (Thackray, 1986), and Courtaulds had fiber manufacturing to retail functions in the UK (Knox & Newton, 1998). Some of the other recent examples of vertical integration are "textile city" in Mexico, Xiqiao Light Textile City in China, and Brandix Apparel City in India. Although it was denim concentration in Mexico, the Chinese city had a comprehensive production chain from product R&D, material production, weaving, dyeing and finishing, fashion design to sale, and the export of finished apparel. The Brandix group produces exclusive casual trousers, inner activewear, textiles, knitted fabrics, sewing and embroidery thread, accessories and hangers, and also offers wet processing and finishing as well as fabric printing. There are

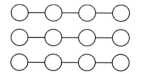
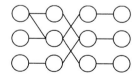
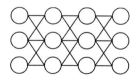

Completely disconnected networks	Partially overlapping network	Completely overlapping network
Highly vertically integrated manufacturers–retailers such as Zara in fashion apparel	The limited vs. branded apparel products from Levis sold through retailers	Private label apparel retailers that source from contract or FOB manufacturers in Southeast Asia

Figure 13.5 Classification of supply chain (Rice & Hoppe, 2001).

several advantages and disadvantages of vertical integration; although there are better control, faster communication, and lower cost, it also leads to limiting competition, less flexibility, and higher risk, and so on.

A V-shape network (refer to Fig. 13.6) starts with limited raw material at the input stage a and wide variety of finished products with product variety determined early in the transformation process. Examples are textile and apparel, metal fabrication, and so on. An A-shape network has numerous raw materials at the input stage a with limited variety of finished products, for example, in aerospace and food retail. A T-shape tries to keep a simple flow path until the latest possible moment before suddenly branching out into a wide variety of finished products, for example, in electronics and home appliances.

The partnership is one of the key pillars behind both the V-shape and A-shape network. The basis of partner selection changed from primarily price-based (adversarial) to collaborative/technology/core competency-based. In apparel and textile supply chains, retailers partnered with the apparel manufacturers, who in turn partnered with trims and accessories suppliers. The essence of partnership is trust, which played a significant role in collaboration. Marks and Spencer's erstwhile partnership with UK manufacturers speak about the unmatchable trust and loyalty achieved in those years (up to the late 20th century).

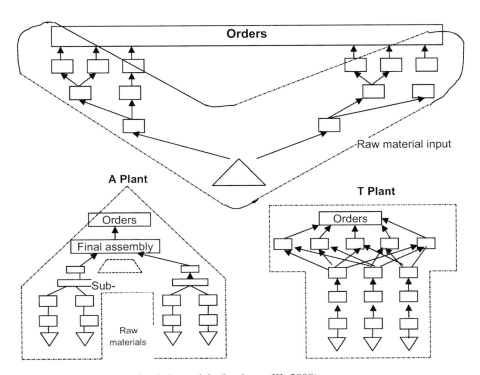

Figure 13.6 V-A-T supply chain models (Lockamy III, 2008).

The T-shape network is an extension of V-shape and favorable from the supply chain management viewpoint since the decisions on final product specification are delayed till the end and thereby support fast response to varying demand of customers (postponement technique). However, due to multisourcing practice apparel and textile supply chain take more of V shape.

13.3 Lean applications in retail supply chain

Traditional mass merchants sought a cost advantage through economies of purchasing scale. Given the limited information on sales, this meant that these retailers purchased large inventories of goods that they would then "push" to consumers, often through price reductions and sales promotions. Wal-Mart is the most well-known of the early lean retailers in apparel. Beginning in the late 1970s, Wal-Mart used cutting edge information technologies to track consumer sales at the checkout counter, monitor its inventory of goods within and across stores, and then supply its stores on an ongoing basis via highly efficient, centralized distribution methods. By capitalizing on "real-time" information on sales and inventory position, Wal-Mart increased its ability to let consumer demand "pull" its orders. To support lean retail the warehousing, distribution, transport, purchasing functions also adopted lean philosophies.

13.3.1 Lean warehousing and transport

The seven steps to embrace lean warehousing concepts are profile analysis, standardize processes, establishing a baseline measure, go to the Gemba, improving the processes, recruiting the right talent who understand "lean" and lastly avoid three deadly sins: travel distance, touches, and paper (Kenco, 2001). Although one can start building a profile of the warehouse by asking questions like how many pallets are coming into the warehouse? are there seasonal shifts in the product mix? are the products single or mixed? The profile creation then decides the labor, equipment, and layout and helps to do zoning and slotting the warehouse properly, which ensures optimization of locating products (minimizing travel distance) and maximize operational efficiency by reducing the processing of paper and touches.

Lean management challenges existing assumptions about size-based cost efficiency, where frequent inbound and outbound movement of smaller batches is considered more expensive than transporting larger lots. This poses a challenge for transportation managers but transporting any more than necessary adds to transportation waste. Along with reducing shipment sizes, inventories, and transportation costs, lean transportation also minimizes lead times. Faster shipments and deliveries lead to higher customer satisfaction, which is a key factor in any organization's success. The transportation cost structure is divided into unit costs and productivity costs (such as trailer utilization, waiting time for vehicles and weighing scales, total distance covered, etc.). The latter provides a larger opportunity for cost reduction according to Lean, so transportation managers started focusing on these.

Optimizing productivity costs instead of focusing on carrier rates allows organizations to minimize the wastage of materials, time, and effort while ensuring that carriers are not running unprofitably. It is essentially a win-win for both.

The selection of the transportation method for e-commerce giant Amazon for a given package is driven, first, by the promised delivery date to the customer. Lower-cost options enter the equation only if they provide an equal probability of on-time delivery. That is a lean principle of transportation (Onetto, 2014).

13.3.2 Lean distribution

Lean management in distribution is based on three main pillars: being able to respond to (the complete) demand, delivering at the promised date and time, and keeping stocked at the minimum possible expense (Landmark Global, 2015). Any inventory arriving late will impact customer service and must be avoided. So, inventory generally arrives early, increasing capital requirement, space, handling, tracking, and so on. If inventory can be moved faster and more reliably, then it could reduce distribution costs and resources, and arrive close to customer needs.

The two principles of Lean—Pull and Linkage—across operations improve total flows across the distribution network. The commute time of any supply vehicle to the store from the warehouse will follow the normal distribution curve (Fig. 13.7). A wide and low profile indicates a highly variable commute time, which requires allowing more time to ensure on-time arrival. The more time allowed, the higher the chance that arrival will be early.

This can be addressed by maximizing the total flow of cars by moving from push to pull system (Zylstra, 2012). During rush hour, speed and spacing between vehicles deteriorate as cars "push" their way onto the highway, reducing the total

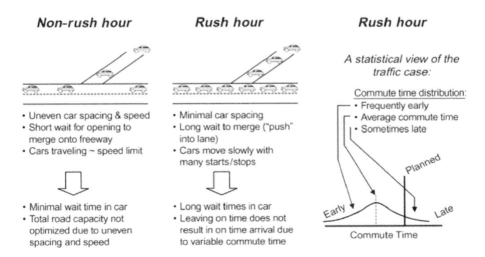

Figure 13.7 Normal commuting time (Zylstra, 2012).

flow. Once on the highway, the variability of each driver's speed reduces spacing, causing the average speed and volume of all vehicles to deteriorate. Improved speed and spacing could be accomplished by using "pull" to bring vehicles onto the highway up to the point where optimal speed and spacing start to deteriorate. Refer to Fig. 13.8, for the "pull" system to work, each vehicle must be ready to depart on fairly short notice (the pull signal) and able to maintain a constant spacing for the given speed (imaginary linkage to the next vehicle).

13.3.3 Lean in e-commerce

Given the business evolution of Amazon from a bookstore to the store for everything, Amazon had to reinvent automation, following the lean principle of "autonomation": keep the humans for high-value, complex work and use machines to support those tasks. Humans are extremely creative and flexible. The challenge of course is that sometimes they are tired or angry, and they make mistakes. From a Six Sigma perspective, all humans are considered to be at about a Three Sigma level, meaning that they perform a task with about 93% accuracy and 7% defects. Autonomation helps human beings perform tasks in a defect-free and safe way by only automating the basic, repetitive, low-value steps in a process. The result is the best of both worlds: a very flexible human being assisted by a machine that brings the process up from Three Sigma to Six Sigma.

Amazon implemented the *Andon* cord principle in customer service. Bezos was enthusiastic about it right away and we implemented it in about 6 months. The process begins when a service agent gets a phone call from a customer explaining that there is a problem with the product he or she has just received from us. If it is a repetitive defect, the customer service agent is empowered to "stop the line," which

Rush hour with lean-based pull

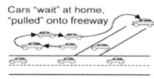

- Even car spacing & speed
- Short wait to merge onto freeway
- Cars traveling ~ speed limit
- Leave home based on Pull

A statistical view of the traffic case:

Distribution comparison:
- Low variation (more consistent commute time)
- Shorter commute time
- Small early arrival time to insure rarely late

- Minimal wait time in car
- Variable window for leave time; predictable arrive time
- Total road capacity optimized by best speed/spacing
- Previous wait time becomes productive "home" time

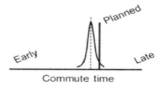

Figure 13.8 Pull transportation (Zylstra, 2012).

means taking the product off the website until we fix the defect. The objective is to start the line again with the defect resolved. Amazon created an entire background process to identify, track, and resolve these defects.

The *Andon* cord has had an amazing impact; it eliminates tens of thousands of defects per year. The other wonderful thing is that the *Andon* cord has empowered frontline workers. The authority to stop the line is an enormous proof of trust for customer service agents, who usually have no real authority to help irate customers over the telephone. With the cord, the agents have been able to tell customers that the product has been placed in the lab for quality problems until the defect can be resolved. At the same time, they offer customers a new product or reimburse them. Customers can see products pulled for quality issues on the website in real-time. This has created incredible energy and motivated the frontline people of Amazon to do great work for the customers and their assessments are found correct 98% of the time: that is, the *Andon* cord is pulled for a real defect.

eBay, one of the world's most successful e-commerce companies, has introduced continuous improvement methods through Lean and Six Sigma several years ago with a focus on improving the customer experience. The company concentrated on process improvement and long-term organizational learning on the front lines of its customer service department. eBay set out to make sure its customer service agents were empowered to apply improvement tools and implement workflows to advance their problem-finding and problem-solving capabilities. Within the first year, one frontline service agent improved the contract subscription process by eliminating a redundant workflow and clearly defining service requirements. His initiative reduced the processing time from 7 days to 2 days, and the single employee was responsible for a revenue gain of $114,000 (Four Principles, 2018).

13.4 Conclusion

While lean tools are popularly used in a manufacturing scenario, they are found to be equally effective in product development, inventory management, logistics, and warehousing functions. Although this chapter has discussed examples of waste minimization, pull system, just-in-time, critical chain, and Andon in some of the nonmanufacturing functions of the supply chain, other lean tools, and other nonmanufacturing functions also can be explored. Lean in marketing and sales is already gaining popularity among consultants and academicians (Ivanov, 2018; Elias & Harrison, 2015). As the improvement potential at the apparel manufacturing domain is bottoming out and the automation in the overall supply chain is increasing, we will see more and more use of lean tools in the supply chain functions in the future.

References

AAMA. (1995). *Bobbin Show International Seminar*. Atlanta: AAMA since renamed as AAFA, url:. Available from http://www.aafa.org.

Abernathy, F., Dunlop, J., Hammond, J., & Weil, D. (1999). *A stitch in time: Lean retailing and the future of manufacturing: lessons from the textile and apparel industries.* New York: Oxford University Press.

Ackere, A., Larsen, E., & Morecroft, J. (1993). Systems thinking and business process redesign: An application to the beer game. *European Management Journal, 11*(4), 412–423.

Alber, K. L., & Walker, W. T. (1998). *Supply chain management principles and techniques for the practitioner.* Falls Church: APICS Educational & Research Foundation.

Bruce, M., Moore, C., & Birtwistle, G. (2004). *International retail marketing: A case study approach.* Butterworth-Heinemann.

Chopra, S., & Meindl, P. (2001). *Supply chain management: Strategy, planning and operation.* Prentice Hall, Inc.

Christopher, M. (2016). *Logistics and supply chain management.* UK: Pearson, 5th ed.).

Cid, F. d, Gordon, R., Kearns, B., Lennick, P., & Sattleberger, A. (2000). *Operations logistics and supply chain management.* Kellogg Graduate School of Management.

Coats. (1999). Technical leaflet on sewing thread: Module No. 22.

Cosgrove, E. (2020). *Supplychaindive.* Retrieved from https://www.supplychaindive.com/news/coronavirus-apparel-fashion-sourcing-suppliers. (Accessed 20 July 2020).

Dapiran, P. (1992). Benetton—Global logistics in action. *International Journal of Physical Distribution & Logistics Management, 22*(6), 7.

Dejonckheere, J., Disney, S., Lambrecht, M., & Towill, D. (2003). Measuring and avoiding the bullwhip effect: A control theoretic approach. *European Journal of Operations Research, 147,* 567–590.

Disney, S., & Towill, D. (2003). The effect of vendor managed inventory (VMI) dynamics of the bullwhip effect in supply chains. *International Journal of Production Economics, 85* (2), 199–216.

Elias, S., & Harrison, R. (2015). Lean competency system. Retrieved from Simon Elias & Richard Harrison. (Accessed 20 July 2020).

Forrester, J. (1961). *Industrial Dynamics.* Boston, MA: MIT Press.

Four Principles. (2018). Retrieved from https://www.fourprinciples.com. (Accessed 13 July 2020).

Freshtex. (2020). *Freshtex.* Retrieved from https://freshtex.com/en/#services. (Accessed 19 July 2020).

Glatzel, C., Helmcke, S., & Wine, J. (2009). Building a flexible supply chain for uncertain times. *McKinsey Quarterly.* Retrieved from https://www.mckinsey.com/business-functions/operations/our-insights/building-a-flexible-supply-chain-for-uncertain-times. (Accessed 20 July 2020).

Harrigan, K. (1983). *Strategies for Vertical Integration.* Lexington, MA: Lexington Books.

Hoek, V. (2001). The rediscovery of postponement a literature review and directions for research. *Journal of Operations Management, 19*(2), 161–184.

Ivanov, B. (2018). *Kanbanize.* Retrieved from https://kanbanize.com/blog/what-is-lean-marketing/. (Accessed 20 July 2020).

Jana, P. (2010). *An investigation into Indian apparel and textile supply chain networks.* Nottingham: The Nottingham Trent University.

Jana, P., Banerjee, P., & Knox, A. (2003). *VMI in apparel manufacturing.* Retrieved from https://www.just-style.com/.

Jana, P., Gupta, S., Joshi, C., & Knox, A. (2005). *Sypply chain dynamics in indian apparel export manufacturing.* Retrieved from www.techexchange.com.

JBA. (1998). *1998 Apparel and footwear industries survey.* JBA.

Kenco. (2001). *Embracing lean warehousing.* Tennessee: Kenco Group.

Knox, A., & Newton, E. (1998). Clothing industry supply chain management & technology. In *International conference on apparel manufacturing: Future scenario*. New Delhi: National Institute of Fashion Technology.

Landmark Global. (2015). Retrieved from https://landmarkglobal.com/trends-insights/lean-management-in-e-commerce-deliveries-2/. (Accessed 10 July 2020).

Lawson, B., King, R., & Hunter, A. (1999). *Quick response managing the supply chain to meet consumer demand*. Chichester: John Willey and Sons.

Lee, H., Padmanabhan, V., & Whang, S. (1997). The bullwhip effect in supply chains. *Sloan Management Review*, *38*(3), 93−102.

Li, L. (2007). *Supply chain management: concepts, techniques and practices: enhancing value through collaboration*. Singapore: World Scientific Publishing Co. Pte. Ltd.

Lockamy, A., III (2008). Examining supply chain networks using V-A-T material flow analysis. *Supply Chain Management*, *13*(5), 343−348.

Mabert, V. A., & Venkataramanan, M. (1998). Special research on supply chain linkages: Challenges for design and management in the 21st century. *Decision Sciences*, *29*(3), 537−552.

Macbeth, D., & Ferguson, N. (1994). *Partnership sourcing: An integrated supply chain approach*. UK: Pitman Publishing.

Mageean, L. (2018). *3D Virtual Sampling: Evolution or Revolution?* Retrieved from https://www.whichplm.com/3d-virtual-sampling-evolution-or-revolution/. (Accessed 19 July 2020).

Mason-Jones, R., & Towill, D. (1999). Using the information decoupling point to improve supply chain performance. *International Journal of Logistics Management*, *10*(2).

Massey, L. (2000). *An investigation into apparel and textile supply chain developments*. Nottingham: The Nottingham Trent University.

Mentzer, J., DeWitt, W., Keebler, J., Min, S., Nix, N., Smith, C., & Zacharia, Z. (2001). Defining supply chain management. *Journal of Business Logistics*, *22*(2), 1−25.

Mintel. (2007). *Retail review*. London: Mintel.

Onetto, M. (2014). *When Toyota met e-commerce: Lean at Amazon*. McKinsey Quarterly. Retrieved from https://www.mckinsey.com/business-functions/operations/our-insights/when-toyota-met-e-commerce-lean-at-amazon. (Accessed 10 July 2020).

Paik, S. K., & Bagchi, P. K. (2007). Understanding the causes of the bullwhip effect in a supply chain. *International Journal of Retail & Distribution Management*, *35*(4), 308−324.

Papahristou, E., & Bilalis, N. (2016). Can 3D virtual prototype conquer the apparel industry? *Journal of Fashion Technology & Textile Engineering*, *4*(1).

Parnell, C. (1999).). Reshaping apparel: SCM & enterprise control. *Apparel Industry Magazine*, 16.

Plumlee, T. M., & Little, T. J. (1998). No-interval coherently phased product development model for apparel. *International Journal of Clothing Science and Technology*, *10*(5), 342−364.

Rice, J. B., & Hoppe, R. M. (2001). Supply chain versus supply chain: The hype and the reality. *Supply Chain Management Review*, *5*(5).

Roser, C. (2020). *AllAboutLean.com*. Retrieved from https://www.allaboutlean.com/dfma-1/. (Accessed 13 July 2020).

Sareen, R. (2008). I want it faster. . . I want it cheaper: I want e-fitting solutions. *StitchWorld*.

Scott, C., & Westbrook, R. (1991). New strategic tools for supply chain management. *International Journal of Physical Distribution & Logistics Management*, *21*(1), 23−33.

Sterman, J. (1989). Modelling managerial behaviour: Misperceptions of feedback in a dynamic decision-making environment. *Management Science*, *35*(3), 32−399.

Thackray, J. (1986). America's vertical cutback. *Management Today*.

Towill, D. (1991). Supply chain dynamics. *International Journal of Computer Integrated Manufacturing, 4*(3), 197–208.

Waller, M., Dabholker, P., & Gentry, J. (2000). Postponement, product customization, and market-oriented supply chain management. *Journal of Business Logistics, 21*(2), 133–159.

Womack, J., & Jones, D. (1996). *Lean Thinking—Banish waste & create wealth in your corporation*. New York: Simon & Schuster.

Womack, J. P., Jones, D. T., & Daniel, R. (1990). *The machine that changed the world*. New York: Simon & Schuster Inc.

Yang, B., Burns, N., & Backhouse, C. (2004). Postponement: A review and an integrated framework. *International Journal of Operations & Production Management, 24*(5), 468–487.

Zylstra, K. D. (2012). *Lean distribution: Applying lean manufacturing to distribution, logistics, and supply chain*. John Wiley & Sons.

Agile manufacturing

14

Riddhi Malviya
VF Corporation, Greensboro/Winston-Salem, North Carolina, United States

> *The best opportunities are "visible, but not seen." Next practices are all about imagining what the future will look like, identifying the mega-opportunities that will arise, and building the capabilities to capitalize on them. — Peter Drucker*

14.1 Introduction

In recent times, lean manufacturing has been evolving to serve goals including, but not limited to, the following:

1. reducing the lead time of processes;
2. improving the overall quality of the end product;
3. eliminating "waste" in the form of unproductive time, wasted effort, wasted raw, material, etc.; and
4. reducing the total manufacturing cost.

In classic lean manufacturing methodology, there are seven types of waste which must be reduced or eliminated (Howard, 2009):

- overproduction (occurs when production should have stopped)
- waiting (periods of inactivity)
- transport (unnecessary movement of materials)
- extra processing (rework and reprocessing)
- inventory (excess inventory not directly required for current orders)
- motion (extra steps taken by employees due to inefficient layout)
- defects (do not conform to specifications or expectations)

As any procurement professional is aware, no matter what product you are buying, there are a few standard Key Performance Indicators (KPIs) that are considered when selecting the manufacturer: how quickly can they meet your demands, how defect-free can they make the product, and how cheap can they sell you the final products while continuing to be sustainable. Lean manufacturing targets improvement of all those KPIs, which increases the efficiency of the production lines.

From all fronts, lean seems like an unquestionably positive concept which all manufacturers should seek to apply. Unfortunately, there are still some risks associated with the implementation of lean, which are discussed in the next section.

Lean Tools in Apparel Manufacturing. DOI: https://doi.org/10.1016/B978-0-12-819426-3.00011-4
© 2021 Elsevier Ltd. All rights reserved.

14.1.1 Lean manufacturing: risks and mitigations

Whenever a leadership team decides to implement lean in a company or a specific department, they focus on the objective, quantitative side of it. This is partly because the proponents of lean describe its value with the use of case studies and fancy figures. They tend to forget that the key facets of lean are the way of thinking, the culture that it represents. Subsequently, lean is implemented using only the tools and methodologies it stands for, with no effort to inculcate lean thinking as a normal way of doing things. This may lead to an early-stage abandonment of lean.

Another common complaint comes from employees who work on the shop floor and are the closest to the everyday issues lean is trying to solve. The managers, often too entangled in the standard approaches described by the lean practitioners, attempt to apply the same standard formula to solve every problem. This not only causes distrust of lean among the floor workers but also causes a lot of work and rework.

Furthermore, a significant number of well-known lean activities, originating from the Toyota Production System (TPS), are answers for explicit issues that Toyota was confronting. They might not apply to all facets of the apparel industry, much less to a particular company. Thus it is usually unwise to implement lean teachings directly. Instead, it is important to study the purpose of each of the teaching and subsequently collect data about the specific problems the company's shop floor is facing. Also, the leadership should create a detailed image of what the final stage should look like after lean has been implemented. They should know, in measurable terms, what they are looking to achieve. Then, the plan to bridge the gap will become easier to think about. This is also important when a company wants to ensure that the solutions that are being implemented are permanent and will not be abandoned easily.

The lean way of thinking expects to lessen costs while upgrading and improving execution. Value stream mapping (VSM) and 5S are the most widely recognized methodologies organizations make on their first strides toward making their association leaner. In an apparel company the lean activities can begin by examining both facets of the shop floor: the SKUs (stock keeping units) as well as the workers. The SKUs usually depend on the orders placed by the company's clients. For the SKUs a company can record past quantity information, the specifications of the SKU, anomalies in the specifications (any special orders received), any recurring issues with quality faced, number of reworks, etc. If any of the above is not being measured, tools should be put in place immediately to start tracking such key KPIs. Next, the team members, starting from the external team including the vendors to the purchaser and in house, accountants should be assessed for their role in the value stream. These are some of the basic steps to ensure that the team has a complete understanding of the current process, before making plans for the next steps. Another part of this process is the FMEA (failure mode impacts examination) to distinguish and avoid any hazard factors. Also, if a department has highly skilled or experienced workers, they should be made part of the VSM process, because they might know some intricacies of the process that the managers miss.

When lean is being embraced by an organization, both leaders and field workers experience a change in their everyday tasks and processes. Thus it is paramount that the leaders of the lean journey are both decisive and determined about the implementation process. Lean specialists suggest creating a lean plan specific to the organization itself, with input and guidance from a few selected managers called the "lean leadership." Also, a group of people may be formed as the "lean team" who is a neutral group with the core intent of managing the lean implementation by measuring the KPIs for all departments and giving necessary instructions to ensure that all departments stay on track. The lean team should also be reporting insights gathered by analyzing the KPIs and the data to the leadership.

14.1.2 Lean paves the way for agility

In recent years, many large-scale apparel manufacturing companies have adopted lean as a way of becoming more efficient in dealing with the high volume of orders they are used to getting.

On the other hand, the rise of consumerism and the growing desire for unique fashion have paved the way for countless clothing brands promoting creativity and innovation in design. For such brands the epitome being Zara, change and adaptability are key, rather than value or process perfection. Thus a methodology like agile manufacturing (AM) is fitting.

Bruce, Daly, and Towers (2004) discuss the perspectives of lean, agile, or leagility (a combination of lean and agile) in the context of the dynamic changes happening in the global apparel industry especially in the context of global sourcing and a higher level of price competition. The researchers emphasize some key characteristics of the apparel industry, including shorter product life cycles, high volatility, low predictability, and high level of impulse purchase, which are making the scenario even more challenging (Bruce et al., 2004).

Küçük and Güner (2014) advocate for adopting AM approaches in the current competitive business environments in this globalized world without any commercial boundaries. There is a need for new processes, methods, ideas to combat fierce competition. AM is competent in achieving the same, as it is an ability to be successful in continuously changing environments. AM is a composition of total quality management, just in time, and lean manufacturing but with a focus on handling smaller quantities catering to individual needs (Küçük & Güner, 2014). The AM can be referred as the fourth stage of manufacturing (Küçük & Güner, 2014) with other three stages of manufacturing being craft-based manufacturing, mass manufacturing, and lean manufacturing (Kasap & Peker, 2009).

14.2 Agile manufacturing

AM represents a unique and intriguing approach to developing a competitive advantage in today's fast-moving marketplace. It places the main focus on rapid response to the customer—turning speed and agility into a key competitive advantage. An agile company is in a much better position to take advantage of short windows of opportunity and fast changes in customer demand (Vorne, 2011).

14.2.1 Agile

The philosophy for agile was initially developed for software development, somewhat similar, or competing with lean. It started as a countermovement to a heavily regulated and inflexible software development approach that was used before, hence the name Agile. At its core is the agile manifest, which resulted in several principles such as "customer focus," "trust," or "communication." One popular tool is used in conjunction with agile. Similar to lean, there is a focus on practical solutions over rigid methods, a focus on working with people, and a focus on value for the customer. One difference is that lean uses more standardization, which is useful if you want to make many identical physical products, whereas agile doesn't have work standards and focuses on flexibly getting results, which is well suited for software development, where every software is different and rigid standards would not help. Another difference is that lean focuses on the production and agile on the product. Agile would be more for the development of a product, lean more for the manufacturing of it (Roser, 2014). According to Naylor, Naim, and Berry (1999), agility means using market knowledge and a virtual corporation to exploit profitable opportunities in a volatile marketplace. According to Gunasekaran et al. (2019), AM is a means of remaining competitive. It focuses on excellence on several competitive objectives and not limited to just achieving quality or low cost. Aiming to address producing leading-edge customized products at the cost of mass manufacturing is one important characteristic of AM. Such products, which are output from an agile enterprise, should provide a competitive advantage as it surpasses customer expectation (Gunasekaran et al., 2019).

14.2.2 Agile manifest

The core philosophy of agile is the agile manifest. It is refreshingly brief, hence listed here in full.

We are uncovering better ways of developing software by doing it and helping others to do it. Through this work we have come to value:

1. individuals and interactions over processes and tools,
2. working software over comprehensive documentation,
3. customer collaboration over contract negotiation, and
4. responding to change over following a plan (Roser, 2014).

14.2.3 Agile manufacturing system

AM systems (AMSs) are characterized by a high level of dynamism, and nonequilibrium functioning regimes are dominating this class of systems. Simultaneously,

Agile manufacturing

AMSs fall in the class of organizational, hierarchical, multifunction, multiple objectives; multiply connected nonlinear dynamic complex systems that consist of subsystems set of different physical nature (structural subdivisions), agents, having own local purposes and containing the man as an active element (Ilyasov, Ismagilova, & Valeeva, 2001). Due to the inherent complexities and agility associated with the modern manufacturing systems, modeling these complex interacting subsystems using common analytical and mathematical approaches has proved to be very difficult (Wang, 2001).

Agility is a quality of timely response to changing conditions. For an organization, it is quickly and successfully adapting to change which may be in any area such as market, regulation, or technological advancement (Vernadat, 2001).

In the year 1991 the Iacocca Institute (in its vision document titled *21st Century Manufacturing Enterprise Strategy* primarily created for the US industry) highlighted the need for an agile enterprise, which could operate efficiently and effectively in a rapid and unpredictable change environment. Such business environments are constantly changing and highly competitive (Bruce et al., 2004). The ultimate objective of the vision document was to encourage a transition from mass production to AM in pursuit of regaining the manufacturing leadership by the US industry (Küçük & Güner, 2014; Nagel & Dove, 1991).

In his book, *Agile Manufacturing: The 21st Century Competitive Strategy*, Gunasekaran (2001) describes AM as the capability of surviving and prospering in a competitive environment of continuous and unpredictable change by reacting quickly and effectively to changing markets, driven by customer-designed products and services. Critical to accomplishing AM are a few enabling technologies such as the standard for the exchange of products, concurrent engineering, virtual manufacturing, component-based hierarchical shop floor control system, and information and communication infrastructure (Gunasekaran, 2001).

According to Christopher and Towill (2001), "Agility is a business-wide capability that embraces organizational structures, information systems, logistics processes and in particular, mindsets." Being flexible is a key characteristic of an agile enterprise. Hence, the roots of agility lie in the concept of flexible manufacturing systems (Christopher & Towill, 2001).

Initially, the manufacturing flexibility was limited in scope, and the key focus was on automation and rapid changeovers only enabling the manufacturing of orders with product mix or volume. In later years the concept of agility (flexibility) was extended to a larger business context (Nagel & Dove, 1991).

14.2.4 The growing popularity of agile in manufacturing

The key reason for its growth in this sector is that it places the consumer in the center stage. An agile company understands the following key characteristics of consumers and fashion:

1. Since a majority of consumers are classified as "fashion followers," it makes sense that their fashion choices are easily influenced by the changes in trends.

2. Every season brings with it fresh trends and fads—including new color pallets, a unique cut or style, blended fabrics.
3. As sophisticated as forecasting algorithms are becoming, they can never accurately truly predict what consumers want. There will always be a need to give them a broad range of choices.

All of the above prompts a hustle in consumers to attain a new wardrobe from time to time, which in turn causes most apparel companies to hustle to keep up with the changing demands. AM has the unique ability to take these complications and turn them into a competitive edge. It's this ability that can help domestic manufacturing in the United States, which would ordinarily be struggling to keep up with the low labor rates provided by overseas manufacturing, to instead emerge successful by focusing on providing what the competitors cannot: speed and quick customization.

14.3   The symbiotic relationship between lean and agile

Both lean and agile were developed as concepts to make manufacturing more efficient and help businesses carve out a competitive edge, even though they take different approaches to do that.

In the early stages of development, and even until this day, both the concepts face a multitude of challenges from traditional manufacturing practitioners. If a company decides to use either of these concepts, the ideal scenario is that such decisions are recognized and taken while the company is still building the manufacturing organization from the ground up—because it's always easier to start something new afresh rather than go back and change something that has been done the same way for years. The best thing about these concepts is how easily they fit into the modern workplace—thanks to their reliance on seamless communication, advanced data analytics, and sense equality amongst hierarchies.

A brilliant way to approach manufacturing is to combine the best parts of both lean and agile and implement them in different departments to increase productivity: lean in the finance and sourcing teams to slash budgets, and agile in the sales and design teams to increase innovation and meet market demands. While such integration will pose some organizational challenges, it will be interesting to see if a company can find that perfect middle ground.

While lean's focus on the reduction and elimination of waste within the factory environment (Ohno, 1988), its application was not extended to other parts of the supply chain where large quantities of the finished product were stock-piled in anticipation of customer orders. It was found that customers would still experience significant delays for delivery of their orders (Fisher, 1997), despite the presence of lean manufacturing facilities in the supply chain where throughput times were being dramatically reduced.

The use of information technology to share data between buyers and suppliers is crucial for agile supply (Harrison, Christopher, & Van Hoek, 1999), which improves visibility of requirements and reduce the amount of stock held in anticipation of predicted and often distorted demand (Hewitt, 1999). While the adoption of lean principles is appropriate for commodity, products where demand can be predicted and agile principles are relevant for innovative products where demand is unpredictable (Childerhouse & Towill, 2000).

14.4 Key advancements needed to support agile in the future

As apparel companies make the move from being complacent manufacturers to reformers, there are a few advancements needed to be made to their working methods.

14.4.1 The digital thread

The digital thread is something that is being used to emphasize that different departments and functions in a company are connected through one common channel—something that is imperative before planning to apply agile (Awad, 2016).

This is also incorporated by the famously AM and retail giant Zara. Zara is one of the only brands of its stature running a largely vertically integrated company, including design, manufacturing, warehousing, and logistics. Although this might not be feasible for all companies, the digital thread can also refer to being connected through one channel to an outsourced company.

14.4.2 Data management and sharing

The key to apply AM is the accurate data. This is must for a precise, accurate, and timely data management and seamless data sharing between different departments of the company and, in some cases, the clients. This is paramount when trying to implement the digital thread, as discussed earlier.

For this, the leadership needs to invest in the development of sophisticated data management platforms and engage managers and employees in comprehensive training programs for the new platforms.

14.4.3 Changing the mindset

Up until now, the process that most apparel manufacturers follow is very straightforward: receive orders from long-standing clients, source material specific to those orders, produce the ordered items, and sell them. While this seems to work for them for the time being, the question is bound to arise—if you are producing garments to the dot, exactly as specified, and adding no other value whatsoever, what is stopping your clients from finding another supplier who can follow orders, at a slightly low price?

The steps involved in an integrative, iterative process are aimed at solving problems:

- Identify the issues.
- Start with small ideas that can be implemented quickly.
- Implement a few ideas in a controlled environment to get an idea about their feasibility.
- Iterate the ideas if needed to reach decisions about which ideas may work.
- If one approach fails, quickly pivot to another line of thinking.
- Scale up the most workable solutions, starting with pilot implementations.

14.4.4 Checking the feasibility of agile

No matter how innovative a concept may seem, it's understandable that there is no one-size-fits-all way to run a business. Thus business leaders spend countless hours strategizing the process before even beginning the implementation phase. Here are some of the questions you should be asking while deciding if agile is the right solution for your company or a specific department in your company:

1. Does at least one of your current products have the potential to be successful as a customized, quick-delivery version of itself?
2. Does your company have the competence to develop a new product, either independently or in collaboration with an external partner, which could potentially benefit from customization and quick delivery?

Even if you can answer one of the above in the affirmative, it doesn't automatically mean that agile needs to be implemented right away. It's still imperative that a comprehensive market, competitive, and financial analysis be done to determine the feasibility of going down the agile path. Agile requires a large amount of investment in terms of both money and time, and unless you are sure about the need for agile, it might be best to wait.

14.5 Lean versus agile

Lean and agile are often lumped together as approaches to increase a firm's efficiency, but they have some inherent differences that should be kept in mind before they are considered as a solution.

14.5.1 Prime integrant

Lean manufacturing is driven by the market and the waste economy. While the market has a definite effect on the trends of production, the major driving force here is the mindset that waste needs to be minimized. It can be used in cases where the end products are predictable to a certain extent, and the assembly line is expected to stay the same for a predetermined amount of time.

Agile, on the other hand, is determined by the customer and the diverse economy. This is driven by the mindset that the company needs to be highly adaptable to the changing needs of the consumer and have the flexibility to meet the consumer demands quickly and efficiently. Thus agile manufacturers accept that it's impossible to predict the end products and work accordingly.

14.5.2 Cornerstones

To successfully implement lean the focus of a company and employees needs to be technology and systems, while for agile, it needs to be human and data. It's true that lean aims to take the functioning of a department to the level of a machine, removing the human error factor. It achieves this by providing clear definitions of each aspect of production—first by assessing whether each process is value or waste, then by creating comprehensive value stream maps and describing how the production is to flow, and finally by implementing technologies such as invoice digitization and automated workflow systems. All these definitions play a part in standardizing the workflow to the dot.

Agile is the opposite—as it needs the human touch as a key part of the system, to quickly grasp the market demand and incorporate it as required. To do that, the data flow is paramount, as is the ability to quickly respond to the data and make decisions that are unique to that exact data point. For example, a manufacturer may realize that there is a particular fabric gaining popularity in a particular area of their target market, and thus the decision may be made to source and produce garments using this fabric as a test, which later may or may not become part of the final portfolio. Such trials and errors are not only forgivable but encouraged in an agile environment also.

14.5.3 External vendors

Whenever a company decides to implement a new system or even make a change to an existing system, it causes a change to not only the internal partners but the external ones as well. External vendors either suffer or rejoice as a result of such changes.

In the case of lean, since we apply a minimalist approach, the number of suppliers is decreased, and the different categories as consolidated as is viable. The chosen few suppliers who remain with the company on its journey toward lean are highly reliable and, usually, have a long history with the company already. They are also expected to provide a high service level.

In the case of agile, there is a need to have a high number of vendors in the mix who often have distinct strengths. The vendors are still expected to be highly reliable, but it's also understood that the relationship between the company and vendor is likely going to be on a short-term basis because changing needs can make a vendor redundant.

14.5.4 Corporate structure and values

A company that wants to apply lean usually needs to apply a top-down structure where the power lies at the higher level of the organization, but a collaborative and teamwork focused culture within the departments is also required there. The workers need to be trained well in lean methodologies before beginning work and then are expected to continue doing what they learned repetitively, as is planned by the executives.

Agile companies put a lot of focus on hiring a talented and motivated workforce because to successfully implement agile, workers at every stage of the process need to be equally invested. Since all stages carry equal importance, they are also given sufficient power to make decisions that will result in the best outcome for their process.

14.5.5 Product

In the case of lean, since the trends are created at the executive level and information largely flows just one way, it makes sense that most of the companies implementing lean are powerhouses—big brands or large retail chains. Also, there is a focus on creating high-quality products—which is not difficult to achieve, since the workers work on the same designs for longer periods.

Agile companies place more focus on expediting every request since the information flow is multiway: first from the customers to the company, then through the departments, then finally to the customers again to advertise the newest creations.

Mason-Jones, Naylor, and Towill (2000) compare lean and agile approaches for apparel supply chains. The agile approach is suitable for fashion goods, which are the products with shorter product life cycles, relatively smaller qualities with volatile market demand (Mason-Jones et al., 2000). Further, the availability of such fashion goods is responsible for the customer drive. This makes customers ready to pay a premium price for the product. The demand patterns of such products are generally unpredictable and may not be dependent on past sales data. Forecast for such products is generally derived in a consultative manner. Because of the same, as indicated by Fisher (1997), it has to be based on an intelligent consultation with inputs from rich marketplace insider sources (Fisher, 1997).

Although agile is perceived as external while lean as internal supply chain, a comparison study of factory lead time (arrival of customer demand to shipment) between two manufacturing enterprise, one following lean and another agile, shows that agile company took less than one-third of the time and around 60% of the operator to complete the production of a personalized shirt (Küçük & Güner, 2014). Table 14.1 shows the distinguishing attributes and elements of lean and agile supply.

14.6 Agile organization

Advocates of agile often say "The only time people in a company should be working in silos is if they are *building* a silo!" On a more serious note, it's true that

Agile manufacturing

Table 14.1 Attributes and elements of lean and agile supply (adapted version from Mason-Jones et al., 2000; Küçük and Güner, 2014).

Differentiating attributes/elements	Lean	Agile
Driving elements	• Market • Waste economy • Predictable products • Conjecture production	• Customer • Diversity economy • Unpredictable products • Job shop production
Focal point	Technology and systems	Human and data
Suppliers	• Less • High-level reliability • Long-term relationship • Collaborator • Buying good	• Multiple choice • High-level reliability • Short-term relationship • Sharing risk/profit • Assigning capacity
Organization	• Teamwork • Low organization chart	• Talented • Authorized
Product	• High possibility to choose • High quality • Low variability • Long life cycle • Commodities	• Convenient to individual requests • Expedient • High variability • Short life cycle • Fashion goods
Process	• Flexible • Automatic • Low-profit margin • Physical costs • Forecasting is algorithmic • Information enrichment is highly desirable	• Fit to different conditions • Knowledge-based • High-profit margin • Marketability costs • Forecasting is consultative • Information enrichment is obligatory
Philosophy	Executive	Leadership

collaboration and cooperation, while often talked about, are seldom implemented in the apparel manufacturing world.

14.6.1 Reasons for a low emphasis on collaboration

Some reasons for a low emphasis on collaboration are:

1. *Low margins in the apparel industry:* As a whole, the manufacturing industry operates with very low margins and very high-pressure situations. This causes all the employees to

focus their energy on just the tasks ahead of them and not have the bandwidth to interact with any other department.

2. *Interdepartmental conflicts:* Most of the departments in an apparel manufacturing company function under the assumption that they are working against each other, and someone else's wins will be their losses.

For example, the design team would generally prefer using unique kinds of trims and accessories, things that stand out, and act as a platform for their creativity. This may make things either more difficult or easier for the marketing team, based on the difficulty of finding a target audience for the new designs. The buying teams will probably have a hard time sourcing the particular trims in larger quantities at a feasible price. The manufacturing team will need to train the employees since they would be working with a new product.

3. *Focus on hierarchy:* Most apparel companies follow a strict system of hierarchy, where ideas always flow from top to bottom. That isn't a system where brainstorming is encouraged because most superiors will not be used to listen that their ideas are wrong.

Thus, if a company decides to use agile, there are a lot of steps that are needed to be taken, not just in the process but in the core mindset and culture of the company.

14.6.2 Ways agile can help a company improve its culture and be more innovative

Some of the ways agile can help a company improve its culture and be more innovative are:

1. *Identifying and removing excess work:* A lot of times, because two teams are not well integrated, there are many instances of duplicate work and gray areas of responsibility. When teams start talking to each other, the process keeps becoming clearer, which also helps establish a clearer division of responsibility. It also increases the level of accountability between people, as what you are doing affects more people.
2. *Helps employees in realizing their value:* The worst thing about working in a silo is that employees lose motivation easily and start thinking about their work as meaningless. When you connect teams, it becomes easy to realize that all the employees are working toward the same company goal, and each of them has a part to play.
3. *Integrate multiple tools and tasks into one:* Agile focuses on using a single, or as few as possible, sources of data sharing between teams, thus making employees' lives easier and decreasing the amount of time they spend checking their emails and other tools for information.

14.6.3 Is bringing outside help imperative to apply agile concepts?

When people start thinking about applying concepts such as lean and agile, they automatically think that they need a large group of consultants to come, evaluate the company, and offer credible solutions. But that need not always be the best option. We have to think of agile not as the implementation of a new process, but as changing the core of how manufacturing decisions are made and implemented.

Agile manufacturing 411

For this reason, it makes sense to invest in forming an in-house team of people who are trained in AM. This team should ideally be a cross-functional mix of people from fields such as data science, finance, design, operations, and supply chain.

14.7 Strategic tools for future agile firms

14.7.1 Understand that clients are more important to the business than investors

Amazon's Chief Executive Officer (CEO), Jeff Bezos, famously declared in his letter to investors in 1997 that investor worth should be the outcome of their efforts, not the objective. This holds for almost all industries, as long as you focus on delivering value to the end customers, the worth of your company will keep growing.

CEO Bezos uses symbolic acts to reinforce customer obsession, such as the famous "empty chair": Bezos is known to leave one seat open at a conference table to inform all attendees that they should consider that seat occupied by their customer, "the most important person in the room." Ultimately, these symbolic gestures are highly calculated and often repeated messages created to reinforce Amazon's leadership principles.

14.7.2 Performance metrics centered around clients

To drive customer fixation, it's important to incorporate customer-centric measurement tools in daily metrics. For example, in most apparel companies, budgeting is done intensely, so all money-related information is stored diligently and is accessible between departments as needed. At the same time, client information is either absent or inconsistent. The firms that want to survive agile implementation need to isolate themselves from a strictly money centered attitude.

14.7.3 Activities report to the association, not the immediate department

In most enormous associations, without client fixation, one can undoubtedly envision that an arranging procedure would rapidly decline into a fight between units battling for assets for "their" exercises and "their" units. This should be prevented at all costs, by harboring a single line reporting structure.

14.7.4 Budgeting is performed by the association, not at the single department level

Often what makes conventional planning so disappointing is that everybody realizes that substantive issues about reason and needs are not being tended to in the budgeting talk: the budgeting fights are an intermediary for uncertain power battles between contending units and the dissimilar perspectives of ranking directors.

14.7.5 Work should be done in small groups working short cycles

As in all adept administrations, most work should be performed by independent multi-disciplinary groups. Such groups should ideally work like semiautonomous innovative nurseries. Protected from the more prominent association's administration, small groups support driven pioneers, give opportunity, and impart a feeling of proprietorship.

14.7.6 Connection to HR

The issues of customary planning are regularly bothered by the company's HR framework, which may reflect and strengthen the siloed and various leveled structure and spending plan, in which client results and effects are not completely in the image. Hence, people inside each authoritative storehouse will, in general, be assessed and remunerated as far as delivering the planned degree of yields inside the given spending envelope for their unit.

Pay structure at agile firms should be centered around incentivizing work done toward the long haul—the end goal that everyone is trying to accomplish.

14.8 New horizons for lean manufacturing

Lean is hardly stationary. Indeed, as senior managers' understanding of lean continues to develop, we expect it to further permeate service environments around the world. In the past few years alone, we've observed lean's successful application to mortgage processing in India, customer-experience improvements in a Colombian pension fund, better and faster processing of political-asylum requests in Sweden, and the streamlining of business services in the United Arab Emirates.

In the years ahead, service and product companies alike will increasingly be able to reach their long-term goal of eliminating waste as defined directly by customers across their entire life cycle or journey with a company. For example, an unprecedented amount of product-performance data is now available through machine telematics. These small data sensors monitor installed equipment in the field and give companies insights into how and where products are used, how they perform, the conditions they experience, and how and why they break down. Various aerospace and industrial-equipment companies are starting to tap into this information. They are learning directly from customer experience with their products—about issues such as the reliability of giant marine engines and mining equipment or the fuel efficiency of highway trucks in different types of weather.

The next step is to link this information back to product design and marketing, for example, by tailoring variations in products to the precise environmental conditions in which customers use them. Savvy companies will use the data to show customers evidence of unmet needs they may not even be aware of and to eliminate product or service capabilities that aren't useful to them. Applying lean techniques to all these new insights arising at the interface of marketing, product development, and operations should enable companies to make new strides in delighting their customers and boosting productivity.

Information about customers won't be coming only from sensors and databases. The understanding of what makes people tick has been improving dramatically; companies are starting, more and more, to apply psychology to their operations. Disney, for example, recognized that visitors in its theme parks respond to different emotional cues at different times of the day and embedded this realization into its operations in precise ways. In the morning, for example, Disney employees are encouraged to communicate in a more inspirational style, which resonates with eager families just starting their day at the park. In the late afternoon (when children are tired and nerves become frayed), employees aim for a more calming and supportive style of communication. The integration of these psychological insights with Disney's operating philosophy allows the company to eliminate waste of a different sort: employee behavior that would not be desired by customers and might inadvertently alienate them at certain times of the day.

Finally, market- and consumer-insight tools (for instance, statistically based regression analysis, as well as advanced pricing- and financial-modeling tools) are creating a far more sophisticated (and much closer to real-time) view of what customers value. The changes may just be getting started. Better-integrated datasets across channels and touchpoints are rapidly enabling companies to get much more complete views of all interactions with customers during the journeys they take as they evaluate, buy, consume, and seek support for products and services. Usage patterns of mobile devices and services are painting a richer picture than companies previously enjoyed.

The result should be more scientific insight into how product and service attributes contribute to customer value; new ways to look at what matters most for classic lean variables, such as lead time, cost, quality, responsiveness, flexibility, and reliability; and new opportunities for cross-functional problem solving to eliminate anything that strays from the customer-defined value (Duncan & Ritten, 2014).

Lean standards have started to assume a key job in many different enterprises beyond just manufacturing, enterprises that are currently looking to grow and adjust to the rapid advancements in technology and markets of the Internet age. As organizations change from the mechanical economy to an advanced one, lean ways of thinking are impacting almost all parts of their business. Lean's central idea is of increasing the focus on client's needs while decreasing the amount of waste produced fitting to our times, as there is a growing need for associations to better understand what customers truly value. In addition, companies need to organize their work activities to efficiently develop and deliver the right products and services and continuously improve customer value and efficiency based on actual marketplace feedback. This "pull" approach is markedly different from the "push" approaches of the past.

Inside the mechanical economy of the past, firms were generally sorted out around push draws near, which appeared to be the correct method to arrange the quickly developing modern age organizations and ventures. While open and private foundations were encountering significant changes, those changes were generally little and unsurprising.

The greatest test was to deal with the developing methods for manufacturing in the most productive manner conceivable. Establishments grasped top-down authoritative structures to help scale their production of products and ventures. This served them well in a moderately reliable reality where similar activities yielded similar outcomes, and models could make generally exact forecasts.

However, the push economy is vanishing amid our complex and regularly advancing world. The present world, with its huge range of segments and their mind-boggling examples of interconnectivity, displays every one of the properties of dynamic, complex frameworks, including flighty and random practices.

An organization that needs to stand separated from its rivals can do so through one of two key ways: by giving a predominant client experience or by offering the most minimal costs. For organizations that go for the former, advanced innovations are the best methods for connecting with clients and giving them an unrivaled at focused expenses.

Nevertheless, providing such exceptional service to clients who are themselves becoming increasingly empowered and digital is getting more grueling. New advancements are hitting the market quicker than any time in recent memory, brand dedication keeps on diminishing and the challenge of moving control from the institution to the individual is growing. Customers have more choices than they did at any time in the past for not only products but the platforms they can use to acquire those products. Also, clients are exploiting all the data they can now effectively access to scan for and distinguish the most ideal qualities. Access to such a significant number of choices puts weight on organizations. A couple of driving, forefront organizations can keep up; however, a larger chunk of organizations is trailing behind. While working more earnestly to turn out to be increasingly proficient and unsurprising, they keep on attempting to actualize innovation into old plans of action. This methodology may have functioned admirably in the environment of the modern economy, yet it won't prevail in the regularly changing advanced economy.

Organizations need to focus more on their digital transformation to keep up with the needs of their inexorably computerized clients. To enable them to do as such, we are seeing lean standards connected to the administration of existing organizations just as new businesses. The lean management enterprise, an ongoing report by the McKinsey Global Institute, records four lean administration controls drilled by the best performing associations they've worked with throughout the years.

"Conveying esteem proficiently to the client" is the principal discipline. The association must begin by understanding what clients genuinely esteem—and where, when, how, and why too, the report states. At that point, it includes three extra administration disciplines:

1. *Empowering individuals to lead and add to their fullest potential:* empower workers to claim their improvement without disregarding them to find it.

This one might be the most difficult in an apparel company because the leadership is so used to the top-down approach that they will probably not be comfortable taking advice from those beneath them. This is where lean leadership comes in and bridges the apparent gap between different hierarchies, by explaining why a collaborative approach is the only way to implement lean.

2. *Finding better methods for working:* the whole company should persistently consider how the present techniques for functioning and overseeing could be improved.

Lean goes hand in hand with continuous improvement. Apparel companies need to move away from the idea that "we only need to fix what's broken" and start thinking along the lines of "this is going well, how can I make it go better?". The generation of such ideas cannot happen solely in the offices of a few authorities. The entire shop floor needs to be involved and invested in making things better.

3. *Connecting strategies, objectives, and meaningful purpose:* keep up an advancing vision of what the company is about, which shapes techniques and goals in manners that offer significance to your everyday work.

This is a golden rule, not just for companies looking to implement lean but for all companies in the present day. People are no longer content just keeping their heads down and doing what they're told, and jobs are no longer simply about money. People are looking to find the purpose behind what they do every day and companies that provide that successfully are the companies that will eventually thrive.

Lean methods of reasoning underline the idea of nonstop improvement, a progressing exertion planned for accomplishing steady changes that, over the long haul, will fundamentally improve items, administrations, and procedures just as the general administration of the association. This endless cycle of trading, testing, and adapting maybe best represents the developing significance of lean standards in the advanced economy (White, 2014).

Bruce et al. (2004) discuss the approach of combining lean and agile, termed as "leagile." The leagile approach advocates the combination of lean and agile concepts at a decoupling point to optimize the supply chain (Bruce et al., 2004). Several researchers have discussed the need of the leagile supply chains (Christopher & Towill, 2001; Martin & Towill, 2000; Naylor et al., 1999; Van Hoek, 2000).

According to Mason-Jones et al. (2000), agility should be applied downstream while the lean approach should be applied upstream from the decoupling point in a supply chain. This may enable improved results in terms of cost-effectiveness as well as improved service (Mason-Jones et al., 2000).

According to Mason-Jones et al. (2000), agility cannot be achieved without experiencing lean. Both lean and agile concepts share a significant amount of similarities. Agile practices based on a solid foundation with lean may help in steepening the learning curve. Further, an agile supply—based transformation requires mastery of each of the processes in the supply chain, and that requires a great deal of process enhancement (achieving excellence). Adopting lean practices surely helps in achieving excellence in the industry environment. A combination of lean and agile (leagile) may result in great yields, as lean focuses on total waste reduction while agile focuses on total flexibility.

Mason-Jones et al. (2000) suggest a model of leagile supply chain. Availability of right information at the right time (effective information-based outputs regarding when the next deliveries to be pulled out from the suppliers) is vital for an agile supply chain, as it helps in enabling the supply chain in quick and accurate response and meeting the demands at very short notice.

To execute the demand orders (which have received at a short notice and need to be delivered with a shorter lead time), the processes need to be capable enough to perform in a rapid response mode. Here, at this decoupling point, the role of lean becomes vital. A lean process that is based on industrial engineering and operations engineering provides a much-needed push to meet the demand in a quick response manner. Here it can be observed that in the execution part lean approach plays a dominant role and agile remains in a supportive role, while in assessing and responding the demand roles of lean and agile become supportive and dominant respectively. Refer to Table 14.2 for an adapted version of the same.

Table 14.2 Leagile supply chain (adapted version from Mason-Jones et al., 2000).

Strategy	Techniques	Role	
		Lean	**Agile**
Industrial engineering interventions	Quick-change over (reduction in setup time or Rapid setup using Single Minute Exchange of Die (SMED) Method improvements (method study and method reengineering) Improvements in product design (reengineering)—Design for manufacture	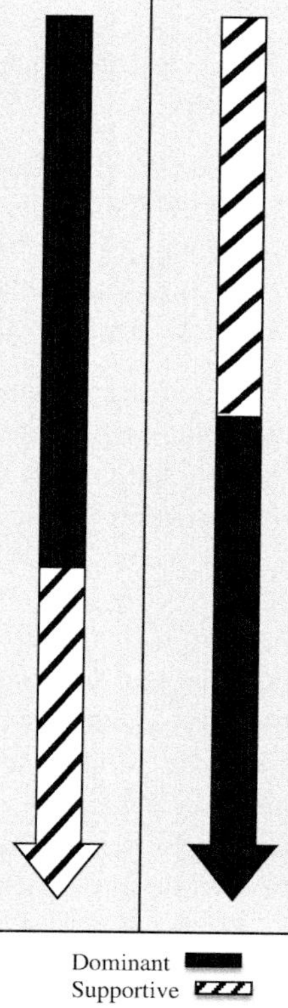	
Operations engineering interventions	Kanban (pull-based demand driven flow) JIT (quick replenishment with minimum or no waiting time)—ability to handle smaller order quantities with variations Shared call off information—accuracy in decision making		
Information technology interventions	Quick and accurate data capturing (enabling quick response supply chain) —promoting technology-based quick solutions Electronic data transfer (EDI)—paperless practices for automatic ordering and replenishment		
Production engineering interventions	Process integration Process sequencing Alternative manufacture		

Dominant
Supportive

Agile manufacturing 417

Combining the best practices of lean and agile philosophies to optimize the efficiency and effectiveness of the supply chains, a road map for the leagile supply chain is shown in Table 14.3.

Van Hoek (2000) also confirmed that in the leagility approach, lean capabilities may contribute to agile performance. Lean and agile are not contradicting each other but complementary, hence don't challenge each other. For an agile supply chain, being lean is a prerequisite (Van Hoek, 2000). Further, it is emphasized that waste elimination is a key feature of a lean approach if responsiveness is incorporated (maybe in the form of mass customization, refer Fig. 14.1), it may result in an agile system. The popular business models that are based on make-to-measure or customized tailoring may be categories as AM models. It is indeed an effort to combine the operational efficiency with responsiveness in a supply chain.

Bruce et al. (2004) advocate the need for leagile supply chains as it helps in effective sequencing the supplies, and efficient manufacturing process to reduce lead times. Such supply chains focus on improved customer-order-demand management with the elimination of waste. This approach is very relevant to the fashion and apparel industry which is highly competitive and generally involves small businesses with limited resource availability (Bruce et al., 2004).

Table 14.3 Road map for a leagile supply chain (adapted version from Mason-Jones et al., 2000).

Market knowledge	Supply-chain knowledge	Optimize for leanness agility	Result	
• Identify product demand variability • Identify product variability • Identify the point of differentiation • Identify lead-time requirements	• Integrate supply-chain material flow • Integrate supply-chain information flow • Strategic positioning of decoupling point • Lead-time compression	• Eliminate waste • Maximize flexibility without addition to waste	• Profit maximization with minimum cost • Enhanced customer service and satisfaction	Lean
		• Design for total flexibility • Minimize waste without restricting flexibility	• Maximized profit and customer service and satisfaction at minimum cost in the volatile environment	Agile

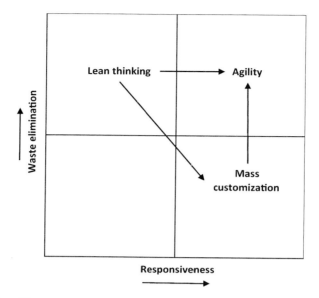

Figure 14.1 Leagility approach (Van Hoek, 2000).

14.9 Case study

14.9.1 Agility: an unlikely savior in trying times

The year 2020 has brought upon unprecedented challenges globally in terms of a pandemic and failing economy. The apparel industry has been hard hit from the beginning since the pandemic initially started growing in the Western countries and resulted in immediate and long-lasting shutdowns of multiple countries in Europe and North America. These countries happen to be the biggest buyers of garments from countries such as India, Bangladesh, and Sri Lanka. The sudden decrease in demand for garments caused a rippling effect and resulted in canceled orders, factory shutdowns, and massive layoffs in many Asian countries as well.

This pandemic exposed a huge inadequacy of agility in the apparel sector. It also exposed how brittle our medical supply chains are. During the COVID-19 pandemic, frontline healthcare workers urgently need more supplies of personal protective equipment (PPE) to keep themselves and their patients safe. While the shortage of face masks, face shields, and hospital gowns had been widely reported, it is no small task for companies to shift gears and manufacture products they don't typically offer. The growing demand for PPE quickly caused supply shortages, price gouging, and left the medical industry in distress.

Fortunately, there were at least some companies who were able to turn things around for themselves and offer much-needed help to the medical industry. Williamson-Dickie Mfg. Co., an apparel manufacturing company based in the United States, is primarily known for its largest brand, Dickies. Answering the call for PPE shortage, the legacy workwear brand Dickies worked for weeks to convert manufacturing facilities in Mexico and Honduras to begin producing FDA-compliant medical isolation gowns. It started with plans to deliver 50,000 gowns to hospitals in the first month and build the capacity to produce up to 3.4 million gowns within a few months (Verdi, 2020).

Agility was a huge factor in making this happen. The brand had one huge advantage in its product mix and worker adaptability. Since it had multiple key products at any given time, the idea of switching to a new product quickly was not unfamiliar to the sourcing teams and the operators. Alix Reyes Martin, Senior Director, Engineering, VF Supply Chain Americas, and the project lead for the isolation gown project, said the process of converting manufacturing during the COVID-19 pandemic had been "a bit of a rollercoaster," with teams in the United States and Latin America working long hours to manage the logistics to switch over manufacturing and establish protocols for factory safety. "One of the virtues of workwear is the great variety of product categories we make. All of our facilities are flexible and can produce different types of products," Reyes said. "We were able to re-balance our capacity and re-purpose machines to build isolation gown production lines at each of our factories."

14.10 Conclusion

According to Professor Nick Vyas, from University of Southern California, Marshall School of Business, here are some of the ways we can see lean manufacturing evolve in the coming years (Vyas, 2020):

1. Up till now, lean manufacturing was going from a closed loop to an open-loop system. We decided to integrate partners, suppliers, to the concept of production planning. It created a collaborative environment for everyone. For example, automatically points were created based on general inventory levels fluctuations.
2. Where is this going to go? Emerging technology has evolved so greatly that it's almost impossible to predict. Let's think about this in terms of tiers, where tier 1 contains suppliers who are directly in contact with the manufacturers, while tier 5 are the suppliers who have four middlemen between them and the original manufacturer. Technologies such as artificial intelligence, machine learning, blockchain truly have the power to integrate tier 1, tier 2 manufacturing networks, and can even penetrate tier 5. Lean is going to be heavily integrated with emerging technology.
3. The fate of lean is energizing. Lean tools and techniques are instrumental in killing the waste and for increasing the value of the end product. The clients describe it in the terms of colossal gains in the volume and nature of the data. To collect such an information or the data about the product, the organizations can accumulate about client conduct, the estimation of the marketing bits of knowledge, and the complexity of the mental bits of knowledge applied according to the client's needs and wants. These advances carry new importance to the exemplary lean proverb "figuring out how to see." The complexity between where organizations are presently and where they'll be 20 years on will appear as distinct as the contrast between a static shading photo and superior quality, three-dimensional video.

To remain competitive and survive, customer focus is a must. Customer focus has to be the ultimate aim and the center of the strategy, irrespective of the type of product or services we are catering to, or the methods or processes we are dealing with. Also, it is not only limited to manufacturing facilities or dealing with the suppliers but also has a wider scope with a coverage of the entire organization and its surrounding businesses (Sharifi & Zhang, 2001). In the context of the fashion and apparel industry, customer focus becomes even more vital where more varieties with low-order quantities are in practice these days. Further, it is worth mentioning that trend of competing with quality and cost has been replaced with competing through speed, service, and innovation (Christopher, 2000). In such a scenario, becoming an agile enterprise is the most suitable strategy.

It's time to move from "Best Practice" to "Next Practice"

References

Awad, Y. (2016). *Agile manufacturing: Not the oxymoron you might think*. InfoQ. Retrieved from https://www.infoq.com/articles/agile-manufacturing-not-oxymoron/. (Accessed November 2019).

Bruce, M., Daly, L., & Towers, N. (2004). Lean or agile: A solution for supply chain management in the textiles and clothing industry? *International Journal of Operations & Production Management, 24*(2), 151–170.

Childerhouse, P., & Towill, D. (2000). Engineering supply chains to match customer requirements. *Logistics Information Management, 13*(6), 337–345.

Christopher, M. (2000). The agile supply chain: Competing in volatile markets. *Industrial Marketing Management, 29*(1), 37–44.

Christopher, M., & Towill, D. (2001). An integrated model for the design of agile supply chains. *International Journal of Physical Distribution & Logistics Management, 31*(4), 235–246.

Duncan, E., & Ritten, R. (2014). *Next frontiers for lean*. McKinsey & Company Retrieved from. Available from https://www.mckinsey.com/business-functions/operations/our-insights/next-frontiers-for-lean#. (Accessed December 2019).

Fisher, M. (1997). What is the right supply chain for your product. *Harvard Business Review, 75*(2), 105–116.

Gunasekaran, A. (2001). *Agile manufacturing: The 21st century competitive strategy*. Elsevier.

Gunasekaran, A., Yusug, Y. Y., Adeleye, E. O., Papadopoulos, T., Kovvuri, D., & Geyi, D. G. (2019). Agile manufacturing: An evolutionary review of practices. *International Journal of Production Research, 57*. (15–16: Special issue: Selected surveys on cutting-edge problems in production research), 5154-5174.

Harrison, A., Christopher, M., & Van Hoek, R. (1999). *Creating the agile supply chain*. Cranfield: School of Management Working Paper, Cranfield University.

Hewitt, F. (1999). Supply or demand? Chains or pipelines? Co-ordination or control? In: *Proceedings from international symposium in the information age* (pp. 785 – 90). Florence.

Howard, W. S. (2009). *An overview of lean manufacturing*. Stability Technology. Retrieved from https://www.stabilitytech.com/lean.html. (Accessed January 2020).

Ilyasov, B., Ismagilova, L., & Valeeva, R. (2001). The control problems of agile manufacturing. In A. Gunasekaran (Ed.), *Agile manufacturing: A 21st century competitive strategy* (pp. 559−581). Elsevier Science.

Kasap, G. C., & Peker, D. (2009). Agile manufacturing: A research of presenting the agility of a primary automobile industry organization. *Elektronik Sosyal Bilimler Dergisi, 8*(27), 57−78.

Küçük, M., & Güner, M. (2014). Agile manufacturing for an apparel product. In: *XIIIth international Izmir textile and apparel symposium* (pp. 343−352).

Martin, C., & Towill, D. R. (2000). Supply chain migration from lean and functional to agile and customised. *Supply Chain Management, 5*(4), 206−213.

Mason-Jones, R., Naylor, B., & Towill, D. R. (2000). Engineering the leagile supply chain. *International Journal of Agile Management Systems, 2*(1), 54−62.

Nagel, R., & Dove, R. (1991). *21st Century manufacturing enterprise strategy.* Iacocca Institute, Leigh University.

Naylor, J. B., Naim, M. M., & Berry, D. (1999). Leagility: Integrating the lean and agile manufacturing paradigms in the total supply chain. *International Journal of Production Economics, 62*(1−2), 1070−1118.

Ohno, T. (1988). *The Toyota productive system: Beyond large scale production.* Portland: Productivity Press.

Roser, C. (2014). Facing change in modern manufacturing systems — The difference between flexible, agile, reconfigurable, robust, and adaptable manufacturing systems. AllAboutLean. Retrieved from https://www.allaboutlean.com/change-in-manufacturing/. (Accessed December 2019).

Sharifi, H., & Zhang, Z. (2001). Agile manufacturing in practice: Application of a methodology. *International Journal of Operations & Production Management, 21*(5−6), 772−794.

Van Hoek, R. I. (2000). The thesis of leagility revisited. *International Journal of Agile Management Systems, 2*(3), 196−201.

Verdi, K. (2020). *Dickies®/VF corporation to produce 3.4 million isolation gowns in support of US COVID-19 response.* Business Wire. Retrieved from https://www.businesswire.com/news/home/20200416005116/en/Dickies%C2%AE-VF-Corporation-Produce-3.4-Million-Isolation. (Accessed May 2020).

Vernadat, F. B. (2001). Enterprise integration and management in agile organizations. In G. A. (Ed.), *Agile manufacturing: The 21st century competitive strategy* (pp. 461−481). Elsevier.

Vorne. (2011). Agile manufacturing. Vorne Industries Inc., LeanProduction. Retrieved from https://www.leanproduction.com/agile-manufacturing.html. (Accessed February 2020).

Vyas, N. (2020). *Evolution of lean manufacturing.* (R. Malviya, Interviewer).

Wang, K. (2001). Computational intelligence in agile manufacturing engineering. In A. Gunasekaran (Ed.), *Agile Manufacturing: The 21st century competitive strategy* (pp. 297−315). Elsevier.

White, G. (2014). The future of lean manufacturing. (BizClik Media Limited). Manufacturing Global. Retrieved from https://www.manufacturingglobal.com/lean-manufacturing/future-lean-manufacturing. (Accessed March 2020).

Index

Note: Page numbers followed by "*f*" and "*t*" refer to figures and tables, respectively.

A

A-B control, 18
A3
  problem-solving, 107−112
  report, 17
  sheet, 107, 108*f*
Acceptable quality level (AQL), 322−325,
    323*t*
  acceptance sampling plan, 323−325
  ensuring success of acceptance sampling
    plans, 325
  Pareto analysis of defects, 326*f*
Additive manufacturing (AM), 366−368
Affinity diagram, 18, 195−198, 197*f*
Agile manufacturing (AM), 306, 401−404
Agile manufacturing systems (AMS), 18,
    402−403
Agile/agility, 402−411, 418−419
AM. *See* Additive manufacturing (AM);
    Agile manufacturing (AM); Autonomous
    maintenance (AM)
AMS. *See* Agile manufacturing systems
    (AMS)
Analytical Hierarchical process, 372−373
Andon, 5, 18−19, 182−188, 233
  Andon boards, 185, 186*f*
  Andon cords, 183, 395
  Andon switch, 183, 184*f*
ANN. *See* Artificial neural network (ANN)
AQL. *See* Acceptable quality level (AQL)
AR. *See* Augmented Reality (AR)
Artificial neural network (ANN), 370
Assembly cell, 19
Assembly line, 295−297
Augmented Reality (AR), 366−368
Automation, 233, 234*t*
Autonomation, 233, 234*t*
Autonomous maintenance (AM), 19, 356

B

Back Propagation algorithm, 370
Baka yoke, 19
Barcode/RFID scanning, 248, 248*f*
Batch setting time (BST), 217
Batch-and-push levels, 19
Batch-and-queue, 19
Baton touch, 20
Blitz (problem-solving tool), 81
Bottleneck, 20
"Bouncing line". *See* Bucket brigade
BST. *See* Batch setting time (BST)
Bucket brigade, 20, 300−305, 301*f*
Buffer, 20
Built-in-quality, 21
Bullwhip effect, 21, 257−258, 382−384
"Bump-back". *See* Bucket brigade

C

Cause-and-effect analysis. *See* Ishikawa
    diagram
Cellular manufacturing, 21, 298
Chaku-Chaku system, 21−22, 300−305,
    303*f*
Chalk Circle, 22
Changeover, 212−215
Changeover time (COT), 22, 211−212
Circle exercise, 22
Clean−Lubricate−Inspect−Tighten (CLIT),
    365
Collaborative product development (CPD),
    382, 386
Constant Work In Process (CONWIP), 22,
    287
Constraint management, 273−275
Contact method, 243−244
Continuous flow, 22
  manufacturing, 259−273

Index

Continuous improvements. *See* Kaizen
Continuous quality improvement (CQI), 316
Conveyance Kanban. *See* Withdrawal Kanban
CONWIP. *See* Constant Work In Process (CONWIP)
COT. *See* Changeover time (COT)
CPD. *See* Collaborative product development (CPD)
CPS. *See* Cyber-physical system (CPS)
CQI. *See* Continuous quality improvement (CQI)
Customer Takt, 23
Cyber-physical system (CPS), 373
Cycle time, 23, 261−265

D

Dandorigae, 23
Dashboard, 23, 198−204, 199*f*
DBR system. *See* Drum−buffer−rope system (DBR system)
Decoupling point, 381
Defects per million (DPM), 24, 326
Define, measure, analyze, improve, and control (DMAIC), 24, 119−121, 314
Delivery-quality-cost (DQC), 21, 24
Demand amplification, 382−384
Design-led changeover, 212
Digital thread, 405
DMAIC. *See* Define, measure, analyze, improve, and control (DMAIC)
Downsizing, 338−339
Downtime, 24
DPM. *See* Defects per million (DPM)
DQC. *See* Delivery-quality-cost (DQC)
Drum−buffer−rope system (DBR system), 23−24, 258, 273−274
  implementation, 274−275, 274*f*

E

E-commerce, lean in, 394−395
Economic order quantity (EOQ), 25
Eliminate-combine-rearrange-simplify (ECRS), 25
Eliminate, Simplify, Externalize, and Execute model (ESEE model), 214−215
EOQ. *See* Economic order quantity (EOQ)
ESEE. *See* Eliminate, Simplify, Externalize, and Execute model (ESEE model)

F

Failure modes and effects analysis (FMEA), 26, 326
FAS. *See* Flexible Assembly System (FAS)
FIFO principle. *See* First-in-first-out principle (FIFO principle)
First time right (FTR). *See* First time through (FTT)
First time through (FTT), 19, 25, 233, 240, 313−314
First-in-first-out principle (FIFO principle), 25, 100−101, 268
First-time pass yield, 101, 101*t*, 103, 106*t*
First-time-yield (FTY), 313
Fishbone diagram. *See* Ishikawa diagram
Five Ss (5Ss), 25−26, 156−181, 169*f*
5-Why analysis, 122, 124−126
Flexible Assembly System (FAS), 360−361
Flexible Manufacturing System (FMS), 360−361
Flow management (FM), 4, 279
Flow manufacturing, 261−266, 262*t*, 267*t*, 271*f*
Flow shop, 26
FM. *See* Flow management (FM)
FMEA. *See* Failure modes and effects analysis (FMEA)
FMS. *See* Flexible Manufacturing System (FMS)
Fordism, 26
Four M (4M), 26
4P model. *See* People, Practice, Process, and Product model (4P model)
FTT. *See* First time through (FTT)
FTY. *See* First-time-yield (FTY)
Fuzzy logic, 373

G

Gantt chart, 27
Gemba, 27, 133−137
Gemba walk, 134−137
Genchi Genbutsu, 27
Genetic algorithm, 370
Genjitsu, 27

H

Hanedashi, 27
Hansei, 28
Harada method, 28, 346−350

Index 425

Heijunka, 28, 272−273
Hitozukuri, 28
HMI. *See* Human−machine interface (HMI)
Holonic manufacturing system, 28
Hoshin Kanri, 29, 65−66
Hourensou, 29
HRM. *See* Human resources management (HRM)
Human Automation. *See* Jidoka
Human resource (HR), 331, 339−346
Human resources management (HRM), 331−332
Human−machine interface (HMI), 368−369

I

I-V-A-T analysis, 390−391
IE practices. *See* Industrial engineering practices (IE practices)
IIoT. *See* Industrial Internet of Things (IIoT)
Industrial engineering practices (IE practices), 112−113
Industrial Internet of Things (IIoT), 366−368
Information waste, 60
Internet of Things (IoT), 371
Inventory management, 386−390, 388*f*
Ishikawa diagram, 38, 122−125, 122*f*, 123*f*, 126*f*

J

Jidoka, 29, 233−242, 237*f*
Jinbou, 29
Jinzai Katsuyou, 29
Jiro Kawakita method (KJ method), 18
JIS. *See* Just in sequence (JIS)
Jishuken, 30
JIT. *See* Just-in-time (JIT)
Jundate, 30
Just in sequence (JIS), 18, 30
Just-in-time (JIT), 2, 4, 30, 73−74, 159−162, 257, 294, 315−316, 315*f*, 389−390, 389*t*

K

Kaikaku Kaizen, 30
Kaizen, 2, 67−72, 81−86, 83*f*, 84*f*, 85*f*
Kaizen Blitz, 30−31
Kaizen event, 67−69, 81
Kanban, 5, 31, 279−286

Kanban card, 106*t*, 279−280
Key Performance Indicators (KPIs), 139, 178, 181*f*, 340−342, 341*f*, 342*f*, 364
Key Result Areas (KRA), 340−342, 341*f*
Kitting, 31
KJ method. *See* Jiro Kawakita method (KJ method)
KPIs. *See* Key Performance Indicators (KPIs)
KRA. *See* Key Result Areas (KRA)

L

LAMDA cycle. *See* Look−ask−model−discuss−act cycle (LAMDA cycle)
Last-in−first-out (LIFO), 25, 32
Lead time, 31−32, 266
Leagile approach, 415
Leagile supply chain, 416*t*, 417, 417*t*
Leagility approach, 417, 418*f*
Lean Enterprise Research Center (LERC), 48−49
Lean journey, 47−75, 343−344
Lean Lexicon, 17
Lean manufacturing, 3−4, 4*f*, 227, 400−401, 412−417
Lean Production system, 299−300
Lean quality management system (LQMS), 316, 327−328
Lean Sigma, 6
Lean supply chain, 382−395
LEGO layout, 305, 305*f*
LEGO Module sewing layout, 306*f*
LERC. *See* Lean Enterprise Research Center (LERC)
LIFO. *See* Last-in−first-out (LIFO)
Line balancing, 32
Little's law, 32
Load leveling, 279
Look−ask−model−discuss−act cycle (LAMDA cycle), 24, 31, 118−119
Lot Tolerance Percent Defective (LTPD), 324
LQMS. *See* Lean quality management system (LQMS)

M

Machine Learning (ML), 366−368
Maintenance performance indicator, 359−360

Index

Make to order (MTO), 32
Make to stock (MTS), 33
Make-through approach, 295
Manufacturing systems, 292−293, 293*f*, 307
Mass customization, 33
Mass production, 33
Mathematical modeling, 372−373
Mean time between failures (MTBF), 24, 34, 360
Mean time to failure (MTTF), 24, 34, 374−375
Mean time to repair (MTTR), 24, 34, 360
Mendomi (Japanese expression), 33, 343−344
Mieruka, 33
Missing weft, 234
ML. *See* Machine Learning (ML)
Modular manufacturing system, 297−306
MTBF. *See* Mean time between failures (MTBF)
MTO. *See* Make to order (MTO)
MTS. *See* Make to stock (MTS)
MTTF. *See* Mean time to failure (MTTF)
MTTR. *See* Mean time to repair (MTTR)
Muda (waste), 48, 55*t*, 332−333
  of inventory, 58−59
  of motion, 57
  of overprocessing, 58
  of overproduction, 59−60
  of rework, 57−58
  of transportation, 57
  of waiting, 57
Mura (unevenness), 53−54
Muri (overburden), 53−54

N

Nagara, 34
Nagare, 35
National Institute of Fashion Technology (NIFT), 9, 9*f*, 11*f*
Necessary non-value added activities (NNVA activities), 35, 48−49, 49*f*, 51*f*, 189−190, 210
Nemawashi, 35
NNVA activities. *See* Necessary non-value added activities (NNVA activities)
Non-value-added activities (NVA activities), 24, 35, 48−49, 49*f*, 51*f*, 81, 189−190
NVA activities. *See* Non-value-added activities (NVA activities)

O

OEE. *See* Overall equipment effectiveness (OEE)
On time in full (OTIF), 257−258
On-time in-full and error-free (OTIFEF), 87
One-piece flow, 35−36
One-Touch Exchange of Dies (OTED), 210
OTED. *See* One-Touch Exchange of Dies (OTED)
OTIF. *See* On time in full (OTIF)
OTIFEF. *See* On-time in-full and error-free (OTIFEF)
Overall equipment effectiveness (OEE), 35, 359−362

P

Pacemaker process, 36, 268, 268*f*
Paired-cell overlapping loops of cards with authorization (POLCA), 287
Pareto principle, 36
PBSs. *See* Progressive bundle systems (PBSs)
PCDs. *See* Planned cut dates (PCDs)
PDCA. *See* Plan−do−check−act (PDCA)
People, Practice, Process, and Product model (4P model), 212, 213*f*
Pitch time, 37, 265
Plan−do−check−act (PDCA), 36, 65, 81, 117−118, 272−273, 314
Plan−do−study−act (PDSA). *See* Plan−do−check−act (PDCA)
Planned cut dates (PCDs), 268
Poka-Yoke, 19, 233, 242−253
POLCA. *See* Paired-cell overlapping loops of cards with authorization (POLCA)
Postponement, 387−388
PPMS. *See* Predictive preventive maintenance system (PPMS)
Predictive preventive maintenance system (PPMS), 372
Process balancing, 258
Process inventory, 106
Process matrix, 96, 96*f*, 152−153
Process wastes, 60
Product leveling, 277−279
Production Kanban, 106, 280
Production leveling, 275−279
Production monitoring system (PMS), 364−365

Index 427

Progressive bundle systems (PBSs), 296–297
Pull system, 37
Push system, 37

Q

QCD + flexibility (QCDF), 24
QCD + morale + safety (QCDMS), 24
Quality, cost, and delivery (QCD), 135
Quality circle (QC), 19, 37, 81, 344
Quality function deployment (QFD), 19, 38, 313–314
Quality management systems (QMSs), 313–314
Quick changeover (QCO), 211–212
Quick response (QR), 294
Quick win problem-solving, 346, 347t

R

R control chart, 318, 319f
Rabbit Chase, 38, 300–305, 302f
Rapid setup, 209–228
Rapid tooling, 211
RCA. *See* Root-cause analysis (RCA)
RDT. *See* Rundown time (RDT)
Red tagging, 170, 171f
Repetitive strain injuries (RSIs), 295
Right-first-time (RFT). *See* First time through (FTT)
Root-cause analysis (RCA), 26, 38, 81, 121–126, 135, 314
RSIs. *See* Repetitive strain injuries (RSIs)
"Rule of tens", 47, 48f
Rundown time (RDT), 216–217

S

SAM. *See* Standard Allowed Minutes (SAM)
Seiketsu–standardize, 172–173
Seiri (sort), 168–170
Seiso (shine), 170–172
Seiton (set in order), 170
Setup time (ST), 209
Seven waste (7 waste), 38–39
Shingo's SMED methodology, 214, 215f
Shitsuke (sustain), 173–175, 175f
Shoujinka, 39
Single Minute Exchange of Die (SMED), 22, 39, 73, 210, 214f
Single-piece flow systems, 296–297

SIPOC. *See* Supplier–input–process–output–customer (SIPOC)
Six Sigma (6S), 24, 39, 175–176, 179f, 326
Smart maintenance, 366–376
SMART methodology. *See* Specific, Measurable, Attainable, Relevant, Time-based methodology (SMART methodology)
SMED. *See* Single Minute Exchange of Die (SMED)
SMED-ZERO, 216, 217t
Spaghetti diagram, 39
Specific, Measurable, Attainable, Relevant, Time-based methodology (SMART methodology), 340–341
SQC. *See* Statistical quality control (SQC)
ST. *See* Setup time (ST)
Standard Allowed Minutes (SAM), 227
Standard work. *See* Standardized work
Standard work-in-process (Standard WIP), 117
Standardized work, 40, 112–117
Statistical quality control (SQC), 311
Supermarket, 40, 282–283
Supplier–input–process–output–customer (SIPOC), 87, 89–90, 91f, 191
Supply chain, 381–382
Supply lead time, 381
System Takt, 40

T

Takt time, 113, 265
Taylorism, 40
Theory of Constraints (TOC), 20, 40–41, 273
Three M (3M), 41
of lean manufacturing, 53–54
Throughput, 41
TOC. *See* Theory of Constraints (TOC)
Tool changeover, 211–216
Total productive maintenance (TPM), 19, 41–42, 73–75, 272–273, 355–376, 356f
Total quality management (TQM), 65
Toyota Production System (TPS), 2, 19, 42, 53, 233, 299
Toyota Sewing System (TSS), 294–295, 299–300, 299f, 300f
TPM. *See* Total productive maintenance (TPM)

TPS. *See* Toyota Production System (TPS)
TQM. *See* Total quality management (TQM)
Triangular Kanban card, 282*f*
TSS. *See* Toyota Sewing System (TSS)
Tunnel Kanban. *See* Kanban
Turtle diagram, 42, 191−194, 192*f*

U
U-line, 42
Unit flow production system, 296
Unit production system (UPS), 261, 296

V
V-A-T supply chain models, 391*f*
V-shape network, 391
Value-added (VA), 43, 49, 49*f*, 51*f*
Value added ratio (VAR), 43
Value addition, 48−53
Value stream design (VSD), 44
Value stream loop (VSL), 44
Value stream mapping (VSM), 5, 17, 43−44, 81, 87−106, 88*f*, 92*f*, 400
Value stream summary matrix, 97*f*
Value stream walk-through, 95
VAR. *See* Value added ratio (VAR)
Vendor managed inventory (VMI), 36, 43, 383, 388−389
Visual control, 131−132, 146, 147*t*, 149*t*, 150
Visual display, 142−145, 147*t*, 150
Visual factory, 137−155
Visual garment factory, 152−153, 152*f*
Visual management system (VMS), 131−132, 148*f*
Visual metrics, 146, 148*t*
Visual workplace, 137
VMI. *See* Vendor managed inventory (VMI)
VMS. *See* Visual management system (VMS)
Voice of Customer (VoC), 91−92, 95*f*, 313
Volatile, uncertain, complex, and ambiguous (VUCA) environment, 1, 18

VSD. *See* Value stream design (VSD)
VSL. *See* Value stream loop (VSL)
VSM. *See* Value stream mapping (VSM)

W
Waste, 54−60, 54*f*
  Muda
    of inventory, 58−59
    of motion, 57
    of overprocessing, 58
    of overproduction, 59−60
    of rework, 57−58
    of transportation, 57
    of waiting, 57
Waste elimination (WE), 4, 417
WCM. *See* World class manufacturing (WCM)
WE. *See* Waste elimination (WE)
Whiplash effect, 21
Whipsaw effect, 21
Why Why Because of Logical Analysis tool (WWBLA tool), 364
Why-Why analysis. *See* 5-Why analysis
WIP. *See* Work in process (WIP)
Withdrawal Kanban, 106, 280, 283−284
Work cell setup, 268−270
Work in process (WIP), 57, 257−258, 295
World class manufacturing (WCM), 18, 44
WWBLA tool. *See* Why Why Because of Logical Analysis tool (WWBLA tool)

X
X-bar chart, 318, 318*f*
X-Matrix, 66, 68*f*

Y
Yamazumi Charts, 5, 44, 188−191, 189*f*
Yokoten, 44

Z
Zero defects (ZD), 19, 45, 313−314
Zoning approach for plant layout, 151−152

Printed in the United States
By Bookmasters